Cooperative Networks

NEW DIMENSIONS IN NETWORKS

Series Editor: Anna Nagurney, *John F. Smith Memorial Professor, Isenberg School of Management, University of Massachusetts at Amherst, USA*

Networks provide a unifying framework for conceptualizing and studying problems and applications. They range from transportation and telecommunication networks and logistic networks to economic, social and financial networks. This series is designed to publish original manuscripts and edited volumes that push the development of theory and applications of networks to new dimensions. It is interdisciplinary and international in its coverage, and aims to connect existing areas, unveil new applications and extend existing conceptual frameworks as well as methodologies. An outstanding editorial advisory board made up of scholars from many fields and many countries assures the high quality and originality of all of the volumes in the series.

Titles in the series include:

Supernetworks
Decision-Making for the Information Age
Anna Nagurney and June Dong

Innovations in Financial and Economic Networks
Edited by Anna Nagurney

Urban and Regional Transportation Modeling
Essays in Honor of David Boyce
Edited by Der-Horng Lee

The Network Organization
The Experience of Leading French Multinationals
Emmanuel Josserand

Dynamic Networks and Evolutionary Variational Inequalities
Patrizia Daniele

Supply Chain Network Economics
Dynamics of Prices, Flows and Profits
Anna Nagurney

Cooperative Networks
Control and Optimization
Edited by Panos Pardalos, Don Grundel, Robert A. Murphey and Oleg Prokopyev

Cooperative Networks

Control and Optimization

Edited by

Panos Pardalos

University of Florida, USA

Don Grundel

Eglin Air Force Base, USA

Robert A. Murphey

Eglin Air Force Base, USA

and

Oleg Prokopyev

University of Pittsburgh, USA

NEW DIMENSIONS IN NETWORKS

Edward Elgar
Cheltenham, UK • Northampton, MA, USA

Published by
Edward Elgar Publishing Limited
The Lypiatts
15 Lansdown Road
Cheltenham
Glos GL50 2JA
UK

Edward Elgar Publishing, Inc.
William Pratt House
9 Dewey Court
Northampton
Massachusetts 01060
USA

A catalogue record for this book
is available from the British Library

Library of Congress Control Number: 2008927955

ISBN 978 1 84720 453 0

Printed and bound in Great Britain by MPG Books Ltd, Bodmin, Cornwall

Editorial Board

Contents

Contributors   ix

Preface   x

1 Token-Based Approach for Scalable Team Coordination
Yang Xu, Paul Scerri, Michael Lewis and Katia Sycara   1

2 Consensus Controllability for Coordinated Multi-Agent Systems
Zhipu Jin and Richard M. Murray   27

3 Decentralized Cooperative Optimization for Systems Coupled through the Constraints
Yoshiaki Kuwata and Jonathan P. How   45

4 Consensus Building in Cooperative Control with Information Feedback
Wei Ren   63

5 Levy Flights in Robot Swarm Control and Optimization
Yechiel J. Crispin   81

6 Eavesdropping and Jamming Communication Networks
Clayton W. Commander, Panos M. Pardalos, Valeriy Ryabchenko, Oleg Shylo, Stan Uryasev and Grigoriy Zrazhevsky   101

7 Network Algorithms for the Dual of the Constrained Shortest Path Problem
Bogdan Grechuk, Anton Molyboha and Michael Zabarankin   113

8 Towards an Irreducible Theory of Complex Systems
Victor Korotkikh   147

9 Complexity of a System as a Key to its Optimization
Victor Korotkikh and Galina Korotkikh   171

10 GRASP with Path-Relinking for the Cooperative Communication Problem on Ad hoc Networks
Clayton W. Commander, Paola Festa, Carlos A. S. Oliveira, Panos M. Pardalos, Mauricio G. C. Resende and Marco Tsitselis   187

11 Optimal Control of an ATR Module - Equipped MAV/Human Operator Team
Meir Pachter, Phillip R. Chandler and Dharba Swaroop   209

12 An Investigation of a Dynamic Sensor Motion Strategy
Nathan P. Yerrick, Abhishek Tiwari and David E. Jeffcoat   253

13 Market Based Adaptive Task Allocations for Autonomous Agents
Jared W. Patterson and Kendall E. Nygard   269

14 Cooperative Persistent Surveillance Search Algorithms using Multiple Unmanned Aerial Vehicles
Daniel J. Pack and George W. P. York   279

15 Social Networks for Effective Teams
Paul Scerri and Katia Sycara   291

16 Optimization Approaches for Vision-Based Path Planning for Autonomous Micro-Air Vehicles in Urban Environments
Michael Zabarankin, Andrew Kurdila, Oleg A. Prokopyev, Ryan Causey, Anukul Goel and Panos M. Pardalos   311

17 Cooperative Stabilization and Tracking for Linear Dynamic Systems
Yi Guo   343

Index   355

Contributors

Ryan Causey, University of Florida, USA
Phillip R. Chandler, Air Force Research Laboratory, USA
Clayton W. Commander, Air Force Research Laboratory, USA
Yechiel J. Crispin, Embry-Riddle Aeronautical University, USA
Paola Festa, University of Napoli, Italy
Anukul Goel, University of Florida, USA
Bogdan Grechuk, Stevens Institute of Technology, USA
Yi Guo, Stevens Institute of Technology, USA
Jonathan P. How, Massachusetts Institute of Technology, USA
David E. Jeffcoat, Air Force Research Laboratory, USA
Zhipu Jin, California Institute of Technology, USA
Galina Korotkikh, Central Queensland University, Australia
Victor Korotkikh, Central Queensland University, Australia
Andrew Kurdila, University of Florida, USA
Yoshiaki Kuwata, Massachusetts Institute of Technology, USA
Michael Lewis, University of Pittsburgh, USA
Anton Molyboha, Stevens Institute of Technology, USA
Richard M. Murray, California Institute of Technology, USA
Kendall E. Nygard, North Dakota State University, USA
Carlos A. S. Oliveira, Princeton Consultants, USA
Meir Pachter, Air Force Institute of Technology, USA
Daniel J. Pack, United States Air Force Academy, USA
Panos M. Pardalos, University of Florida, USA
Jared W. Patterson, IBM Corporation, USA
Oleg Prokopyev, University of Pittsburgh, USA
Wei Ren, Utah State University, USA
Mauricio G. C. Resende, AT&T Labs Research, USA
Valeriy Ryabchenko, University of Florida, USA
Paul Scerri, Carnegie Mellon University, USA
Oleg Shylo, University of Florida, USA
Dharba Swaroop, Texas A&M University, USA
Katia Sycara, Carnegie Mellon University, USA
Abhishek Tiwari, California Institute of Technology, USA
Marco Tsitselis, University of Napoli, Italy
Stan Uryasev, University of Florida, USA
Yang Xu, University of Pittsburgh, USA
Nathan P. Yerrick, Air Force Research Laboratory, USA
George W. P. York, United States Air Force Academy, USA
Michael Zabarankin, Stevens Institute of Technology, USA
Grigory Zrazhevsky, University of Florida, USA

Preface

Cooperative networks have gained a tremendous amount of attention over the last several years because of their existence not only in biological, social and economic arenas, but because of direct applications in many more areas such as communications, robotics and military sciences. While the benefits of networks in general have been recognized for quite some time, the idea of cooperative networks has several novel implications. Because distributed processing by heterogeneous nodes promises significant increases in system capability, performance and efficiency, research has intensified with a breadth of promising applications. Much of the research has focused on cooperation and optimization within the networks given real-life challenges such as limited bandwidth, noisy communication, hazardous environments, highly mobile nodes, protocol restrictions and scalability. Examples of cooperative control networks might include: robots operating within a manufacturing cell, unmanned aircraft in search and rescue operations or military surveillance and attack missions, arrays of micro satellites that form distributed large aperture radar, employees operating within an organization, software agents, etc. Economists first delved into the analysis of the economics of cooperative networks in the early 1980s. Today we talk about cooperative financial networks, cooperative markets and cooperative decision making.

This volume reflects a cross fertilization of ideas from a broad set of disciplines and creativity from a diverse array of scientific and engineering research. Coverage of topics include areas such as networks of unmanned vehicles in uncertain environments, networks subject to eavesdropping and jamming, optimal node task-allocation, network complexity analysis, cooperative search involving multiple unmanned vehicles, and cooperative communications in ad-hoc networks, optimal control and optimization in man-in-the-loop scenarios.

Many of the chapters in this volume were presented at the 6th International Conference on Cooperative Control and Optimization, which took place on February 1-3, 2006 in Gainesville, Florida. The three-day event was sponsored by the Air Force Research Laboratory and the Center of Applied Optimization of the University of Florida.

Don Grundel, Rob Murphey, Panos Pardalos, Oleg Prokopyev
June 2007

1 Token-Based Approach for Scalable Team Coordination

Yang Xu, Paul Scerri, Michael Lewis and Katia Sycara

Summary. Efficient coordination among large numbers of heterogeneous agents promises to revolutionize the way in which some complex tasks, such as responding to urban disasters can be performed. However, state of the art coordination algorithms are not capable of achieving efficient and effective coordination when a team is very large. Building on recent successful token-based algorithms for task allocation and information sharing, we have developed an integrated and efficient approach to effective coordination of large scale teams. We use *tokens* to encapsulate anything that needs to be shared by the team, including information, tasks and resources. The tokens are efficiently routed through the team via the use of local decision theoretic models. Each token is used to improve the routing of other tokens leading to a dramatic performance improvement when the algorithms work together. We present results from an implementation of this approach which demonstrates its ability to coordinate large teams. By comparing with the market-based approach, we demonstrated how the token-based approach has made a good trade off between team performance and communication cost.

1 Introduction

Efficient and flexible coordination among large numbers of robots, agents and people promises to revolutionize the achievement of complex and distributed tasks. In domains such as disaster response [13], the military [29] and business organizations [9], decentralized cooperative coordination can dramatically reduce costs and improve efficiency while lowering risks and improving safety. In these applications, a large number of heterogeneous agents need to coordinate in a dynamic, uncertain environment and adjust their activities according to the status of the team and their teammates. Typically, coordination requires tasks including plan monitoring, information delivery, role allocation and resource sharing.

Previous work on coordination has typically focused on only one specific coordination task, e.g. role allocation [24] or planning [11], precluding the use of knowledge from other aspects of coordination being used to improve the performance of that algorithm. For example, results of the task allocation process have not been used to guide resource allocation, although intuitively they will improve the search.

On the other hand, even algorithms designed for single coordination tasks do not scale well to very large teams. The major challenges include communication limitation, decentralized control and incomplete knowledge for decision support. Existing approaches, which are successful in small-team coordination when apply to large teams [14, 28], are incapable of making decisions under incomplete team knowledge with the communication limitations. Decentralized communication decision approaches require accurate models of all team members which is infeasible for agents in a large team [22, 34]. Algorithms that are scalable, often rely on swarm-like behavior that, while robust, can be very inefficient [6]. Other approaches, e.g., using an auctioneer [11], require some degree of centralization which is not always desirable. Although rapid progress has been made in developing coordination algorithms [10], teamwork algorithms that scale to large numbers of agents while remaining efficient, distributed and flexible are not yet available.

In this chapter, we present an integrated and scalable approach to coordinating a large number of heterogeneous agents. Three novel ideas underlie this approach. The first idea is to encapsulate *all* coordination interactions, including information, assignable tasks and sharable resources within *tokens*. The agent holding the token has exclusive control over whatever is represented by that token, hence tokens provide a type of access control. Agents either keep tokens or pass them to teammates. For example, an agent holding a resource token has exclusive access to the resource represented by that token and passes the token on to transfer access to that resource. The resulting movement of tokens implements the coordination by distributing information, resources and tasks with low communication overhead.

The second novel idea is for agents to use local decision theoretic models to determine when and where to pass tokens. When an agent passes a token to another agent, that exchange is used to refine local models of the team. These models are used in a decision theoretic way to determine whether and where to forward any token the agent currently holds, so as to maximize the expected utility of the team. Informally, agents will try to pass tokens to where they help team performance the most by inferring from their local models which team member will either have use for the information, resource or task represented by the token or be in the best position to know who will. A logical static network across the team, limits agents to forwarding tokens to their neighbors in this network. As a result an agent directly receives tokens from only a small number of neighbors in the network and can thus build better models of those agents. By ensuring that the network has a small world property [16], i.e., the distance between any two nodes in the network is small, the effect of these better models outweighs the additional number of "hops" a token might need to take to get where it is required.

The third novel idea in this work is to leverage *all* available information for creating models of the team, specifically using the movement of one token to inform the movement of other tokens. This synergistically integrates the execution of key coordination algorithms in a way not done before. For example, tokens representing resources useful for a particular task should be passed to the same agent as the token representing that task was. Intuitively, making use of the relationship between tokens, each coordination task becomes more efficient because it focuses its search based on the progress of other coordination tasks.

In the remainder of this chapter we describe how tokens are routed around the network to maximize the expected utility of the team. Specifically, we begin with a Markov Decision Process (MDP) model based on the full observation of team state then make a series of approximations to develop efficient, local reasoning for routing the tokens. To test our approach, there are two groups of experiments. In the first group, the results show that the local routing models lead to a dramatic improvement in coordination performance and excluding any type of token from the development of the local reasoning models decreases performance building on previously described individual token-based algorithms [24, 12, 27, 31]. In the second group of experiments, we systematically and scientifically compare our token-based coordination with the market-based coordination, a popular, centralized algorithm to find the optimal coordination solution. Our experiment results meet the hypothesis based on the nature of each approach. Auctions are focused on maximizing overall utility taking into account the *bids* of all team members [2]. Token-algorithm is focused on scalability, hence it minimizes communication, sometimes at the expense of overall utility.

2 Large Scale Coordination

In this section, we provide a detailed model of the organization and coordination problem for the team.

2.1 Problem Description

Coordination is required between a team $A = \{\alpha_1, \alpha_2, ..., \alpha_n\}$ of agents that share a top level common goal G (as in [28]). Achieving G requires achieving a number of sub-goals $\{g_1, g_2, ..., g_i, ...\}$. When sub-goal g_i is satisfied the team receives a reward $reward_i$. For example, sub-goals of a high level goal to respond to a disaster might be to extinguish fires and provide medical attention to injured civilians. To satisfy sub-goals, the team follows plan templates $Plan = \{plan_1, plan_2, ..., plan_i, ...\}$ represented in a library. Each template i includes four parts and is written as $plan_i = <g_i, conditions_i, roles_i, reward_i>$. The first element is the sub-goal g_i; the second is the conditions under which it is applicable, $conditions_i = event_1 \cap event_2 \cap ... \cap event_l$; the third element is the individual roles $roles_i = \{r_1, r_2, ..., r_k\}$ which are required to achieve g_i and the last part, the $reward_i$ is to be received by the team on successful satisfaction of g_i. Each role

$r_i =< task_i, ability_i, resource_i >$ is represented by its task, i.e., a description of the actual thing to be done, the capabilities required to perform that task and the resources needed to perform the role.

For example, a fire fighting template can be defined as: $< plan_{fire}=$(Fight fire at location X), (Fire alarm at X $\cap$ Smoke at X), $\{r_1, r_2, r_3\}$, (100) $>$. This template requires two conditions before it is initiated: a fire alarm and smoke. After this plan is initiated, three roles, $\{r_1, r_2, r_3\}$ need to be assigned and a reward 100 will be credited to the team. The three roles in this template are: driving the fire truck, fighting the fire and searching for victims, i.e., $r_1=<$(Driving the fire truck), (Skillful in driving truck), (Fire truck)$>$, $r_2=<$(fighting the fire), (Have training in fire fighting), (Hose, water)$>$ and $r_3=<$(Searching for victims), (None), (Breathing equipment)$>$. To perform r_1, an agent is required to be able to drive and have access to a fire truck which is an exclusive resource.

2.2 Acquaintance Network

It has been observed that in a human group, members typically maintain a small number of acquaintances but can rapidly transmit information to any member of the group in a series of hops, a phenomenon known as a *small world effect* [16]. The most popular manifestation of this phenomenon is the *six degrees of separation* concept [19]. Milgram concluded that there is a path of acquaintances with typical length six between any two people in the United States. By using very vague (and often incorrect) information about other members of the population, people will pass a message to someone better placed to find the intended recipient until the information reaches the desired recipient.

Inspired by such social networks, we arrange the team as an *acquaintance network*. The *acquaintance network* is a graph $G = (A, N)$, where A is the team of agents and N is the set of links between agents. Specifically, for $\alpha_i, \alpha_j \in A$, $< \alpha_i, \alpha_j >\in N$ denotes that α_i and α_j are acquaintances and are able to exchange tokens directly. $n(\alpha)$ is defined as all the acquaintances of agent α. Note that $n(\alpha) << |A|$. We additionally require that the acquaintance network be a *small world network*, which means that a relatively small number of links separate any two agents in comparison to a regular grid network. Previous work has shown that such networks lead to better performance of token-based algorithms [31]. A subset of a typical acquaintance network for a large team is shown as Figure 1. In the Figure, each node represents a team member and when pairs of agents are connected by a line, they can exchange tokens with each other directly.

3 Tokens for Coordination

Token-based algorithms for specific tasks have been developed by us and others and have been shown to be effective for specific coordination tasks [31, 24].

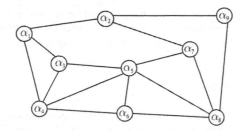

Fig. 1. An example of a subset of a typical acquaintance network.

Control information is included in the token to help the agents determine what to do within the token. For example, for information sharing the control information is the number of "hops" a token can move before stopping [31]. For task allocation, the control information is the minimum capability an agent must have to accept a task [24]. However, while these algorithms share the important common feature of being based on tokens, they operate separately. In this chapter, we generalize and integrate token-based approaches to make a complete approach to coordination.

Let $\Gamma = \{\Delta_1, \Delta_2, ..., \Delta_m\}$ be all types of coordination tokens. These tokens can be classified into three basic types: information tokens, roles tokens and resources tokens. Each token Δ_i is defined as a tuple with four elements: $\Delta_i = < Type, Coordination, Path, Threshold >$. $Type = \{inf, role, res\}$ denotes the type of coordination; it contains information, role or resource, respectively. *Coordination* captures the specific coordination element represented by this token. In the case of an information token, it is the information to be shared. In the case of a resource token, it is a description of the resource to which this token grants exclusive access. In the case of a role token, it is a description of the task for which the accepter of this token is responsible. *Path* records the route the token has taken through the network. $\Delta.path$ is also used as stop condition for information and role tokens when $|\Delta.path| > TTL$ where TTL is empirically set to be the maximum number of "hops" that Δ is allowed to be passed. *Threshold* generalizes the *control information* for resource and role tokens but is not required for information tokens. An agent may keep a resource if its need for that resource is greater than the token's *threshold*. Determining an agent's requirement for a resource is outside of the scope of this chapter. While an agent holds a resource token Δ, $\Delta.threshold$ slowly increases up to some maximum. When the token is passed, $\Delta.threshold$ is decreased to avoid the token being passed indefinitely. This mechanism ensures that resources can flow through the team resource. For example, to coordinate a team of unmanned aerial vehicles (UAVs) holding tokens representing airspace, might be forced to relinquish tokens to the longest held regions as their thresholds increased unless the airspace was critical. Similarly, a role token Δ will be accepted by an agent whose capability is greater than $\Delta.threshold$ and its $threshold$ will be decreased if it has not been accepted. The role allocation algorithm is described in detailed in [24].

4 Routing Tokens

Token-based coordination is a process by which agents attempt to maximize the overall team reward by moving tokens around the team. If an agent were to know the exact state of the team, it could use an MDP to determine the expected utility maximizing way to move tokens. Unfortunately, it is infeasible for an agent to know the complete state, however, [22] it is illustrative to look at how tokens would be passed if it were feasible. Then, by dividing the monolithic joint activity into a set of actions that can be taken by individual agents, we can decentralize the token routing process where distributed agents, in parallel, make independent decisions of where to pass the tokens they currently hold. Thus, we effectively break a large coordination problem into many small ones.

4.1 MDP Model for Complete Team State

The basic decision model of agent α for a token Δ can be written as an MDP $< S, Action_\alpha, T, R >$. S is the state space and its specific value in time t defined as $s(t)$, $Action_\alpha$ is the action space of α, $T : S \times A \rightarrow S$, is the transition function that describes the resulting state $s(t+1) \in S$ when executing $\chi \in Action_\alpha$ in $s(t)$. $R : S \rightarrow \mathbb{R}$ defines the instantaneous reward for being in a specific state. This model can be applied to any agent and any token.

In this case, the state $s(t)$ at time t is modelled as the locations of all tokens across the team and is written as: $s(t) = << Tokens(\alpha, t), H_\alpha(t) >,$ $< Tokens(\alpha_1, t), H_{\alpha_1}(t) >, < Tokens(\alpha_2, t), H_{\alpha_2}(t) >, ... > .$ $Tokens(\alpha, t)$ are all the tokens currently held by α and $H_\alpha(t)$ records all the incoming and out-going tokens of α before t. For notation convenience, we write $Tokens(\alpha, t)$ as $Tokens(\alpha)$ and $H_\alpha(t)$ as H_α when there is no ambiguity. Figure 2 shows an example of specific team state $s(t) = << Tokens(\alpha), H_\alpha >, < Tokens(\alpha_1), H_{\alpha_1} >, < Tokens(\alpha_2), H_{\alpha_2} >>$, where $Tokens(\alpha) = \{\Delta, \Delta_2\}$, $H_\alpha = \{\Delta_3\}$, $Tokens(\alpha_1) = \{\Delta_5\}$, $H_{\alpha_1} = \{\Delta\}$, and $Tokens(\alpha_2) = \{\Delta_4\}$, $H_{\alpha_2} = \emptyset$. Since the tokens represent resources, roles and information, $s(t)$ unambiguously defines who is doing what, with what resources and what information.

$Action_\alpha : S \rightarrow (n(\alpha) \cup \alpha)$ is to move Δ to one of $n(\alpha)$ or keep it for itself. For notation convenience, $\chi \in Action_\alpha$ can be written as $move(\Delta, b)$ where $b \in (n(\alpha) \cup \alpha)$. Keeping a token for agent α itself applies when the agent accepts the role, the resource or the information that the token encapsulated.

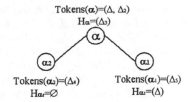

Fig. 2. An example showing part of a team. Agent α holds Δ, Δ_2 and has previously had Δ_3 ; α_2 holds Δ_4 while α_1 holds Δ_5 and has previously had Δ.

In general, we define a function $Acceptable(\alpha, \Delta)$ to determine whether Δ should be kept by agent α.

$R(s(t)) > 0$ when at $s(t)$, a sub-goal g_i are achieved. The team will be credited an instant rewards value of $R(s(t)) = reward_i$.

The utility of state S under a policy π is defined as

$$v^\pi(s) = \sum_{t=0:\infty} (d^t \times R(s(t)) - t \times commcost)$$

where $commcost$ is the communication cost and $d < 1$ is a predefined discount factor. $v^*(s)$ allows the agent to select actions according to optimal policy

$$\pi^*(s(t)) = argmax_{\chi \in Action_\alpha} v^*(s(t+1))$$

By value iteration, $v^*(s(t)) = argmax_{\chi \in Action_\alpha}[R(s(t)) - commcost + d \times v^*(s(t+1))]$. This policy tells the agent where to move resources information and roles to maximize the team's expected utility.

We define a matrix V where each element $V[s(t), b] = R(s(t)) - commcost + d \times v^*(s(t+1))$ when $\chi = move(\Delta, b)$. Then $V[b]$ represents the expected utilities vector for α to send token Δ to b at each different state $s(t)$.

4.2 Local POMDP Model

Knowing the complete team state is only feasible for small teams. In large teams, agents must make token coordination decisions based on a more limited view of the team. Thus the reasoning must be modelled as a Partially Observable Markov Decision Process (POMDP). Standard POMDP techniques such as [16] and [15] could be used to solve the POMDP to determine optimal token routing. However, for fast routing of tokens, while this local POMDP does tell the agent the optimal action, the computational complexity is still too high for practical applications. However, the POMDP model does provide important hints for how to do a heuristic approach.

The POMDP model is defined as $< S, Action_\alpha, T, \Theta_\alpha, O, R >$. In this case, the observations of agent α are defined as $\Theta_\alpha = < Tokens(\alpha, t), H_\alpha(t) >$ to include not only the tokens the agent currently holds but also all the previously incoming and out-going tokens (in $H_\alpha(t)$). The observation function is defined as $O : \Theta_\alpha \times S \rightarrow \Omega_\alpha$. *Belief state* Ω_α is a discrete probability distribution vector over the team state $s(t)$ inferred from current local state Θ_α. For example, if $S = \{s_1,\ s_2,\ s_3\}$ and $\Omega_\alpha = [0.6, 0.2, 0.2]$, α estimates that the probability of $s(t)$ being s_1 is 0.6 and being s_2 and s_3 are 0.2.

One way of solving a POMDP is via a Q-MDP [15]. Agent α makes use of V, see above, to calculate the expected reward vector $EU(\Omega_\alpha) = \Omega_\alpha \times V$. For example, if α has acquaintances b, c, and d and $EU(\Omega_\alpha) = [5, 10, 6, 4]$, then $EU(\Omega_\alpha, b) = 10$ represents the expected utility to send Δ to b according to the Q-MDP. The locally-optimal policy $\pi^{**}(\Omega_\alpha)$ is $argmax_{\chi \in Action_\alpha} EU(\Omega_\alpha, c)$. This is the action the agent should take to maximize expected utility, given that it has an incomplete view of the team state. As in the previous example, passing Δ to b is the best choice because $EU(\Omega_\alpha, b) = 10$ is the maximum value of $EU(\Omega_\alpha)$.

5 Local Heuristic Approach

In this section, we provide a heuristic approach for token-based team coordination inspired by the local POMDP. The resulting approach allows fast, efficient routing decisions, without requiring accurate knowledge of the complete state. In the next section, we show that this approach is effective for improving token routing.

5.1 Local Model

P_α is the decision matrix agent α uses to decide where to move tokens. Each row $P_\alpha[\Delta]$ in P_α represents a vector that determines the decision where to pass a token Δ to one of its acquaintances. Specifically, each value $P_\alpha[\Delta, b] \rightarrow [0, 1], b \in n(\alpha)$ represents α's decision that the probability of passing token Δ to an acquaintance b would be the action that maximize team reward. Then our policy π^{***} for this local model is to choose action χ to $argmax_{\chi \in Action_\alpha} P_\alpha[\Delta, c]$ where $\chi = move(\Delta, c)$. Figure 3 shows an example where $P_\alpha[\Delta] = [0.6, 0.1, 0.3]$ and agent α has three acquaintances $\alpha_1, \alpha_2, \alpha_3$. $P_\alpha[\Delta, \alpha_1] = 0.6$, $P_\alpha[\Delta, \alpha_2] = 0.1$, $P_\alpha[\Delta, \alpha_3] = 0.3$ and π^{***} will choose the action $move(\Delta, \alpha_1)$ to pass Δ to α_1. The key to this distributed reasoning lies in how the probability model P_α for each agent α is updated. If the action indicated by P_α matches the optimal policy π^* from the MDP model, then the team will act optimally.

Initially, agents do not know where to send tokens, but as tokens are received, a model can be developed and better routing decisions made. That is, the model, P_α is based on the accumulated information provided by the receipt of previous tokens. For example, when an agent sends a role to an acquaintance that has previously rejected a similar role, the team is potentially hurt because this acquaintance is likely to reject this role too and thus communication bandwidth has been unnecessarily wasted.

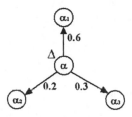

Fig. 3. Agent α's local model for where to send Δ. The probability that α_1 is the best to send Δ to is 0.6 and α will pass Δ to α_1 according to π^{***}

From this view point, P_α can only depend on α's history of received tokens, H_α. The update function $Update(P_\alpha[\Delta], \Delta_i)$ for $P_\alpha[\Delta]$ defines the calculation of the probability vector of where to send Δ based on previously received token Δ_i in H_α. It will be explained in detail in next section.

Algorithm 1 shows the reasoning of agent α when it receives incoming tokens from its acquaintances via function $getToken(sender)$ (line 2). For each incoming token Δ, function $Acceptable(\alpha, \Delta)$ determines whether the token will be kept by α (line 4). When a resource is kept, its threshold is raised (line 6). If α decides to pass Δ, it will add itself to the path of Δ (Line 9) and $Update(P_\alpha[\Delta], \Delta_i)$ will update how to send Δ according to each previously received token Δ_i in α's history (line 11). If Δ is a resource or role token, its threshold will be decreased (line 14). Then α will choose the best acquaintance to pass the token to according to $P_\alpha[\Delta]$ (line 16) and record Δ in its history, H_α (line 18).

Algorithm 1: Decision process for agent α to pass incoming tokens

```
 1: while true do
 2:    Tokens(α) ← getToken(sender);
 3:    for all Δ ∈ Tokens(α) do
 4:       if Acceptable(α, Δ) then
 5:          if Δ.type == Res then
 6:             Increase(Δ.threshold);
 7:          end if
 8:       else
 9:          Append(self, Δ.path);
10:          for all Δᵢ ∈ Hα do
11:             Update(Pα[Δ], Δᵢ);
12:          end for
13:          if (Δ.type == Res)||(Δ.type == Role) then
```

14:            $Decrease(\Delta.threshold)$;
15:        **end if**
16:        $acquaintance \leftarrow Choose(P_\alpha[\Delta])$
17:        $Send(acquaintance, \Delta)$;
18:        $AddtoHistory(\Delta)$;
19:      **end if**
20:    **end for**
21: **end while**

5.2 Model Update Function

The effectiveness of the token-based approach depends on how well agents maintain their local models so that tokens are routed to where they lead to the highest gain in expected reward. In this section, we describe an algorithm to update the localized decision model by utilizing previously received tokens. The key is to make use of relationships between tokens, which we refer to as *relevance*.

Deciding where to send one token based on the receipt of another relies on knowing something about the relationship between the tokens. We quantify this relationship as the *Relevance* and define the relationship between tokens Δ_i and Δ_j as $Rel(\Delta_i, \Delta_j)$. $Rel(\Delta_i, \Delta_j) > 1$ indicates that an agent with use for Δ_i will often also have use for Δ_j, while $Rel(\Delta_i, \Delta_j) < 1$ indicates that an agent, which has use for Δ_i is unlikely to have use for Δ_j. If $Rel(\Delta_i, \Delta_j) = 1$ then nothing can be inferred. Details about how *relevance* is computed to ensure appropriate behavior will be explained in the next section. The update function of $P_\alpha[\Delta_j]$ according to H_α, written as $Update(P_\alpha[\Delta_j], \Delta_i)$ where $\Delta_i \in H_\alpha$ is found by using Bayes' Rule as follows:

$$\forall b \in n(\alpha), \ \forall \Delta_i \in H_\alpha, \ d = first(n(a), \Delta_i.path)$$

$$Update(P_\alpha[\Delta_j, b], \Delta_i) = \begin{cases} P_\alpha[\Delta_j, b] \times Rel(\Delta_i, \Delta_j) & \text{if } \Delta_i \neq \Delta_j, b = d \\ P_\alpha[\Delta_j, b] & \text{if } \Delta_i \neq \Delta_j, b \neq d \\ P_\alpha[\Delta_j, b] \times \varepsilon & \text{if } \Delta_i = \Delta_j, b \in \Delta_j.path \cap n(\alpha) \end{cases}$$

where $Update(P_\alpha[\Delta_j, b], \Delta_i)$ is to update the $P_\alpha[\Delta_j, b]$ in $P_\alpha[\Delta_j]$ according to Δ_i and $first(n(a), \Delta_i.path)$ extracts from the recorded path of the token the acquaintance of agent α that had the token Δ_i earliest. The first case in this function is the most important. The probability that the sender of previous token Δ_i is the best agent to receive the token Δ_j is updated according to $Rel(\Delta_i, \Delta_j)$. The second case in the equation changes the probability of sending that token to agents other than the sender in a way that ensures the subsequent normalization has the desired effect. Finally, the third case encodes the idea that α should typically not pass a token back from where it came. $P_\alpha[\Delta_j]$ is subsequently normalized to ensure that $\sum_{b \in n(\alpha)} P_\alpha[\Delta_j, b] = 1$.

To see how the updating function works, consider the following example. Supposed agent α has five acquaintances $\{a, b, c, d, e\}$ and $P_\alpha[\Delta_j] = $

$[0.1, 0.4, 0.2, 0.2, 0.1]$. Moreover, $H_\alpha = \{\Delta_i, \Delta_k\}$, $rel(\Delta_i, \Delta_j) = 1.2$ and $rel(\Delta_k, \Delta_j) = 0.4$. $\Delta_i.path = \{b, ..\}$; $\Delta_k.path = \{c, ..\}$; $\Delta_j.path = \{e, ..\}$. If currently α holds Δ_j, by applying our updating function to $P_\alpha[\Delta_j]$, we get the result as $P_\alpha[\Delta_j] = [0.12, 0.56, 0.09, 0.23, \varepsilon]$ and Δ_j will be most likely passed to acquaintance b.

5.3 Token Similarity

When an agent receives two tokens that are relevant to one another they are more likely to be usable in concert to obtain a reward for the team. While infeasibly complex, the POMDP model can suggest how relevance should be defined. If the local policy π^{***} always matches with π^{**}, $P_\alpha[\Delta]$ will be the normalization of $EU(\Omega_\alpha)$.

$$\forall b \in n(\alpha), \quad P_\alpha[\Delta, b] = \frac{EU(\Omega_\alpha, b)}{\displaystyle\sum_{c \in N(\alpha)} EU(\Omega_\alpha, c)}$$

That is, the largest expected utility for sending a token to an acquaintance should result the highest probability. Following the previous example, if $EU(\Omega_\alpha) = [5, 10, 6, 4]$, then for optimal behavior $P_\alpha[\Delta] = [0.2, 0.4, 0.24, 0.16]$.

Now we are in a position to see how the receipt of a token affects the locally optimal policy for routing token Δ and hence determine how to compute relevance. Suppose that the state estimation of agent α just before a token Δ_{pre} arrives is Ω_α while after it arrives, the state estimation is changed to Ω'_α because α gains additional knowledge from this token. Thus, according to Q-MDP, before Δ_{pre} is received, the expected reward of α is $EU(\Omega_\alpha) = \Omega_\alpha \times V$ while after the arrival of Δ_{pre}, $EU(\Omega'_\alpha) = \Omega'_\alpha \times V$. Moreover, agent α's local model will also be updated according to Δ_{pre}. Suppose Δ_{pre} comes from acquaintance b and $P_\alpha[\Delta, b]$ is the probability that α will send Δ to b before Δ_{pre} comes while $P'_\alpha[\Delta, b]$ is the updated probability after the arrival of Δ_{pre}. According to our assumption that policy π^{***} (according to the P_α model) and π^{**} (according to the POMDP model) will choose the same action which is to send Δ to b. Thus, we have

$$\frac{P'_\alpha[\Delta, b]}{P_\alpha[\Delta, b]} = \frac{EU(\Omega'_\alpha, b)}{EU(\Omega_\alpha, b)} = \frac{[\Omega'_\alpha \times V]_b}{[\Omega_\alpha \times V]_b}$$

Where $[\Omega'_\alpha \times V]_b$ is the value of the component in vector of $[\Omega'_\alpha \times V]$ according to acquaintance b. It is the same vector as $EU(\Omega'_\alpha, b)$.

According to the update function $Update(P_\alpha[\Delta, b], \Delta_{pre})$, we should get $P'_\alpha[\Delta, b] = Rel(\Delta, \Delta_{pre}) \times P_\alpha[\Delta, b]$. Thus, the relationship between Rel and our POMDP model is:

$$Rel(\Delta, \Delta_{pre}) = \frac{[\Omega'_\alpha \times V]_b}{[\Omega_\alpha \times V]_b}$$

From this equation, we can conclude that a received token changes the agent's estimation of the probability distribution of the team's state, which in turn directly influences the decision of where to send related tokens. If we know a little bit of how the probability distribution changes for an agent after it has passed a token, we can use this to predict how this agent updates its distributed decision model and therefore define the relevance between tokens. A heuristic that captures this relationship will approximate the locally optimal policy and hence lead to good behavior. In this chapter, we simply estimate this value based on the similarity between tokens. Intuitively, if two tokens are similar, receiving one token allows an agent to update its estimation of the team state and infer where to pass the similar tokens. For example, receiving a role token from a particular acquaintance, tells the agent that it is relatively less likely that similar role tokens will be accepted in the part of the network accessible via that acquaintance; receiving an information token with information about Pittsburgh tells the agent that some agents in that part of the network must currently be in Pittsburgh.

The similarities between tokens come from the *coordination* they carry and the calculation depends on the domain knowledge of applications. We assume that, from $\Delta_i.coordination$ and $\Delta_j.coordination$, we can deduce the similarity between two tokens as $sim(\Delta_i, \Delta_j)$. $sim(\Delta_i, \Delta_j) > 1$ if Δ_i and Δ_j are a pair of similar tokens. For example, if two tokens both reference Pittsburgh, we deem them similar because both are involved with the same location; we deem two tokens which require driving a specific machine as similar because they need the same kind of capacity; two tokens that are both preconditions of the same plan would also be considered similar.

We distinguish the relationship between relevance and similarity of two tokens as positively related or negatively related. For two similar tokens Δ_i and Δ_j, if an agent previously received a token from an acquaintance and would prefer to send a similar token to that acquaintance, similar tokens are positively related to each other and $Rel(\Delta_i, \Delta_j) = sim(\Delta_i, \Delta_j)$. Otherwise, if this agent is less likely to send the similar token to that acquaintance similar tokens are negatively related to each other, so $Rel(\Delta_i, \Delta_j) = \frac{1}{sim(\Delta_i, \Delta_j)}$

The similarity between different types of tokens potentially influences agents' estimation in different ways. As we have shown in the previous example, receipt of role tokens discourages sending similar tokens to agents along the role tokens' paths because the previous token senders refused the role token and are incapable of accepting the role, therefore are less likely to be interested in the information, tasks or resources that similar tokens carry. Thus, a previous role token is negatively related to its similar tokens. Similarly, receipt of an information token will indicate that agents along the information tokens' paths are more likely to work on things related to that information and are interested in other similar tokens. Hence, a previous information token is positively related to its similar tokens.

If the *threshold* of a resource token Δ_i is greater than its initial value (*init*) upon arrival to current agent, this means that the resource has been used by the agents previously holding Δ_i and that those agents are potentially engaged in tasks requiring the resource. Therefore, if the current agent gets similar tokens, it will be more likely to send them to the part of the network where the previous token has been passed. In this case, the previous resource token is positively related to similar tokens. Alternatively, if $\Delta_i.threshold$ is lower than its initial value ($init$), it means that agents passing the token did not need it. In such a case, the previous resource token is negatively related to similar tokens.

Supposing Δ_i is a previously received token, we can summarize the calculation of $Rel(\Delta_i, \Delta_j)$ according to $sim(\Delta_i, \Delta_j)$. No matter what $\Delta_j.Type$ is, this function only depends on the type of previously incoming token:

$$Rel(\Delta_i, \Delta_j) = \begin{cases} sim(\Delta_i, \Delta_j) & \text{if } \Delta_i.Type = inf \\ sim(\Delta_i, \Delta_j) & \text{if } \Delta_i.Type = res, \ \Delta_i.threshold > init \\ \frac{1}{sim(\Delta_i, \Delta_j)} & \text{if } \Delta_i.Type = role \\ \frac{1}{sim(\Delta_i, \Delta_j)} & \text{if } \Delta_i.Type = res, \ \Delta_i.threshold < init \end{cases}$$

6 Evaluation

In this section, we describe an empirical evaluation of our approach. Tests were conducted using an abstract simulation called CoordSim [33] configured to simulate a group of 400 distributed UAVs searching a hostile area. The network topology was that of a small world network where each UAV had, on average, four acquaintances. Simulating automatic detection rates, 200 pieces of information were randomly sensed by UAVs and passed around the team. Fifty plans instances, each with four independent preconditions, were given to the team. After a plan was initiated, tokens for the four roles needed to realize the plan were circulated through the acquaintance network. To accept a role an agent must be close to the region the role requires and have access to resource tokens for airspace at the role allocation. Airspace over the hostile area was divided into fifty regions. Each of these regions was duplicated in three resource tokens allowing a maximum of three UAVs to simultaneously access that airspace. Each UAV needed to obtain the resource for the region related to its task before it be performed. If all four roles of a plan were successfully executed, a reward of 10 units was credited to the team. A maximum reward of 500 units (10 units x 50 plan instances) was possible. Results for each experiment are based on one hundred trials.

The first experiment investigated the algorithm's performance in enhancing overall team reward. Reward obtained was recorded for each tick of the simulation which corresponded to the time taken for a token to move from agent to agent. Five configurations of the algorithm were compared. In the first configuration, agents passed tokens randomly if they did not keep them. In the next three configurations, local reasoning model updating is applied to

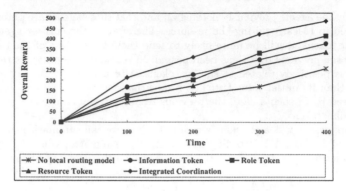

Fig. 4. The team gets more reward (y-axis) over time (x-axis) when all token types are used to update local models.

only one type of token, i.e., information, resource or role, with no updating by the other two types. The fifth configuration provided integrated coordination using tokens of each type to update agents' local model for routing tokens. The results are shown in Figure 4. Use of any previous tokens for token routing improved team reward. This benefit was most pronounced when all token types were used. In fact reward was almost doubled over random token movement. Notice that early on using plan instantiation alone was more effective than using other types of tokens but later role allocation tokens were the most effective type. We hypothesize that as roles are allocated they become most useful but before then it is critical to know who is initiating plans.

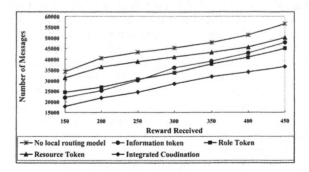

Fig. 5. The team needs to send less messages (y-axis) to coordinate when all token types are used to create local models.

The second experiment investigated the effect of our algorithm on communication. In this experiment we compared the number of messages needed for a team to gain a particular level of reward. A message was credited to each transfer of a token from an agent to its acquaintance. The same five configurations (random, three with coordination for single token-types, and integrated) were employed. As shown in Figure 5 configurations using token coordination algorithm performed better. They used fewer communications to attain the same level of reward as the random configuration. Once again, complete integrated token routing was superior to the partial token-based algorithm in attaining the same results with substantially fewer messages. Notice also that the total number of references is quite low, even for a large team.

The third experiment examined in more detail the scalability of our algorithm to larger teams. In this experiment teams of 200 to 800 agents were run under conditions otherwise identical to the first two experiments, however, only two configurations were used: no use of previous tokens to improve agents' local reasoning model versus use of all types of previous tokens to improve agents' local reasoning model to routing tokens. Performance was measured using *average number of messages per agent*. In Figure 11, the top configuration is to get a team reward of 200 units while the bottom one is to get a reward of 400. Our results show that integrated token-based routing produced lower message overload under all conditions. For both 200 and 400 reward levels observed message overhead was lower for teams of 800 agents using the integrated algorithm than for 200 agent teams using the random one.

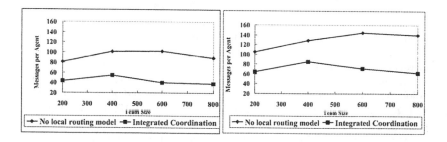

Fig. 6. The average number of messages per agent (y-axis) stays relatively constant as the team size is increased (x-axis). This holds when tokens are used for integrated model and when they are not.

7 Comparing with Market-based Coordination

In addition to testing the token-based approach itself, we compare the token-based coordination with the market-based coordination, a popular centralized algorithm. Although this algorithm is good at finding the optimal coordination solution, it ignores the communication costs, which are critical to large teams. The objective for this comparison is to verify how the token-based coordination approach makes the trade off between team utilities and communication cost.

Our implementation of market-based approach was based on TraderBots [8]. One agent acts as auctioneer and both tasks and resources are treated as merchandize. Agents bid for either single items or combinatorial sets of items in order to maximize their own utilities. The auctioneer maximizes its utility by "selling" their "merchandize". Because of the centralized position of the auctioneer, it develops a complete knowledge of how agents will use a task or resource if allocated and the auctioneer can perform assignments that maximize the team utility. Notice that several constraints also apply to this approach. The auction should last for a fixed period of time and early determination is infeasible; Agents are allowed to bid for resources after tasks have been allocated. Moreover, to prevent deadlock in resource allocation, agents are only allowed to bid for resources for their *first* pending task.

We designed a hybrid approach which is to combine the two different approaches. In the hybrid approach, the auctioneer algorithm runs exactly as before, except that instead of broadcasting announcements for auctions an *auction token* is created. Each auction token is allowed to exist from the start of the auction to the end of the auction. The auctioneer has a probabilistic model of the team state, just as all agents do in the token-based approach. The auction token is then intelligently routed to the agents most likely to be able to submit the best bids. The token stops moving after the auction it presents is closing or has visited a fixed number of teammates.

In those experiments, we only focused the comparison on task and resource allocation which can be performed by both token and auction. The basic experiment settings are configured as follows. There are 100 agents to perform 50 tasks with 50 resources. Each task requires only one resource which was interchangeable with four others. In the default setup, there is only one type of capability required and all agents have non-zero value for this capability, i.e., all agents are at least somewhat capable of all tasks. Auctions are held open for 40 time steps, and the task tokens and resource tokens are allowed to move unless accepted. The initial threshold on a task token is 100, meaning that the task will not be accepted by an agent until it can get a reward of more than 100 by performing this task. We measured two key statistics required to support or refute our hypothesis about the algorithms. "Reward" is the sum of reward received by each agent. "Messages" is the number of times agents communicated, either between themselves or with the auctioneer. The "messages" count indicates messages sent to perform sensor fusion, plan initiation and information sharing. Simulation runs for 2000 time steps. The experiment results below are based on 100 runs.

7.1 Heterogeneous Team

In the first experiment, we examined team performance by varying team composition and the capabilities required to perform tasks. For example, in an emergency response experiment some agents might only be able to fight fires while others could only provide medical treatment. As capabilities grew more varied fewer agents were available to perform particular tasks. In this experiment, we varied the number of capabilities from 1 to 46 where in the most heterogeneous condition, only two agents on average are capable of performing a task.

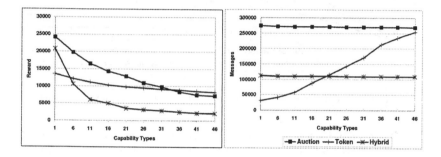

Fig. 7. The average reward for heterogeneous teams dramatically decreases in auction and hybrid approaches. Token-based approach maintains constant reward but requires more messages.

The experimental results in Figure 7 show that for heterogeneous teams, auction and hybrid approaches earn less reward as the team becomes more heterogeneous because there are fewer agents able to compete for the more specialized tasks. The advantages of team wide maximization of utility by the auctioneer decrease as there are progressively fewer feasible alternative bids. In contrast, reward for the token-based approach remain almost flat with increasing specialization. We propose two reasons. One is that token-based approach greedily finds a reasonable solution rather than searching for the optimal.

The other reason is that by passing a higher number of tokens around the network and making use of the relevance between them, the intelligent routing algorithm gains a better knowledge of how to route tokens. This results in a higher number of messages but an equal amount of reward.

7.2 Time Critical Tasks

In the second experiment, we investigated team performance when many tasks needed to be performed within a short period of time. To increase their reward, teams were required to perform tasks and allocate resources as rapidly as possible. In this study we varied the number of tasks the teams were required to finish from 20 to 182. After 2000 time steps, the accumulated reward and message count were recorded as shown in Figure 8.

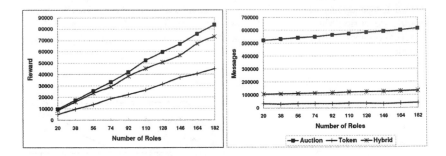

Fig. 8. Reward and messages increase dramatically with the number of tasks.

All three approaches performed more tasks in order to get higher reward. As expected, the auction approach attained higher reward than the hybrid or token-based approaches. Considering both reward and messages, however, the hybrid approach performs well by almost matching the reward obtained by the auction at just a quarter of the communication cost. The reason the hybrid approach achieves such good performance with so little communication overhead is that the intelligent routing algorithm limits communication to a small number of agents while high bidders must always be informed in auctions.

7.3 Competitive Resources

The third experiment used 200 tasks each requiring an average of four resources with no interchange possibilities. As available resources are increased from 4 to 40, competition for them declines and they become less likely to be a bottleneck.

The experiment was stopped after 1000 time steps. Figure 9 shows that the reward for the auction based approach increased rapidly with increases in resources. Both the token-based and hybrid approaches remained flat with the token-based approach earning the highest reward at all levels of scarcity while the hybrid approach yielded very limited reward.

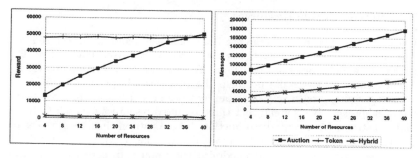

Fig. 9. Reward decreases with scarcity in auction approach and is very low in hybrid approach, but with token-based approach is uniformly high.

We hypothesized that because resource contention in this experiment was high the centralized control of the auction and hybrid approaches would often force agents to either bid for all four resources together or miss the task while the distributed token-based approach weakened this constraint. If our hypothesis were true, auction and hybrid approaches would get more reward if we either increased the simulation length or reduced the length of the auction to weaken the constraint. Figure 10 shows the effect of shortening auction length from 40 to 20 steps (a) and increasing session length from 2000 and 4000 steps (b). The token-based approach continues to produce its constant level of reward while the hybrid approach obtains very little higher. The auction-based approach, however, improves with increasing resources, exceeding the token-based approach at most levels in the two alternative experiments.

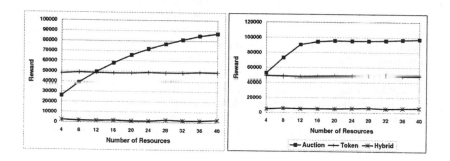

Fig. 10. Auction approach gets more reward when auction last length is 20 than when it is 40 (a) and when experiment lasts from 2000 time points to 4000 (b).

7.4 Interchangeable Resources

In the fourth experiment, there were 100 tasks each requiring three resources. The number of interchangeable resources was varied from one to five. Experiments were stopped at 1000 steps. Results are shown in Figure 11. Interchangeable resources did not help the token-based approach, helped the auction-based approach very little but substantially increased reward for the hybrid approach. We contend that three required resources for each task is a high constraint for a centralized auction. The constraint has been weakened in the market-based approach because this experiment lasts long enough for auctioneer to search bids and maximize the reward. In contrast, this constraint is higher in the hybrid approach because within a limited number of moves, all resource tokens for a role are required to visit the agent who has this task pending. This is also a reason why the hybrid approach gained such a low reward. Moreover, interchangeable resources led to dramatic increases in the number of messages for auction based approach because every agent could participate in every resource auction leading to the submission of a large number of multiple bids. For example, when interchangeable resources are 5, an agent should submit 5^3 resource bids.

7.5 Conclusion

The experiment results above support our hypothesis. Auctions are focused on maximizing overall utility by taking into account the *bids* of all team members [2], but at the expense of higher communications. Token-algorithms are well focused on scalability and make the trade off between team performance and communication. Therefore, sometimes at the expense of overall utility. More subtly, the performance advantage of an auction should be most pronounced when small changes in allocations lead to big differences in performance, i.e., typically highly constrained cases, while the token algorithms should maximize their communication advantage when the probabilistic models they rely on are most advantageous, i.e., weakly constrained cases.

The hybrid algorithm only performs well under restricted circumstances. If the problem is so tightly constrained that the auctions need to see many bids to make good allocations then using tokens to solicit bids only adds overhead. Conversely, if the coordination is so underconstrained that the tokens can reliably and accurately target the best agents, then the auction only adds unnecessary overhead.

8 Related Work

Multi-agent coordination is an extensively studied area of multi-agent systems, but most of the existing work does not scale well to very large teams. STEAM combined the joint intention model and the shared plan model and developed

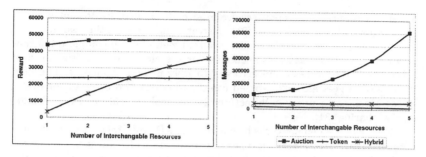

Fig. 11. Reward for hybrid approach increases rapidly with more interchangeable resources.

a general model of teamwork for persistent coordination in an uncertain, complex, and dynamic environment [28]. Unfortunately, its coordination requires a precise model of each individual agent. Distributed constraint-based algorithms [17, 20] have high communication requirements that get dramatically worse as the team size is increased. Swarm-inspired approaches [6] have been used for large-scale coordination, but the behavior can be inefficient.

Centralized coordination approaches typically make improper assumptions for scalable team coordination: complete communication with central agent, e.g., message board [5], or central information agents which have complete knowledge of the team [7]. Combinatorial auctions [11] have an exponential number of possible combinations of bids, and frequently use centralized auctioneers that can become severe bottlenecks. Generalized Partial Global Planning (GPGP) and its associated hierarchical task network representation TAEMS (Framework for Task Analysis, Environment Modelling, and Simulation) are focused on optimization to maximize the overall team utility. Unfortunately, this approach requires a centralized reasoning framework which is computationally hard to coordinate a moderate or scalable team [14].

Decision theoretic approaches, such as MDPs and POMDPs, have been used for team coordination, but the MDP model requires agents to have a complete view of team states. Decentralized POMDP model is more realistic for handling uncertainty in scalable teamwork; however Bernstein provided the mathematical evidence that decentralized decisions to find the optimal joint policy is NEXT-complete [4]. Xuan and Pynadath [10, 22, 34] recast the decentralized POMDP as a communication problem [22] by considering that sending messages to teammates with a belief that they did not observe, the team can get higher reward. But such approaches are known to be intractable in general [22], and so are of limited use in large scale applications.

The token-based approach was first introduced from networking design [1]. This idea was introduced to multi-agent coordination research by Wagner [27]. He renamed tokens as "keys" and applied them to the coordination of dynamic readiness and repair service in aircraft simulation. Recent work focusing on scalable coordination [21] illustrates that exponential search spaces, excessive communication demands, localized views, and incomplete information of agents pose major problems for large scale systems. Initial work on token-based approaches promises a way to address these challenges. The effectiveness of large-scale, token-based coordination has also been demonstrated in the Machinetta proxy architecture [25] for task allocation [24] and information sharing [31].

Research on social networks began in physics [1, 16]. [18] showed that social network structures in team formation can dramatically affect team abilities to complete cooperative tasks. In particular, using scale-free network structures for agent teams facilitates team formation by balancing the number of skill-constrained paths available in the agent organization with the effects of potential blocking. [12] compared the merits of small world networks and scale-free networks in the application of emergent coordination.

9 Conclusion and Future Work

This chapter presented a novel integrated token-based algorithm for team coordination. By utilizing relationships between tokens we were able to use the execution of one type of coordination algorithm to improve the performance of the others. Our experiments show that our approach is scalable and efficient. This approach opens the possibility to develop a range of new executing applications of heterogeneous agents not possible with existing approaches. By comparing with the market-based approach, we showed that token-based approach makes a good trade off between quality of allocation and use of communication. In some specific circumstances, its performance may outperform the auction which tries to find the optimal solution.

While the results presented here represent a step forward, they also point to significant challenges and exciting questions. We plan to address some of these issues in the near future. From a scientific perspective, the effect of the underlying acquaintance network on the coordination is clearly important but poorly understood. We will investigate the effect of the properties of the network on the team's performance, specifically looking at how the small world nature of the network is important. This work represented a novel attempt at integrating coordination algorithms into a unified approach and showing how by working together the overall performance can be improved. However, the individual algorithms were designed without thought to future integration. A key question is whether knowing that we will be integrating the token algorithms allows us to build algorithms that work better with other algorithms. Finally, but critically, we will use the token-based approach in more realistic domains to understand its utility in the real world. Specifically, we are currently developing large scale coordination applications for rescue response and unmanned aerial vehicle applications.

Acknowledgements

This research has been sponsored in part by AFRL/MNK Grant F08630-03-1-0005 and AFOSR Grant F49620-01-1-0542.

References

1. Token ring access method. In *IEEE. 802.5*.
2. N. Kalra, B. Dias, R. Zlot and A. Stentz, Market-based multirobot coordination: A survey and analysis. Technical report, CMU-RI-TR-05-13, Robotics Institute, Carnegie Mellon University, 2005.
3. A. Barabasi and E. Bonabeau, Scale free networks, *Scientific American*, pp. 60–69, May 2003.
4. D. S. Bernstein, S. Zilberstein and N. Immerman, The complexity of decentralized control of markov decision processes, in *Proceedings of the Sixteenth Conference on Uncertainty in Artificial Intelligence (UAI-00)*, pp. 32–37, 2000.
5. M. H. Burstein and D. E. Diller, A framework for dynamic information flow in mixed-initiative human/agent organizations, *Applied Intelligence on Agents and Process Management*, 2004.
6. V. A. Cicirello and S. F. Smith, Wasp nests for self-configurable factories, in *Proceedings of the Fifth International Conference on Autonomous Agents*, 2001.
7. K. Decker, K. Sycara, A. Pannu and M. Williamson, Designing behaviors for information agents, in *Procs. of the First International Conference on Autonomous Agents*, 1997.
8. B. Dias and A. Stentz, Traderbots: A market-based approach for resource, role, and task allocation in multirobot coordination, Technical report, CMU-RI-TR-03-19, Robotics Institute, Carnegie Mellon University, 2003.
9. D. Goldberg, V. Cicirello, M. B. Dias, R. Simmons, S. Smith and A. Stentz, Market-based multi-robot planning in a distributed layered architecture, in *Multi-Robot Systems: From Swarms to Intelligent Automata: Proceedings from the 2003 International Workshop on Multi-Robot Systems*, vol. 2, pp. 27–38, Kluwer Academic Publishers, 2003.
10. C. Goldman and S. Zilberstein, Optimizing information exchange in cooperative multi-agent systems, in *Proceedings of the Second International Joint Conference on Autonomous Agents and Multiagent Systems*, 2003.
11. L. Hunsberger and B. Grosz, A combinatorial auction for collaborative planning, In *Proceedings of the Fourth International Conference on Multi-Agent Systems*, pp. 151–158, 2000.
12. J. Pujol J. Delgado and R. Sanguesa, Emergence of coordination in scale-free networks, *Web Intelligence and Agent Systems*, pp. 131–138.
13. H. Kitano, S. Tadokoro, I. Noda, H. Matsubara, T. Takahashi, A. Shinjoh and S. Shimada, Robocup rescue: Search and rescue in large-scale disasters as a domain for autonomous agents research, in *Proc. 1999 IEEE Intl. Conf. on Systems, Man and Cybernetics*, vol. VI, pp. 739–743, October 1999.
14. V. Lesser, K. Decker, T. Wagner, N. Carver, A. Garvey, B. Horling, D. Neiman, R. Podorozhny, M. NagendraPrasad, A. Raja., R. Vincent, P. Xuan and X. Q. Zhang, Evolution of the gpgp/taems domain-independent coordination framework, *Autonomous Agents and Multi-Agent Systems*, 9(1), 2004.

15. M. Littman, A. Cassandra and L. Kaelbling, Learning policies for partially observable environments: scaling up, in *International Conference on Machine Learning*, 1995.
16. M. L. Littman, Probabilistic propositional planning: Representations and complexity, in *Proceedings of the 14th National Conference on Artificial Intelligence (AAAI-97)*, pp. 748–761, 1997.
17. R. Mailler and V. Lesser, Solving distributed constraint optimization problems using cooperative mediation, in *AAMAS'04*, 2004.
18. J. Simmons, M. E. Gaston and M. desJardins, Agent-organized networks for dynamic team formation, in *2005 International Conference on Autonomous Agents and Multi-Agent Systems*, 2005.
19. S. Milgram, The small world problem, in *Psychology Today*, 22, pp. 61–67, 1967.
20. P. J. Modi, W. Shen, M. Tambe and M. Yokoo, An asynchronous complete method for distributed constraint optimization, in *Proceedings of Autonomous Agents and Multi-Agent Systems*, 2003.
21. R. Vincent, P. Scerri and R. Mailler, Comparing three approaches to large scale coordination, in *Proceedings of AAMAS'04 Workshop on Challenges in the Coordination of Large Scale MultiAgent Systems*, 2004.
22. D. Pynadath and M. Tambe, Multiagent teamwork: Analyzing the optimality and complexity of key theories and models, in *First International Joint Conference on Autonomous Agents and Multi-Agent Systems*, 2002.
23. T. Sandholm, Algorithm for optimal winner determination in combinatorial auctions, in *Artificial Intelligence*, 135, 2002.
24. P. Scerri, A. Farinelli, S. Okamoto, and M. Tambe, Allocating tasks in extreme teams, in *Proceedings of Fourth International Joint Conference on Autonumous Agents and Multiagent Systems*, 2005.
25. P. Scerri, D. V. Pynadath, L. Johnson, P. Rosenbloom, N. Schurr, M Si and M. Tambe, A prototype infrastructure for distributed robot-agent-person teams, in *The Second International Joint Conference on Autonomous Agents and Multiagent Systems*, 2003.
26. P. Scerri, Yang Xu, E. Liao, J. Lai and K. Sycara, Scaling teamwork to very large teams, in *Proceedings of AAMAS'04*, 2004.
27. V. Guralnik T. Wagner and J. Phelps, A key-based coordination algorithm for dynamic readiness and repair service coordination, in *Proceedings of Second International Joint Conference on Autonomous Agents and Multiagent Systems*, 2003.
28. M. Tambe, Towards flexible teamwork, *Journal of Artificial Intelligence Research*, 7, pp. 83–124, 1997.
29. A. Vick, R. M. Moore, B. R. Pirnie, and J. Stillion, *Aerospace Operations Against Elusive Ground Targets*, RAND Documents, 2001.
30. D. Watts and S. Strogatz, Collective dynamics of small world networks, *Nature*, 393, pp. 440–442, 1998.
31. Y. Xu, M. Lewis, K. Sycara and P. Scerri, Information sharing in very large teams, in *In AAMAS'04 Workshop on Challenges in Coordination of Large Scale MultiAgent Systems*, 2004.
32. Y. Xu, P. Scerri, B. Yu, S. Okamoto, K. Sycara and M. Lewis, An integrated token-based algorithm for scalable coordination, in *Proceedings of Fourth International Joint Conference on Autonumous Agents and Multiagent Systems*, 2005.

33. Y. Xu, P. Scerri, M. Lewis and K. Sycara, Comparing Market and Token-Based Coordination, in *Proceedings of Fifth International Joint Conference on Autonomous Agents and Multiagent Systems*, 2006.
34. P. Xuan, V. Lesser and S. Zilberstein, Communication decisions in multi-agent cooperation: Model and experiments, in *Proceedings of the Fifth International Conference on Autonomous Agents*, 2001.

2 Consensus Controllability for Coordinated Multi-Agent Systems

Zhipu Jin and Richard M. Murray

Summary. A coordinated multi-agent system is consensus controllable if we can control all agents to reach an arbitrary consensus state. We consider consensus controllability of a homogenous coordinated multi-agent system by minimizing the number of agents that have access to the final consensus value. A directed graph is used to describe the topology of the interaction among agents and a reduction process based on the concept of strong component is proposed. The necessary and sufficient conditions on consensus controllability over general directed graphs are presented. Moreover, simulation results are presented to verify our conclusions.

1 Introduction

The consensus problem is currently used to investigate collective behaviors of coordinated multi-agent systems in control community. It has many applications for formation control. For example, it can by used in multiple UAV systems, automatic highway systems, satellite cluster formations, etc. It can also be useful in non-formation cooperative control problems, such as cooperative task assignment, air traffic control, and congestion control of internet. Moreover, benefit can be obtained from current consensus theory to understand animal aggregation behaviors and design protocols for sensor networks.

In the consensus problem, the pattern of information exchanging is a central topic. Vicsek et al. [1] proposed a simple but popular model for coordinated multiple agents in which each agent updates its headings based on the average of its own heading and its neighbors. Using simulation results, they showed that all agents move in the same direction eventually. Based on that, a couple of popular discrete time consensus protocols were presented [2, 3]. On the other hand, continuous time consensus protocols were also developed and analyzed [2, 4, 5, 6]. Conditions for consensus seeking under general interactive topology or switching topology were studied as well [3, 6]. For the consensus convergence speed, Xiao and Boyd [7] treated a consensus process as an optimal linear iteration problem and increased the convergence speed by finding the optimal weights associated with each edge. Olfati-Saber [3] presented a "random rewiring" procedure to boost the convergence speed for large scale graphs. Jin and Murray [8] proposed a multi-hop relay consensus protocol to increase the speed without physically changing the topology.

The consensus problem can be treated as a synchronization problem for coupled dynamical systems when agents' dynamics are considered. Wu and Chua [9] employed Lyapunov's direct method to discuss the stability issue. Also, sufficient conditions for multiple dynamical systems synchronization over general connected, directed graphs were discussed recently [10]. Fax and Murray [4] used Laplacian matrix decomposition method to find the necessary and sufficient conditions for the stability of coordinated multi-agent systems.

However, it is not clear how to determine the final agent state value when a consensus is achieved, especially when the interaction topology is directed and weakly connected. Moreover, no results were presented so far to discuss how to control that value. In this chapter, we introduce a graph reduction process and the concept of consensus controllability so that we can describe the final value of each agent explicitly. The necessary and sufficient conditions are given in order to control a multi-agent system convergence to an arbitrary consensus value asymptotically.

The remainder of this chapter is organized as follows. In Section 2, the concept of consensus controllability is proposed and the graph reduction process is presented. Also, the conditions on consensus controllability with simple agent dynamics are discussed. We extend this result to agents with general LTI dynamics in Section 3. Examples and simulation results are provided in Section 4 and conclusions are listed in Section 5.

2 Consensus Controllability with Simple Agent Dynamics

For a coordinated homogenous multi-agent system, each agent can exchange information with its direct neighbors. Given the topology of the information flow, we investigate how we can control the whole group to reach a certain consensus state value without delivering this global information to all agents. In this section, we assume that the dynamics of each agent is represented by a simple first-order integrator.

2.1 Formulation of Consensus Controllability

We use a directed graph $\mathcal{G} = (\mathcal{V}, \mathcal{E})$ to represent the interaction topology where $\mathcal{V}$ is a set of vertices and $\mathcal{E} \subseteq \mathcal{V}^2$ is a set of edges. Each agent is depicted by one vertex v_i, which has a unique index number $i \in \mathcal{S} = \{1, 2, \cdots, n\}$, and n is the number of agents, which is also called the *order* of the graph. Each edge of the graph is denoted by (v_i, v_j) with $i, j \in \mathcal{S}$ when agent v_i has access to the states of agent v_j. For any edge (v_i, v_j), we call v_i the *head* and v_j the *tail*. The directed graph $\mathcal{G}$ is called *symmetric* if, whenever $(v_i, v_j) \in \mathcal{E}$, then $(v_j, v_i) \in \mathcal{E}$ also. In some literatures, a symmetric graph is also called *undirected*.

In consensus problems, the directions of edges are important. In a directed graph, the number of edges with the same head v_i is called *out-degree* of node v_i; the number of edges with the same tail v_i is called *in-degree* of node v_i. If edge $(v_i, v_j) \in \mathcal{E}$, then v_i is one of the *parent vertices* of v_j. The set of neighbors of vertex v_i is denoted by $N(v_i) = \{v_j \in \mathcal{V} : (v_i, v_j) \in \mathcal{E}\}$.

A *strong path* in a directed graph is a vertex sequence $[u_0, \cdots, u_r]$ so that $(u_{i-1}, u_i) \in \mathcal{E}$ for $i = \{1, \ldots, r\}$. A *weak path* is a sequence $[u_0, \cdots, u_r]$ of distinct vertices so that either (u_{i-1}, u_i) or (u_i, u_{i-1}) belongs to $\mathcal{E}$.

A directed graph is *weakly connected* if any two vertices in the graph can be joined by a weak path, and is *strongly connected* if any two vertices can be joined by a strong path. If a strongly connected directed graph is symmetric, then it is *connected and symmetric*. For a vertices set, Figure 1 reveals the relationships between those concepts.

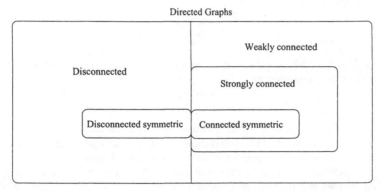

Fig. 1. Classification for directed graphs.

An *adjacency* matrix $\mathcal{A} = \{a_{ij}\}$ of a directed graph $\mathcal{G}$ with order n is a $n \times n$ matrix defined as

$$a_{ij} = \begin{cases} 1, & (v_i, v_j) \in \mathcal{E} \\ 0, & \text{otherwise.} \end{cases}$$

More generally, a *weighted adjacency* matrix $\mathcal{A}$ of a *weighted directed graph* $\mathcal{G}$ is defined as

$$a_{ij} = \begin{cases} w_{ij}, & (v_i, v_j) \in \mathcal{E} \\ 0, & \text{otherwise} \end{cases}$$

where w_{ij} is the weight associated with edge (v_i, v_j). For a weighted directed graph, the out-degree of node v_i is the sum of the weights of the edges whose head is v_i and the in-degree of node v_i is the sum of the weights of the edges whose tail is v_i. Let D be the diagonal matrix with the out-degree of each vertex along the diagonal, then the *Laplacian* matrix L is defined by

$$L = D - \mathcal{A}$$

From now on, we assume that all the weights are 1 unless specifically noted.

Let x_i denote the state of agent i which is represented by node v_i in the graph. The dynamics of each agent is represented by

$$\dot{x}_i = u_i. \tag{1}$$

For some applications, it is necessary to add offsets between different agent states to achieve desired inter-agent spacing. We assume those state offsets are constant and then equation (1) represents the error dynamics. A multi-agent system reaches a *consensus* if $x_i = x_j$ for any $i, j \in S$. This common value is called the *consensus state*, which is depicted by η.

For any agent $i \in S$, we use a uniform distributed control protocol to describe the interactive effect

$$u_i = - \sum_{j \in N(v_i)} (x_i - x_j). \tag{2}$$

where $N(v_i)$ is the neighbor set of agent i. Thus, the entire system with n agents can be presented by

$$\dot{X} = -LX \tag{3}$$

where $X = [x_1, \cdots, x_n]'$.

A multi-agent system is said to be *consensus controllable* if and only if it is possible, by means of distributed inputs, to make each agent asymptotically converge to arbitrary consensus state from any initial state.

This definition means that not only can the system reach a consensus state η, but also we can control the value of η, which is the global information. Obviously, if every single agent has direct access to η, consensus controllability is reduced to controllability of each single agent. In this chapter, we are interested in the consensus controllability with the distributed control protocol (2) when only a limited number (smaller than n) of agents can access the global information.

2.2 Consensus Controllability over Strongly Connected Graphs

Suppose the directed graph $\mathcal{G}$ is strongly connected. It has been shown [5] that rank$(L) = n - 1$ and $\lambda = 0$ is a simple eigenvalue. The right eigenvector associated with $\lambda = 0$ is $\mathbf{1}_n = [1, \cdots, 1]'$. We call a vector *positive* if every element in the vector is positive. Then we have the following Lemma:

Lemma 1. *Let $\mathcal{G}$ be a strongly connected directed graph and L be the Laplacian matrix. There exists a positive left eigenvector $\mathbf{b} = [b_1, \cdots, b_n]$ associated with $\lambda = 0$.*

Proof. Since 0 is a simple eigenvalue of L, then there exists a left eigenvector b so that $b \cdot L = 0$. Since $-b$ is also a left eigenvector, we actually need to prove that all elements in b are non-zero and have same signs. Table 1 lists four possible types of b and we show that only type 1 is possible.

Table 1. Possible left eigenvector associated with $\lambda = 0$.

	No zero elements	Has zero elements
Uniform signs	type 1	type 2
Non-uniform signs	type 3	type 4

First, let us prove type 2 is impossible. Assume that there is only one zero element in b. Without losing generality, we assume that $b_n = 0$ and let l_n denote the last column of L. According to the definition of L, the non-diagonal elements of L are either -1 or 0. So the first $n - 1$ elements of l_n must be zeros in order to satisfy the equation $b \cdot l_n = 0$. However, that means the in-degree of node v_n is zero, which violates the definition of strongly connected graphs. Then we assume that there are k zeros in b where $1 < k < n$. Using the similar argument, we can show that there exists a subgraph composed of k nodes where there are no edges connecting other nodes to this subgraph. That is also impossible for a strongly connected graph.

Second, we prove type 3 is impossible. Suppose there exist k negative elements and $n - k$ positive elements in b. After re-indexing vertices, we can move the positive elements in the front of b and get $b = [b_+, b_-]$ where b_+ is a positive vector with $n - k$ elements and b_1 is a negative vector with k elements. We have

$$vL = [b_+, b_-] \begin{bmatrix} L_{11} & L_{12} \\ L_{21} & L_{22} \end{bmatrix} = \begin{bmatrix} b_+ \cdot L_{11} & b_+ \cdot L_{12} \\ b_- \cdot L_{21} & b_- \cdot L_{22} \end{bmatrix}.$$

It is obvious that $b_- \cdot L_{21}$ is a vector with $n - k$ nonnegative elements and $b_+ \cdot L_{12}$ is a vector with k nonpositive elements. Because $\mathcal{G}$ is strongly connected, then the sum of all elements in $b_- \cdot L_{21}$ is positive and the sum of all elements in $b_+ \cdot L_{12}$ is negative. Moreover, the sum of all elements in $b_- \cdot L_{21}$ and $b_- \cdot L_{22}$ is zero since all rows of L are zero-sum. Then the sum of elements in $b_- \cdot L_{22}$ is also negative. However, since $b \cdot L = 0$, the sum of all elements in $b_+ \cdot L_{12}$ and $b_- \cdot L_{22}$ should be zero which is impossible. So b cannot be type 2.

Combining previous arguments, it is easy to show that b cannot be type 4 either. Thus, b must be type 1.

This Lemma is also true for any strongly connected weighted directed graph with positive weights.

For system (3), the state vector X can be reorganized as

$$Z = V^{-1}X = \begin{bmatrix} b_1 & [b_2, \cdots, b_n] \\ \mathbf{1}_{n-1} & -I_{n-1} \end{bmatrix} \cdot X = \begin{bmatrix} \sum b_i x_i \\ x_1 - x_2 \\ \vdots \\ x_1 - x_n \end{bmatrix}. \tag{4}$$

Without losing generality, we assume that $b_1 = 1$. Using the property of block matrices, we have

$$V = \begin{bmatrix} (\sum b_i)^{-1} & [b_2, \cdots, b_n](\mathbf{1}_{n-1}[b_2, \cdots, b_n] + I_{n-1})^{-1} \\ (\sum b_i)^{-1}\mathbf{1}_{n-1} & -(\mathbf{1}_{n-1}[b_2, \cdots, b_n] + I_{n-1})^{-1} \end{bmatrix}$$

$$= (\sum v_i)^{-1} \begin{bmatrix} 1 & b_2 & \cdots & b_n \\ 1 & (b_2 - \sum b_i) & \cdots & b_n \\ \vdots & \vdots & \vdots & \vdots \\ 1 & b_2 & \cdots & (b_n - \sum b_i) \end{bmatrix}. \tag{5}$$

It is easy to show that V^{-1} is a full rank matrix and

$$V^{-1}LV = \begin{bmatrix} 0 & \mathbf{0} \\ \mathbf{0} & \hat{L} \end{bmatrix} \tag{6}$$

where $\hat{L}$ is a full-rank $(n-1) \times (n-1)$ matrix and all the eigenvalues of $\hat{L}$ are also eigenvalues of L. Thus, the system converges a consensus state η since all eigenvalues of $\hat{L}$ have positive real parts. Because $\dot{z}_1 = \sum b_i \dot{x}_i = 0$ and $\lim_{t \to \infty} x_i(t) = \eta$ for all i, we have $\sum b_i \eta = \sum b_i x_i(0)$. Thus, the consensus state

$$\eta = \sum b_i x_i(0) / \sum b_i. \tag{7}$$

When the directed graph is balanced, i.e., for any vertex, the in-degree equals to the out-degree, $\boldsymbol{b} = \mathbf{1}_n$ and $\eta = \sum x_i(0)/n$.

According to the analysis above, it is clear that the value of η is determined by the initial values of agent states and the left eigenvalue v. For a strongly connected digraph $\mathcal{G}$, the contribution of each initial state to η is indicated by v_i.

Suppose we want to control the consensus state convergence to a certain value α, and only agent v_1 knows this global information. For agent v_1, we add a local feedback besides the consensus protocol:

$$\dot{x}_1 = k_1(x_1 - \alpha) - \sum_{j \in N(v_1)} (x_1 - x_j) \tag{8}$$

where $k_1 < 0$. Then the whole system is:

$$\dot{X} = -LX + k_1 PX - k_1 P \cdot \alpha \mathbf{1} \tag{9}$$

where $P = \text{diag}([1, 0, \cdots, 0])$. We need to show that the equilibrium point is changed to the point $\alpha \cdot \mathbf{1}_n$.

Since $V^{-1}(-L + k_1 P)V$ has the same eigenvalues as $-L + k_1 P$, we have

$$V^{-1}(-L + k_1 P)V = k_1 V^{-1} PV - V^{-1} LV$$

$$= \frac{k_1}{\sum b_i} \begin{bmatrix} b_1 & \cdots & b_n \\ \vdots & \vdots & \vdots \\ b_1 & \cdots & b_n \end{bmatrix} - \begin{bmatrix} 0 & \mathbf{0} \\ \mathbf{0} & \hat{L}_{22} \end{bmatrix}. \tag{10}$$

The rank of the first matrix is 1 and the rank of the second matrix is $n - 1$. However, $[b_1, \cdots, b_n]$ is linear independent of the row space of the second matrix. So $(-L + k_1 P)$ is full-rank. Moreover, according to Gersgorin disk theorem, all of the eigenvalues have negative real parts. Then the system converges to $\alpha \cdot \mathbf{1}_n$ asymptotically, i.e., the system is consensus controllability.

Theorem 1. *A multi-agent system with simple agent dynamics as $\dot{x}_i = u_i$ and distributed control protocol (2) over a strongly connected interaction topology is consensus controllable if and only if there exists at least one agent which has access to the global information.*

Proof. For the sufficient condition: if there exists one agent who knows the global information α, according to previous analysis, the additional local feedback $k_1(x_i - \alpha)$ can make the system consensus controllable. If there exist $l \leq n$ agents who knows α, then the matrix P in equation (9) has l ones and $n - l$ zeros along the diagonal and $-L + k_1 P$ is still full-rank and stable. So the system is consensus controllable.

For the necessary condition: if there does not exist any agent who has access to α, then the consensus state η is $\sum v_i x_i(0) / \sum v_i$ which is independent of α.

According to this theorem, we know that single local feedback on any single agent can completely overrule the consensus state. We can control the whole system by just control any single agent.

However, if there are $0 < l \leq n$ agents which have different reference values, we have

$$\dot{x}_i = k_1(x_i - \alpha_i) - \sum_{j \in N(v_i)} (x_i - x_j) \tag{11}$$

where $i \in [1, \cdots, l]$. The whole system is:

$$\dot{X} = -LX + k_1 PX - k_1 P \cdot \alpha \tag{12}$$

where

$$P = \begin{bmatrix} I_l & 0 \\ 0 & 0 \end{bmatrix} \tag{13}$$

and $\alpha = [\alpha_1, \cdots, \alpha_l, 0, \cdots, 0]$ after we re-assign the agent index properly. The equilibrium point is

$$X^* = k_1 \left(k_1 P - L\right)^{-1} P \cdot \begin{bmatrix} \alpha_1 \\ \vdots \\ \alpha_l \\ \mathbf{0} \end{bmatrix} \tag{14}$$

which is not a consensus vector normally. So in that case, the system can not reach a consensus, but it is still stable since $k_1 P - L$ is full-rank and stable.

2.3 Consensus Controllability over Weakly Connected Graph

When the directed graph $\mathcal{G}$ is weakly connected, the distributed control protocol (2) may not be able to make the system to reach a consensus. We introduce some new concepts in graph theory before we discuss the consensus controllability over weakly connected graphs.

A *rooted directed spanning tree* is a subgraph $\mathcal{G}_r = (\mathcal{V}, \mathcal{S})$ where $\mathcal{S}$ is a subset of $\mathcal{E}$ which connects, without any cycle, all vertices in $\mathcal{G}$ so that each vertex, except the root, has one and only one outcoming edge. (Some literature defines this concept in the opposite direction, i.e., each vertex, except the root, has one and only one incoming edge.) A *strong component* of a directed graph is an induced subgraph that is maximal, subject to being strongly connected. Since a vertex is strongly connected, it follows that each vertex lies in a strong component. Decomposing a directed graph into maximally strongly connected subgraphs is a standard problem in graph theory that can be solved in linear time by using depth-first search (DFS) or breath-first search (BFS). Edges between any two strong components, if any, are uniformly directional, i.e., these edges' heads belong to one component and tails belong to another. Otherwise, these two components compose one bigger strong component.

We can reduce a directed graph by replacing each strong component with a vertex. Edges inside each component are dismissed and edges between any two components are replaced by a single edge. Figure 2 shows an example of this reduction. Part (a) is the original directed graph; part (b) is one of the rooted directed spanning trees for part (a); part (c) indicates that there are three strong components in this graph and part (d) is the result of the reduction. Also part (d) is a root directed tree. If the graph $\mathcal{G}$ is strongly connected, then it can be reduced to a single vertex.

After finding the strong components, the Laplacian matrix L can be transferred to a block lower triangular matrix by a permutation matrix, i.e., there exists a permutation matrix Q so that

$$Q^T L Q = \begin{bmatrix} l_1 & & & \\ \vdots & \ddots & & \\ * & \cdots & l_{k-1} & \\ * & \cdots & * & l_k \end{bmatrix} \tag{15}$$

where k is the number of strong components. Obviously, L is reducible if the graph is weakly connected.

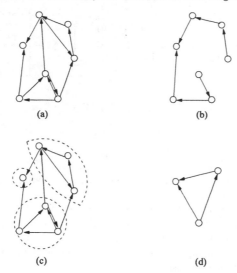

Fig. 2. Root directed spanning tree and reduction of a directed graph.

Lemma 2. *The distributed control protocol (2) can force the system to reach a consensus from any initial value if and only if the directed graph can be reduced to a root directed tree.*

Proof. According to the definition of the reduction process, a weakly connected directed graph $\mathcal{G}$ can be reduced to either a root directed tree or a connected directed *forest* with multiple roots. If a strong component can be reduced to a root, then it is called a *root strong component*.

When $\mathcal{G}$ can be reduced to a root directed tree $\mathcal{T}$ with k vertices, the Laplacian matrix L can be written as a block lower triangular matrix as in equation (15) where $l_1 = L_1$ is the Laplacian matrix of the root strong component.

Other diagonal block l_i can also be presented by $l_i = L_i + P_i$ where L_i is the Laplacian matrix of the corresponding strong component and P_i is a diagonal matrix whose diagonal elements are non-negative and at least one element is positive. According to the analysis in the previous subsection, any l_i for $i \in [2, \cdots, k]$ is full-rank and all the eigenvalues have positive real parts. So $rank(L) = n - 1$ and 0 is a simple eigenvalue.

Suppose the first strong component has n_1 vertices and $[b_1, \cdots, b_{n_1}]$ is a left eigenvector of L_1 which is associated with $\lambda = 0$, the states vector X can be reorganized as

$$Z = V^{-1}X = \begin{bmatrix} b_1 & [b_2, \cdots, b_{n_1}] & 0 \\ \mathbf{1}_{n_1-1} & -I_{n_1-1} & 0 \\ \mathbf{1}_{n-n_1} & 0 & -I_{n-n_1} \end{bmatrix} \cdot X = \begin{bmatrix} \sum_{i=1}^{n_1} b_i x_i \\ x_1 - x_2 \\ \vdots \\ x_1 - x_n \end{bmatrix} \tag{16}$$

and we have:

$$\dot{Z} = -V^{-1}LVZ \tag{17}$$

and

$$V^{-1}LV = \begin{bmatrix} 0 & 0 \\ 0 & \hat{L} \end{bmatrix}. \tag{18}$$

According to previous analysis, we know that $-\hat{L}$ is stable, i.e., the system converges to a consensus.

When $\mathcal{G}$ can be reduced to a connected directed *forest* with multiple roots, the Laplacian matrix L of $\mathcal{G}$ can also be written as a block lower triangular matrix. However, the diagonal blocks corresponding to those root strong components are totally decoupled from other diagonal blocks and they will reach to different "local" consensuses.

Lemma 3. *If the system with a weakly connected graph reaches a consensus, then the consensus state η is:*

$$\eta = \sum_{i=1}^{n_1} b_i x_i(0) / \sum_{i=1}^{n_1} b_i \tag{19}$$

where $[b_1, \cdots, b_{n_1}]$ is the left eigenvalue of l_1.

The lemma is easy to prove. Next, we give out the necessary and sufficient conditions on the consensus controllability over a weakly connected graph.

Theorem 2. *The multi-agent system with simple agent dynamics as $\dot{x}_i = u_i$ and distributed control protocol (2) over a weakly connected interaction topology, whose Laplacian matrix has rank $n - q$, is consensus controllable if and only if there exists at least one agent in each root strong component that has access to the global information.*

Proof. First of all, we need to show that the Laplacian matrix of a weakly connected interaction graph has rank $n - q$ if and only if it can be reduced to a connected directed forest with q roots. According to the definition of the Laplacian matrix and the reduction process, it is clear that this conclusion is true.

Then, we need to show that the whole system reaches a consensus if and only if the agents in the strong components which correspond to the q roots reach a consensus. Suppose the directed graph has u strong components and we know that the Laplacian matrix L can be described in the following form with proper agent indices.

$$L = \begin{bmatrix} l_1 & & & & & \\ & \ddots & & & & \\ & & l_q & & & \\ * & * & * & l_{q+1} & & \\ & & & & \ddots & \\ * & * & * & * & & \\ * & * & * & * & * & l_u \end{bmatrix}. \tag{20}$$

The first q diagonal blocks are Laplacian matrices of the q root strong components and each of them reaches "local" consensus states individually. The agent dynamics in other $u - q$ strong components evolve similarly to equation (11) with $k_1 = -1$. So a consensus can be reached only when the q root strong components reach the same consensus state.

Finally, we know that for each root strong component is consensus controllable if and only if at least one agent has access to the global information. So over a weakly connected graph with q root strong components, the system is consensus controllable if and only if each root strong component has at least one agent which has access to the global information.

This theorem gives the minimum number of agents who should have local feedbacks based on the global reference so that the system can reach a controllable global consensus. Also, how to distribute this global information is clearly stated in terms of the root strong components.

3 Consensus Controllability with LTI Agent Dynamics

In this section, we extend the agent dynamics from a single integrator to general LTI dynamics. Suppose that each agent is represented by

$$\dot{x}_i = Ax_i + Bu_i \tag{21}$$

where $x_i \in \mathcal{R}^m$ and (A, B) is controllable. The distributed control protocol is

$$u_i = K \cdot \sum_{j \in N(v_i)} (x_i - x_j) \tag{22}$$

where $A + BK$ is stable. Then the entire system with n agents can be presented by

$$\dot{X} = (I_n \otimes A + L \otimes BK) \cdot X$$
$$= (I_n \otimes A + (L \otimes B)(I_n \otimes K)) \cdot X \tag{23}$$

where $\otimes$ represents the Kronecker product.

The states vector of system (23) also can be reorganized as

$$Z = V^{-1} \otimes I_m \cdot X = \begin{bmatrix} b_1 & [b_2, \cdots, b_n] \\ 1 & -I_{n-1} \end{bmatrix} \otimes I_m \cdot X = \begin{bmatrix} \sum b_i x_i \\ x_1 - x_2 \\ \vdots \\ x_1 - x_n \end{bmatrix}. \tag{24}$$

The system (23) is changed as:

$$\dot{Z} = V^{-1} \otimes I_m \left(I_n \otimes A + L \otimes BK\right) V \otimes I_m \cdot Z$$
$$= \left(I_n \otimes A + (V^{-1}LV) \otimes BK\right) Z. \tag{25}$$

Remember that the similarity transformation of Laplacian matrix is:

$$V^{-1}LV = \begin{bmatrix} 0 & 0 \\ 0 & \hat{L} \end{bmatrix}. \tag{26}$$

Thus, we have

$$\dot{Z} = \left(I_n \otimes A + \begin{bmatrix} 0 & 0 \\ 0 & \hat{L} \end{bmatrix} \otimes BK\right) Z. \tag{27}$$

Lemma 4. *For the system (23) over a strongly connected interaction topology, the distributed control protocol (22) forces the system reach a consensus asymptotically if and only if K simultaneously stabilizes the following $n-1$ subsystems*

$$\dot{x} = Ax + \lambda_i Bu \tag{28}$$

where λ_i are the non-zero eigenvalues of L. Moreover, the consensus state η satisfies

$$\lim_{t \to \infty} \left(\eta(t) - \frac{\sum b_i x_i(0)}{\sum b_i} e^{-At}\right) = 0. \tag{29}$$

Proof. For a strongly connected directed graph, all the eigenvalues of $\hat{L}$ have positive real parts. Using the Schur theorem, $\hat{L}$ can be transferred to an upper triangular matrix with the non-zero eigenvalues of L along the diagonal. Then $[z_2, \cdots, z_n] = [x_1 - x_2, \cdots, x_1 - x_n]$ asymptotically converges to zero if and only if all subsystems (28) are stable, i.e., all agent states converge to a consensus. Also, since z_1 evolves according to $\dot{z}_1 = Az_1$, the result follows.

Because of the dynamics, the sum of agent states $z_1 = \sum b_i x_i$ is not a constant any more. Also, in order to reach a consensus, more constraints are needed on K.

Next, we want to discuss the controllability of the consensus state. Assume $P = \text{diag}([p_1, \cdots, p_n])$ where

$$p_i = \begin{cases} 1, & \text{agent } i \text{ can access the global reference } \alpha \\ 0, & \text{otherwise,} \end{cases}$$

the distributed feedback control protocol for error dynamics is changed to:

$$u_i = p_i K(x_i - \alpha) + K\left(\sum_{j \in N(v_i)} (x_i - x_j)\right). \tag{30}$$

Please note that we omit the feed forward inputs here. For the consensus controllability, we have the following theorem.

Theorem 3. *A multi-agent system with agent dynamics (21) and distributed control (30) over a strongly connected interaction topology is consensus controllable if and only if*

- *at least one agent has access to the reference;*
- *the local controller K simultaneously stabilizes the following n subsystems*

$$\dot{x} = Ax + \lambda_i Bu \tag{31}$$

where λ_i are the eigenvalues of $P + L$.

Proof. Let $e_i = x_i - \alpha$ and the error dynamics of the system is

$$\begin{aligned}\dot{E} &= (I_n \otimes A + (P + L) \otimes BK) \cdot E \\ &= (I_n \otimes A + ((P + L) \otimes B)(I_n \otimes K)) \cdot E\end{aligned} \tag{32}$$

where $E = [e_1, \cdots, e_n]'$. All eigenvalues of $P + L$ have positive real parts when $P \neq 0$. Then the errors converge to zero if and only if K satisfy the second condition.

For a weakly connected graph, we have the following results:

Theorem 4. *A multi-agent system with agent dynamics (21) and distributed control protocol (30) over a weakly connected interaction topology is consensus controllable if and only if*

- *in each root strong component, there exists at least one agent which has access to the global reference;*
- *the local controller K simultaneously stabilizes the following n subsystems*

$$\dot{x} = Ax + \lambda_i Bu \tag{33}$$

where λ_i are the eigenvalues of $P + L$.

Agents' dynamics makes the condition on consensus controllability more sophisticated. However, how to identify the minimum "informed" agents is still determined by the topology of the interactions.

4 Examples and Simulation Results

In this section, we present a couple of examples on consensus controllability. Assume a multi-agent system with six agents, Figure 3 describes two different interaction topologies. Suppose the dynamics of each agent are identical simple integrators and the initial state is $[1, 2, 3, 4, 5, 6]$. For the strongly connected interaction topology, the first plot in Figure 4 shows the system converges to a consensus $\eta = \sum b_i x_i(0) / \sum b_i = 3.7917$. The second plot in Figure 4 shows that, when only agent 2 has access to a global reference $\alpha = 2$, the system converges to a new consensus state $\eta = \alpha$.

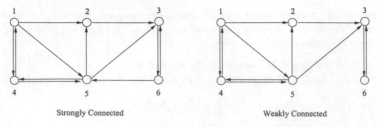

Strongly Connected   Weakly Connected

Fig. 3. Interaction topologies of the simulations.

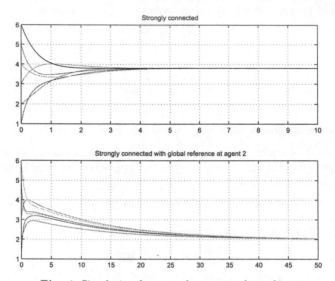

Fig. 4. Simulation for strongly connected topology.

For the weakly connected topology, the first plot in Figure 5 shows the system converges to a consensus $\eta = 4.5$ which equals the average value of initial states of agent 3 and 6 because these two agents compose the root strong component. The second plot in Figure 5 shows the states when only agent 2 has access to a global reference α. Since agent 2 is not in the root strong component, the system does not converge to the reference. However, when agent 3 knows the reference, the system converges to α shown in the third plot in Figure 5.

For a system with unstable agent dynamics, we choose $A = 1.6$, $B = 1$, and $K = -2$. Then, for the weakly connected topology, first plot in Figure 6 shows the system converges to a consensus, but the value of η diverges. When only agent 3 has access to a global reference α, in order to make the system stable, we have to choose $K < -1.6/0.382$ since the smallest eigenvalue of

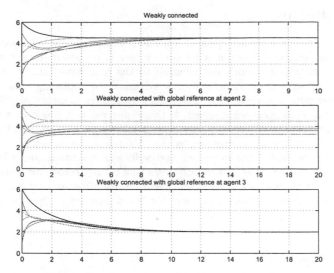

Fig. 5. Simulation for weakly connected topology.

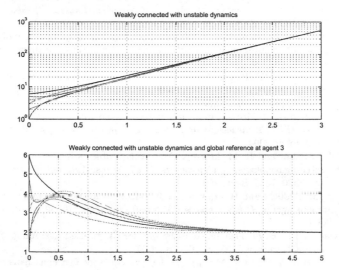

Fig. 6. Simulation for weakly connected topology with unstable agent dynamics.

$L + P$ is 0.382. Second plot in Figure 6 shows that the system converges to α when $K = -7$.

5 Conclusions

In this chapter, we have investigated consensus controllability for coordinated multi-agent systems. A graph reduction process is proposed so that we can clearly identify each agent's contribution to the final consensus state. This reduction process can reduce any directed graph by replacing each strong component with a singe vertex. Those root strong components play important roles in consensus achievement. If there only exists a single root strong component, a normal consensus protocol can achieve a consensus and the final consensus value is only determined by the root strong component.

We also presented the necessary and sufficient conditions on consensus controllability. The whole system converges to a certain consensus state value if and only if there exists at least one vertex in each root strong component that is directly controllable. This condition is independent of the dynamics of the agent.

References

1. T. Vicsek, A. Czirok, E. B. Jocob, I. Cohen and O. Schochet, Novel type of phase transitions in a system of self-driven particles, *Physical Review Letters*, vol. 75, pp. 1226–1229, 1995.
2. A. Jadbabaie, J. Lin and A. S. Morse, Coordination of groups of mobile autonomous agents using nearest neighbor rules, *IEEE Trans. Automat. Contr.*, vol. 48, no. 6, pp. 988–1001, June 2003.
3. W. Ren and R. W. Beard, Consensus seeking in multiagent systems under dynamically changing interaction topologies, *IEEE Trans. Automat. Contr.*, vol. 50, no. 5, pp. 655–661, May 2005.
4. J. A. Fax and R. M. Murray, Information flow and cooperative control of vehicle formations, *IEEE Transactions on Automatic Control*, vol. 49, no. 9, pp. 1465–1476, September 2004.
5. R. Olfati-Saber and R. M. Murray, Consensus problems in networks of agents with switching topology and time-delays, *IEEE Trans. Automat. Contr.*, vol. 49, no. 9, pp. 1520–1533, Sept. 2004.
6. Z. Lin, M. Broucke and B. Francis, Local control strategies for groups of mobile autonomous agents, *IEEE Trans. on Automatic Control*, vol. 49, no. 4, pp. 622–629, 2004.
7. L. Xiao and S. Boyd, Fast linear iterations for distributed averaging, *Systems and Control Letters*, vol. 53, pp. 65–78, 2004.
8. Z. Jin and R. M. Murray, Multi-hop relay protocols for fast consensus seeking, 45th IEEE Conference on Decision and Control, San Diego, CA, USA, submitted, Dec 13-15 2006.
9. C. W. Wu and L. O. Chua, Synchronization in an array of linearly coupled dynamical systems, *IEEE Trans. Circuits Syst.*, vol. 42, no. 8, pp. 430–447, August 1995.
10. C. W. Wu, Synchronization in networks of nonlinear dynamical systems coupled via a directed graph, *Nonlinearity*, vol. 18, pp. 1057–1064, 2005.

11. R. Olfati-Saber, Ultrafast consensus in small-world networks, in *Proceedings of 2005 American Control Conference*, June 2005, pp. 2371–2378.
12. Z. Jin and R. M. Murray, Double-graph control strategy of multi-vehicle formations, in *Proceedings of 43rd IEEE Conference on Decision and Control*, vol. 2, December 2004, pp. 1988–1994.
13. W. Ren, R. W. Beard and E. M. Atkins, A survey of consensus problems in multi-agent coordination, in *Proceedings of 2005 American Control Conference*, Protland, OR, USA, June 2005, pp. 1859–1864.

3 Decentralized Cooperative Optimization for Systems Coupled through the Constraints

Yoshiaki Kuwata and Jonathan P. How

Summary. Motivated by recent research on cooperative unmanned aerial vehicles (UAVs), this chapter introduces a new decentralized optimization approach for systems with independent dynamics but coupled constraints. The application examples include formation control of a fleet of UAVs, vehicle avoidance maneuvers, and multi-vehicle path planning under line-of-sight constraints. The primary objective is to improve the performance of the entire fleet by solving local optimization problems. The issue here is how to coordinate the vehicles without reproducing the global optimization problem for each agent. To achieve cooperation, the approach exploits the sparse structure of active couplings that is inherent in many trajectory optimization problems. This enables each local optimization to use a low-order parametrization of the other agents states, thereby facilitating negotiation while keeping the problem size small. The key features of this approach include (i) no central negotiator is required; and (ii) it maintains feasibility over the iterations, so the algorithm can be stopped at any time. Furthermore, the local optimizations are shown to always decrease the overall cost. Simulation results are presented to compare the distributed, centralized and other (non-cooperative) decentralized approaches in terms of both computation and performance.

1 Introduction

Cooperative distributed control has many areas of application, especially in the field of multi-vehicle control such as formation flying spacecraft, unmanned ground vehicles (UGVs), and unmanned aerial vehicles (UAVs) [1]. To enable fleet-level cooperation, the controller must properly capture the complex interactions between these vehicles, at both the high (tasks) and low (trajectory) levels. One approach is to solve this problem globally, but centralized algorithms typically scale very poorly with the fleet size because of the computational effort involved. This chapter presents a new decentralized optimization approach with specific application to trajectory optimization. Much of the current research on decentralized trajectory optimization uses a setup where each vehicle solves a local problem and communicates this intent information to its neighbors [2, 3, 4, 5, 6, 7]. The challenge in this case is how to achieve fleet level cooperation. For example, previous work by Inalhan [2] softens the constraints and achieves fleet level agreement by iteratively increasing the penalty on the constraint violation, but this iteration process could take a long time before it even reaches a feasible solution. The approach by Richards [4] ensures the feasibility of the entire fleet under the action of disturbances, but there is no consideration of the overall performance. Similarly, in the "communication-based approach" [8], each vehicle optimizes only for its own control and exchanges information only with neighboring subsystems. These decentralized algorithms typically lead to a Nash equilibrium [9] or a Pareto optimal surface [10, 2], which is not necessarily the globally optimal/cooperative solution, because the communication-based approaches do not use the information about the cost functions of the other subsystems.

In order to achieve the cooperative solution, an iterative decentralized scheme has been recently proposed [8] for systems coupled through dynamics. A dual decomposition approach has been proposed for systems coupled through objectives [11]. In this chapter, we focus on UAVs with independent dynamics but with coupling constraints. The application examples include formation control of a fleet of UAVs, vehicle avoidance maneuvers, and multivehicle path planning under line-of-sight constraints. The proposed algorithm minimizes the global cost by solving local optimizations while satisfying all the constraints. This new approach avoids the complexity of global optimization by using a low order model for other systems while retaining the key couplings. In particular, it exploits the problem structure by parameterizing other vehicles' decisions using the active coupling constraints. This approach is suitable for trajectory optimization because trajectory optimization problems tend to have only a few active couplings.

The chapter is organized as follows. First, Section 2 introduces the overall problem and two straightforward approaches. In Subsection 3.1, a simple form of the proposed algorithm is presented first to highlight the implication of this approach. Subsection 3.2 presents the complete decentralized cooperative algorithm. Section 4 analyzes some properties of the algorithm. Finally, Section 5 shows the simulation results and compares the algorithm with other available approaches in terms of performance and computation time.

2 Problem Statement

2.1 Problem Statement

The problem of interest is the general optimization for multi-vehicle systems, with a particular emphasis on the path planning. A fleet of n_v vehicles are assumed to have independent dynamics. Different types of constraints are imposed, but they can be divided into (a) local constraints such as speed bounds, input saturation, and obstacle avoidance; and (b) coupling constraints such as vehicle avoidance, inter-vehicle communication range, and line-of-sight between vehicles.

The system dynamics are in discrete time, and an optimization is performed from the initial states $x_i(0)$ to obtain the optimal input for each vehicle over N steps into the future:

$$\forall i = 1, \ldots, n_v, \quad \forall k = 0, \ldots, (N-1):$$

$$x_i(k+1) = f_i\big(x_i(k), u_i(k)\big) \tag{1}$$

$$g_i\big(x_i(k), u_i(k)\big) \leq 0 \tag{2}$$

$$A_i x_i(k) + A_j x_j(k) \leq b_{ij} \tag{3}$$

$$\forall j = i+1, \ldots, n_v$$

where $f_i\big(x_i(k), u_i(k)\big)$ represents the nonlinear dynamics of vehicle i, and $g_i\big(x_i(k), u_i(k)\big)$ represents all local constraints imposed on the states and the inputs of vehicle i. The pair-wise constraints Eq. 3 capture the couplings between vehicles and are assumed to be a combination of linear constraints, which can express various types of constraints including 1-norm, 2-norm, and ∞-norm bounds and polyhedral constraints.

The objective function for the entire fleet is the sum of individual costs, which could be in conflict with each other.

$$\min_{u(\cdot)} \sum_{i=1}^{n_v} J_i(x_i, u_i) \tag{4}$$

$$\text{s.t.} \quad J_i = \sum_{k=0}^{N-1} l_i\big(x_i(k), u_i(k)\big) + F\big(x_i(N)\big), \quad \forall i \tag{5}$$

where $l_i\big(x(\cdot), u(\cdot)\big)$ is a stage cost and $F_i\big(x_i(N)\big)$ is the terminal penalty.

2.2 Notation

We define the term *neighbor* of vehicle i as a set of vehicles that have any coupling constraint with vehicle i. In particular, if $\mathcal{N}(i)$ denotes this neighbor set for vehicle i, then there exist coupling constraints between i and $j \in \mathcal{N}(i)$. Furthermore, let $\mathcal{A}(i)$ denote a set of vehicles that have active coupling constraints with vehicle i. Figure 1 shows an example of a graphical representation of a vehicle fleet. Each node represents a vehicle, and the arc connecting two nodes shows that there is a coupling constraint between the two vehicles. The shaded nodes in the figure are the neighbors of vehicle i, i.e. $j_1, \ldots, j_4 \in \mathcal{N}(i)$. In general, not all of the coupling constraints are active. In this example, the thick lines show the existence of active couplings, so $j_1, j_2 \in \mathcal{A}(i)$ but $j_3, j_4 \notin \mathcal{A}(i)$.

For notational simplicity, let z_i denote the decision variable of the ith vehicle, i.e. $z_i \triangleq \big[\, u_i(0)^T, \cdots, u_i(N-1)^T \,\big]^T$. Then, with some abuse of notation, a compact form of the optimization Eqs. 1 to 5 can be written as

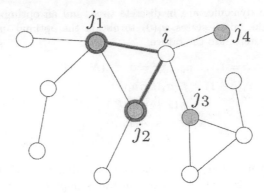

Fig. 1. Node i's neighbor set $\mathcal{N}$ and active coupling neighbor set $\mathcal{A}$.

$$\min_{z_1,\dots,z_{n_v}} \sum_{i=1}^{n_v} J_i(z_i) \qquad (6)$$

$$\text{subject to} \quad \forall i: \quad g(z_i) \leq 0$$

$$h(z_i, z_j) \leq 0, \quad j \in \mathcal{N}(i)$$

where $g(z_i)$ represents the local constraints for vehicle i and $h(z_i, z_j)$ represents the coupling constraints between vehicles i and j.

2.3 Centralized Approach

The centralized approach directly solves the full (and potentially large) optimization given in Eq. 6. This approach produces the globally optimal solution; however, it scales poorly because the optimization becomes very complex for large fleets for certain problem types (i.e. quadratic programming and mixed-integer linear programming) [12].

2.4 Decentralized Non-cooperative Approach

One decentralized approach is to decompose the centralized problem Eq. 6 into smaller subproblems. Figure 2 shows the procedural flow. Similar to Gauss-Seidel iteration [13], the approach sequentially solves the local subproblems and sends the solutions to other vehicles. In a subproblem, each vehicle i freezes the other vehicles' decision variables and solves for its own optimal input $z_i{}^*$. The local optimization for vehicle i can then be written as

$$\min_{z_i} \ J_i(z_i) \qquad (7)$$

$$\text{subject to} \quad g(z_i) \leq 0$$

$$h(z_i, \bar{z}_j) \leq 0, \quad j \in \mathcal{N}(i).$$

Vehicle ID

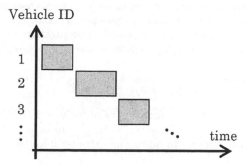

Fig. 2. Decentralized sequential planning. Each gray region represents that the vehicle is computing during that time.

The decision variables of the other vehicles are assumed to be constant, which is denoted as $\bar{z}_j$. The solution $z_i{}^*$ of this optimization is sent to the other vehicles, and the optimization of the next vehicle $i + 1$ starts after receiving $z_i{}^*$.

The main advantage of this approach is that the subproblem has a smaller decision space (approximately $1/n_v$ of the centralized approach), with much fewer constraints. As a result, the computation time is much smaller and the algorithm scales much better than the centralized approach. Furthermore, all the constraints are satisfied while cycling through the vehicles. However, since each vehicle does not account for the objectives of the other vehicles, the resulting solution is coordinated but non-cooperative. This non-cooperative solution is called a Nash equilibrium, where no vehicle can improve its local cost by changing only its own decision, and has to be avoided in the cooperative approach. These benefits and limitations of this approach are illustrated in the examples in Section 5.

3 Decentralized Cooperative Optimization

This section presents a new decentralized cooperative algorithm. We first describe the approach for a simple version of the algorithm with only two vehicles i and j, one coupling constraint, and no local constraints.

3.1 Simple Version

Similar to the decentralized non-cooperative approach, each vehicle solves a local optimization problem in a sequential way by freezing the other vehicle's decision when the coupling constraints are inactive. The key difference is that when there is an active coupling constraint, the new approach recognizes that each vehicle should consider sacrifices to its local performance if it is possible to reap a larger benefit to the overall team performance.

More formally, when vehicle i solves the optimization after vehicle j, the coupling constraint $a_i z_i + a_j \bar{z}_j \leq b_{ij}$ is modified, if active, in the following way

$$a_i z_i + a_j \bar{z}_j \leq b_{ij} - \beta. \tag{8}$$

The parameter $\beta \geq 0$ tightens the constraint for vehicle i, which could make the local performance worse. However, vehicle i can account for the potential benefit to the other vehicle j by adding an extra term to the objective function

$$\min_{z_i, \beta} J_i(z_i) - \lambda \beta \tag{9}$$

where $\lambda \neq 0$ is a Lagrange multiplier of the coupling constraint that is obtained from the previous solution of the optimization by vehicle j. This Lagrange multiplier represents the amount of improvement the optimization of vehicle j can obtain given some change in the right hand side of the coupling constraint $a_i \bar{z}_i + a_j z_j \leq b_{ij}$ [14]. In the hierarchical setup, the Lagrange multiplier could be used to represent the "price" of each vehicle's solution that the centralized negotiator can use to obtain a coordinated solution [15]. Note that this approach is meaningful only when the coupling constraint is active and hence $\lambda \neq 0$.

The decision variables of vehicle i's optimization in Eq. 9 are its local decisions z_i and the negotiation parameter β for the other vehicle. The parameter β allows vehicle i to sacrifice its local cost if that leads to more benefit to the other vehicle j, which is a cooperative behavior. This is essentially minimizing the global performance by solving the local decentralized problems.

3.2 Full Version

This section generalizes the idea introduced in Subsection 3.1 to cases with multiple coupling and local constraints. When there are multiple active coupling constraints, simply including the effect $\lambda \beta$ for each active coupling constraint could doubly count the benefit that the other vehicle can obtain. Also the scope of the simple problem must be expanded because it is possible to have vehicle i tightening its coupling constraint, as in Eq. 8, to enlarge the operating region of vehicle j, but a local constraint of vehicle j prevents it from using that extra region and obtaining the expected benefit $\lambda \beta$.

The simple version used the negotiation parameter β in vehicle i's optimization as an implicit decision variable for vehicle j. In the full version, vehicle i makes an explicit decision for vehicle j, which is denoted by δz_j. Then, vehicle i's optimization is

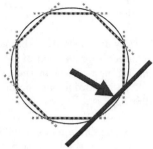

Fig. 3. A two norm constraint approximated by a combination of linear constraints. The arrow shows a relative position vector. The thick line shows the only one active coupling constraint.

$$\min_{z_i,\delta z_j} J_i(z_i) + J_j(\bar{z}_j + \delta z_j) \tag{10}$$

$$g(z_i) \leq 0, \tag{11}$$

$$g(\bar{z}_j + \delta z_j) \leq 0 \tag{12}$$

$$A_i z_i + A_j(\bar{z}_j + \delta z_j) \leq b_{ij} \tag{13}$$

where δz_j is related to β through $A_j \delta z_j \Leftrightarrow \beta$. The modified local optimization in Eq. 10 appears similar to the one solved in the centralized approach, but the key point is that, for the problems of interest in this work, the decision space can be reduced significantly, as discussed below.

Sparse active coupling

The approach avoids reproducing the global optimization problem for each agent by exploiting the structure of active couplings that is typical of trajectory optimization. For example, the vehicle avoidance constraints are imposed over all the time steps of the plan, but they are typically active only at one or two time steps. Communication range limit constraints could be expressed as a nonlinear two-norm constraint on the relative position, but a combination of several linear constraints can approximate it (Figure 3). In such a case, only a few of the many existing constraints are active. The algorithm exploits this sparse structure of the active coupling constraints to reduce the size of the optimization problem.

Low-order parameterization

Without loss of generality, the upper rows of the coupling constraints Eq. 13 can be regarded as active and the lower rows as inactive. Then, Eqs. 8 and 13 are related by

$$\begin{bmatrix} A_j^{\text{active}} \\ \hline A_j^{\text{inactive}} \end{bmatrix} \delta z_j = \begin{bmatrix} \beta^{\text{active}} \\ \hline * \end{bmatrix}$$

We focus on changing β^{active} by δz_j, because a change in these active coupling constraints can lead to the direct improvement of the other vehicle's cost. This corresponds to focusing on the non-zero λ in the simple version. In order to address the change in β^{active}, a low-order parameterization of δz_j can be used because $\dim(\beta^{\text{active}}) \ll \dim(\delta z_j)$.

Let m denote the row rank of A_j^{active}, which is also a number of elements in β^{active} that any δz_j can change independently. Therefore, a new variable $\alpha_j \in \mathbb{R}^m$ could replace δz_j, where in the trajectory optimization problem the dimension m of α_j is significantly smaller than the dimension of δz_j. Let $\breve{A}$ denote a matrix composed of the m independent row vectors extracted from A_j^{active}. Then, δz_j is parameterized by α_j as

$$\delta z_j = \breve{A}^T \left(\breve{A} \breve{A}^T \right)^{-1} \alpha_j \triangleq T_j \alpha_j. \tag{14}$$

The inverse in this equation exists because the product $\left(\breve{A} \breve{A}^T \right)$ is a matrix of full rank m. Thus, the parameterization matrix T_j also exists.

With this new variable α_j, the local optimization can be rewritten as

$$\min_{\substack{z_i \\ \alpha_j, j \in \mathcal{A}(i)}} \left\{ J_i(z_i) + \sum_{j \in \mathcal{A}(i)} J_j(\bar{z}_j + T_j \alpha_j) \right\} \tag{15}$$

subject to

$$g(z_i) \leq 0$$
$$g(\bar{z}_j + T_j \alpha_j) \leq 0, \quad j \in \mathcal{A}(i)$$
$$h(z_i, \bar{z}_j) \leq 0, \quad j \in \mathcal{N}(i), \, j \notin \mathcal{A}(i)$$
$$h(z_i, \bar{z}_j + T_j \alpha_j) \leq 0, \quad j \in \mathcal{A}(i)$$
$$h(\bar{z}_k, \bar{z}_j + T_j \alpha_j) \leq 0, \quad k \in \mathcal{N}(j), k \notin \mathcal{A}(i)$$
$$h(\bar{z}_{j_1} + T_{j_1} \alpha_{j_1}, \bar{z}_{j_2} + T_{j_2} \alpha_{j_2}) \leq 0, \quad j_1, j_2 \in \mathcal{A}(i).$$

Note that the parameterization is based on the active coupling constraints, but the optimization includes both the active and inactive constraints. The first two constraints are the local constraints for vehicle i and for its active coupling neighbors. The next four equations express different types of couplings shown in Figure 4. Type I is between vehicle i and its neighbors with no active couplings; Type II is between vehicle i and its neighbors with active couplings; Type III is between vehicle i's active coupling neighbors and their neighbors; Type IV is between vehicle i's two active coupling neighbors. Note that some constraints in $h(z_i, \bar{z}_j + T_j \alpha_j) \leq 0$ could be omitted if α_j has no impact on them because of the row rank deficiency of T_j.

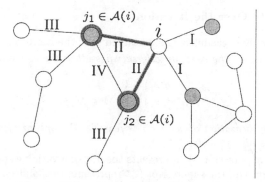

Fig. 4. Different types of coupling constraints.

The key points of this algorithm are 1) it does not freeze the other vehicle's plan, so that it can avoid Nash equilibrium; and 2) it reduces the decision space of the other vehicles, so that the complexity of each local optimization remains low. For the problem of interest with a relatively large decision space (e.g., $\dim(z_i) = 20$, $n_v = 5$) and sparse active couplings (2 active coupling neighbors, $\dim(\alpha) = 2$), the reduction of the decision space of each optimization would be a factor of approximately 4.

The algorithm iterates over the vehicles and the complete flow is summarized below:

1. For each vehicle i
   a) Find the active coupling with this vehicle: $\mathcal{A}(i)$
   b) Calculate the parameterization matrix T_j, $j \in \mathcal{A}(i)$
   c) Solve local optimization Eq. 15 and obtain the solution $(z_i{}^*, \alpha_j{}^*)$
   d) Send the solution to other vehicles. Each vehicle updates the plan

$$z_i := z_i{}^*$$
$$z_j := z_j + T_j\alpha_j{}^*, \quad j \in \mathcal{A}(i)$$

2. Terminate if the maximum number of iterations is reached. Otherwise, increment the vehicle counter $i := i + 1$ and go to Step 1.

Note that in the step 1c) the vehicle i makes a decision on itself and its neighbors. Then, the step 1d) ensures that all of the vehicles have the same values for the decision variables z_i, z_j. The simulation results in Section 5 show that 2 iterations over the fleet produces a good performance that is comparable to the centralized approach.

4 Algorithm Properties

This section discusses two key properties of this algorithm.

4.1 Feasibility Over the Iteration

First, we show that feasibility is maintained over the iteration.

Theorem 1. Assume the fleet initially satisfies all of the constraints

$$\forall j: \quad g(z_j) \leq 0$$
$$h(z_j, z_k) \leq 0, \quad k \in \mathcal{N}(j).$$

Then, if the optimization of vehicle i is feasible, the optimization of the next vehicle $i + 1$ is also feasible.

Proof. Let the superscript $(\cdot)^\circ$ represents the decision variables prior to vehicle i's optimization, and the superscript $(\cdot)^*$ represents the solution optimized by i. Initially,

$$\forall j: \quad g(z_j^\circ) \leq 0$$
$$h(z_j^\circ, z_k^\circ) \leq 0, \quad k \in \mathcal{N}(j).$$

The feasible solution of vehicle i's optimization obtained in step 1c) satisfies the constraints

$$g(z_i{}^*) \leq 0$$
$$g(z_j^\circ + \delta z_j^*) \leq 0, \quad j \in \mathcal{A}(i)$$
$$h(z_i^*, z_j^\circ) \leq 0, \quad j \in \mathcal{N}(i), \ j \notin \mathcal{A}(i)$$
$$h(z_i^*, z_j^\circ + \delta z_j^*) \leq 0, \quad j \in \mathcal{A}(i)$$
$$h(z_k^\circ, z_j^\circ + \delta z_j^*) \leq 0, \quad k \in \mathcal{N}(j), k \notin \mathcal{A}(i)$$
$$h(z_{j_1}^\circ + \delta z_{j_1}^*, z_{j_2}^\circ + \delta z_{j_2}^*) \leq 0, \quad j_1, j_2 \in \mathcal{A}(i).$$

After communicating the solutions and updating them in step 1d), the variables satisfy all the constraints

$$g(\bar{z}_i) \leq 0, \quad \forall i$$
$$h(\bar{z}_i, \bar{z}_j) \leq 0, \quad \forall j \in \mathcal{N}(i) \tag{16}$$

where

$$\bar{z}_i = z_i^*$$
$$\bar{z}_j = z_j^\circ + \delta z_j^*, \quad j \in \mathcal{A}(i)$$
$$\bar{z}_j = z_j^\circ, \quad j \notin \mathcal{A}(i).$$

Given these results, the constraints for the problem for the next vehicle in the optimization sequence $(i + 1)$ are:

$$g(z_{i+1}) \leq 0$$
$$g(\bar{z}_j + \delta z_j) \leq 0, \quad j \in \mathcal{A}(i + 1)$$

$$h(z_{i+1}, \bar{z}_j) \leq 0, \quad j \in \mathcal{N}(i+1),\ j \notin \mathcal{A}(i+1)$$
$$h(z_{i+1}, \bar{z}_j + \delta z_j) \leq 0, \quad j \in \mathcal{A}(i+1)$$
$$h(\bar{z}_k, \bar{z}_j + \delta z_j) \leq 0, \quad k \in \mathcal{N}(j), k \notin \mathcal{A}(i+1)$$
$$h(\bar{z}_{j_1} + \delta z_{j_1}, \bar{z}_{j_2} + \delta z_{j_2}) \leq 0, \quad j_1, j_2 \in \mathcal{A}(i+1). \tag{17}$$

By comparing Eqs. 16 and 17, it can be shown that reusing the solution from vehicle i's optimization

$$z_{i+1} = \bar{z}_{i+1},$$
$$\delta z_j = 0, \quad \forall j \in \mathcal{A}(i+1)$$

provides a feasible solution to the optimization of $i+1$. ∎

Note that because the feasibility of the entire fleet is maintained over the iteration, the algorithm could be terminated at any time. Some previous work [4, 6] also maintains this property but not the following property about the performance.

4.2 Monotonically Decreasing Global Cost

This subsection shows that the global objective value is monotonically decreasing along the iteration.

Theorem 2. The global cost function defined by

$$J(z) = \sum_{i=1}^{n_v} J_i(z_i)$$

is monotonically decreasing over the iteration.

Proof. The local optimization of vehicle i results in

$$J_i(z_i^o) + \sum_{j \in \mathcal{A}(i)} J_j(z_j^o) \geq J_i(z_i^*) + \sum_{j \in \mathcal{A}(i)} J_j(z_j^o + \delta z_j^*)$$

Let $J(z^o)$ denote the global cost prior to the optimization by vehicle i and $J(\bar{z})$ denote the global cost prior to the optimization by the next vehicle $i+1$. Then,

$$J(z^o) = J_i(z_i^o) + \sum_{j \in \mathcal{A}(i)} J_j(z_j^o) + \sum_{j \notin \mathcal{A}(i)} J_j(z_j^o)$$
$$\geq J_i(z_i^*) + \sum_{j \in \mathcal{A}(i)} J_j(z_j^o + \delta z_j^*) + \sum_{j \notin \mathcal{A}(i)} J_j(z_j^o)$$
$$= J_i(\bar{z}_i) + \sum_{j \in \mathcal{A}(i)} J_j(\bar{z}_j) + \sum_{j \notin \mathcal{A}(i)} J_j(\bar{z}_j)$$
$$= J(\bar{z})$$

Therefore, each local optimization decreases the global cost. Because this algorithm maintains feasibility, it monotonically decreases the global cost over the iteration. ∎

5 Simulation Results

5.1 Simulation Setup

In the simulation, all n_v vehicles are assumed to have the same linear dynamics which are described by a simple double integrator model: $\forall i = 1, \ldots, n_v$

$$x_i(k+1) = \begin{bmatrix} I & I \\ O & I \end{bmatrix} x_i(k) + \begin{bmatrix} 0.5I \\ I \end{bmatrix} u_i(k),$$

$$x_i(k) = \begin{bmatrix} r_i(k)^T & v_i(k)^T \end{bmatrix}^T.$$

Constraints are imposed on the position, the speed, and the control input of each vehicle at each time step $k = 0, \ldots, N$

$$\|r_i(k)\|_\infty \le 1, \quad \|v_i(k)\|_2 \le 0.35, \quad \|u_i(k)\|_2 \le 0.18.$$

The systems are coupled with two neighbors through the following position constraints.

$$\|r_i(k) - r_{i+1}(k)\|_2 \le 0.8, \quad i = 1, \ldots, (n_v - 1)$$
$$\|r_{n_v}(k) - r_1(k)\|_2 \le 0.8$$

These two-norm constraints are expressed as a combination of linear constraints as shown in Figure 3. The cost direction for the ith vehicle is

$$c_i = \begin{bmatrix} \cos\left(\frac{i-1}{n_v}\right) & \sin\left(\frac{i-1}{n_v}\right) \end{bmatrix}^T.$$

The overall cost function to minimize is

$$\sum_{i=1}^{n_v} \sum_{k=0}^{N-1} \left\{ x_i(k)^T R_1 x_i(k) + u_i(k)^T R_2 u_i(k) \right\} + c_i^T r_i(N) + r_i(N)^T H r_i(N)$$

where the weights on the states R_1 and inputs R_2 in the stage cost are chosen to be much smaller than the weight H on the terminal position. Both the centralized and the local optimization is a quadratic programming problem, and CPLEX 9.1 is used as a solver.

5.2 Simple Two Vehicle Case

The first example involves two vehicles i and j that can move on a two dimension plane. The terminal position of the vehicle i has its local minimum at coordinates $(0.7, 0)$, i.e.,

$$\begin{bmatrix} 0.7 \\ 0 \end{bmatrix} = \arg \min_{r_i(N)} \left\{ c_i^T r_i(N) + r_i(N)^T H r_i(N) \right\},$$

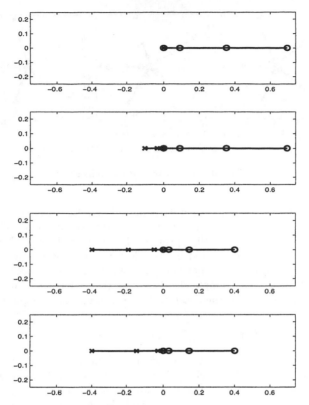

Fig. 5. The evolution of plans over the iteration for the simple two vehicle example.

and that of the vehicle j is at $(-0.7, 0)$. Because the two vehicles must satisfy the separation constraint of 0.8, their separate objectives are conflicting. The planning horizon is three steps for both vehicles.

Figure 5 shows the evolution of plans over two iterations. The plans of vehicle i are marked with $\circ$, and those of vehicle j are marked with $\times$. Originally, both vehicles are at the origin. First, vehicle i solves its local optimization. Because no coupling constraints are active at this point, it plans to reach its local minimum $(0.7, 0)$. Vehicle j then solves its optimization, but given the separation constraint, this vehicle can only plan to move to $(-0.1, 0)$, as shown in the second part of the figure.

The vehicle i solves the next optimization, but since a coupling constraint has become active, it uses a parameterized decision for j with a variable α_j of dimension $m = 1$. The bottom figure shows the plans after two iterations. These final plans are the same as the globally optimal centralized solution.

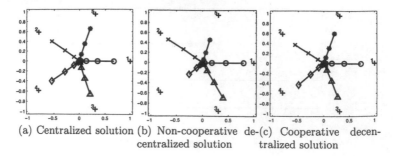

(a) Centralized solution (b) Non-cooperative de-(c) Cooperative decen-
                      centralized solution   tralized solution

Fig. 6. Final plans for five vehicles

If the decentralized non-cooperative algorithm in Eq. 7 were used, it would produce a Pareto optimal solution shown in the second figure of Figure 5, which is clearly not the globally optimal solution in the bottom. Note that if the vehicle j plans first followed by vehicle i, the non-cooperative algorithm results in a symmetric Pareto optimal solution, which again is not the globally optimal solution. This example clearly shows the performance improvement of the new approach over the decentralized non-cooperative approach.

5.3 Five Vehicle Case

Figure 6 shows a more complex case with five vehicles. In this example, the local minimum for each vehicle is located on a unit circle centered on the origin. The planning horizon is three steps for all vehicles, and the planning order was $1 \rightarrow 3 \rightarrow 5 \rightarrow 2 \rightarrow 4$ to highlight the effect of the planning order on the performance.

Two other algorithms described in Section 2 are used as benchmarks. These are (a) the centralized approach Eq. 6 that provides the globally optimal solution, and (b) the decentralized non-cooperative approach Eq. 7 that produces a locally optimized solution.

As shown in Figure 6(b), the decentralized non-cooperative approach produced a suboptimal solution, because the vehicles that plan earlier are less constrained and have more advantages. The decentralized cooperation algorithm produced the plans shown in Figure 6(c) that is similar to the centralized solution shown in Figure 6(a).

5.4 Performance and Computation

Figure 7 compares the global objective value and the cumulative computation time of three algorithms for the five vehicle example. Different lengths of planning horizon $N = 4, 6, 8$ were considered to investigate the scalability of the algorithms.

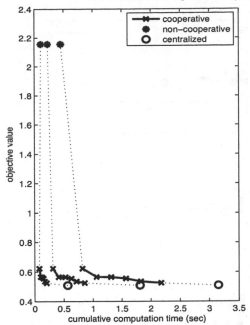

Fig. 7. Comparison of three algorithms in terms of performance and computation.

The solutions of the decentralized non-cooperative approach are marked with ∗. Although the computation time is small, the cost is fairly high. The centralized (and hence globally optimal) solutions are marked with ○. The lines with × show the evolution of the global cost of the decentralized cooperation algorithm. The plot starts from the end of the first iteration when every vehicle has its solution and continues to the end of the second iteration. This proposed algorithm has objective values comparable to those of the centralized solution but scales better than the centralized solution when the problem size increases.

Figure 8 shows cases with more vehicles ($n_v = 5, 7, 10, 15$). The decentralized non-cooperative approach has much higher cost and is out of the range of the plot. For the centralized and the proposed approach, the differences in the computation time scale up significantly for larger fleets. Note that in all the plots of Figures 7 and 8, the lines of the proposed approach are monotonically decreasing, which validates the result in Subsection 4.2 by simulation.

6 Conclusions and Future Work

This chapter presented a decentralized cooperation algorithm for systems coupled through the constraints. By exploiting the sparse structure of the active

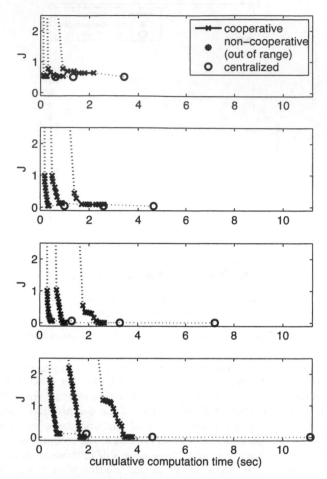

Fig. 8. Trade off between the performance and the computation time. From the top to the bottom, the number of vehicles are 5, 7, 10, and 15.

coupling constraints of trajectory optimization, the algorithm uses low-order parameterization of the neighbors' decisions. Simulation results showed that the proposed algorithm scales much better than the centralized approach and the performance is much better than that of the non-cooperative approach. Over the iteration, it is shown to maintain feasibility and monotonically decrease the global cost.

Future work includes combining this technique with our Robust MPC path planner that uses mixed-integer linear programming; and investigating how it extends to very complex path planning problems using MILP.

7 Acknowledgment

This work was supported by grant AFOSR Grant #FA9550-04-1-0458.

References

1. R. Saber, W. Dunbar, and R. Murray, Cooperative control of multi-vehicle systems usiing cost graphs and optimization, in *Proceedings of the IEEE American Control Conference*, 2003.
2. G. Inalhan, D. M. Stipanovic and C. J. Tomlin, Decentralized optimization, with application to multiple aircraft coordination, in *Proceedings of the IEEE Conference on Decision and Control*, December 2002.
3. E. King, M. Alighanbari, Y. Kuwata and J. How, Coordination and Control Experiments on a Multi-vehicle Testbed, in *Proceedings of the IEEE American Control Conference*, pp. 5315–5320, Boston, MA, 2004.
4. A. Richards and J. How, Decentralized Algorithm for Robust Constrained Model Predictive Control, in *Proceedings of the IEEE American Control Conference*. Boston, MA: IEEE, 2004.
5. W. Dunbar and R. Murray, Receding horizon control of multi-vehicle formation: A distributed implementation, in *Proceedings of the IEEE Conference on Decision and Control*, 2004.
6. T. Schouwenaars, J. How and E. Feron, Decentralized Cooperative Trajectory Planning of Multiple Aircraft with Hard Safety Guarantees, in *Proceedings of the AIAA Guidance, Navigation, and Control Conference*, August 2004.
7. Y. Kuwata, A. Richards, T. Schouwenaars and J. How, Decentralized Robust Receding Horizon Control for Multi-vehicle Guidance, in *Accepted to appear in the proceedings of American Control Conference*, 2006.
8. A. Venkat, J. Rawlings and S. Wright, Stability and optimality of distributed model predictive control, in *IEEE Conference on Decision and Control, and the European Control Conference*, 2005.
9. U. Ozguner and W. R. Perkins, Optimal control of multilevel large-scale systems, *International Journal of Control*, vol. 28, no. 6, pp. 967–980, 1978.
10. P. Heiskanen, Decentralized method for computing Pareto solutions in multiparty negotiations, *European Journal of Operational Research*, vol. 117, pp. 578–590, 1999.
11. R. Raffard, C. Tomlin and S. Boyd, Distributed optimization for cooperative agents: Application to formation flight, in *Proceedings of the IEEE Conference on Decision and Control*, 2004.
12. A. Richards and J. How, Decentralized Model Predictive Control of Cooperating UAVs, in *Proceedings of the IEEE Conference on Decision and Control*, December 2004.
13. D. Bertsekas and J. Tsitsiklis, *Parallel and Distributed Computation: Numerical Methods*, Athena Scientific, 1997.
14. D. P. Bertsekas, *Nonlinear Programming*, Athena Acientific, 1995.
15. W. Findeisen, F. N. Bailey, M. Brdys, K. Malinowski, and A. Wozniak, *Control and coordination in hierarchical systems*, Vol. 9, A. Wiley, Interscience Publication, London, 1980.

4 Consensus Building in Cooperative Control with Information Feedback

Wei Ren

Summary. In this chapter, we study the problem of consensus building in multi-vehicle systems with information feedback. We show how information feedback can be incorporated into the consensus building process so as to improve the robustness and situational awareness of the whole team. We detail the strategies of introducing feedback to the consensus building process through information flow, external feedback terms, state-dependent weights and reference states. Application examples are also given to illustrate the information feedback strategies.

1 Introduction

Autonomous vehicle systems are expected to find potential applications in military operations, search and rescue, environment monitoring, commercial cleaning, material handling, and homeland security. While single vehicles performing solo missions can yield some benefits, greater benefits will come from the cooperation of teams of vehicles.

As an inherently distributed strategy for multi-vehicle coordination, consensus algorithms have recently been studied extensively in the context of cooperative control of multi-vehicle systems [1, 2, 3, 4, 5, 6, 7, 8, 9, 10]. Those algorithms require only local neighbor-to-neighbor information exchange between the vehicles. The basic idea for information consensus is that each vehicle updates its information state based on the information states of its local (possibly time-varying) neighbors in such a way that the final information state of each vehicle converges to a common value. This basic idea can be extended to deal with the case that each vehicle's information states converge to desired relative deviations or to incorporate different group behaviors into the consensus building process. Consensus algorithms have applications in multi-vehicle rendezvous [11, 12], formation control [2, 13], flocking [14, 15], attitude alignment [16, 17], decentralized task assignment [18], sensor networks [19, 20, 21], etc.

Most consensus algorithms studied in the literature do not take into account vehicle performance, environmental information, and sensor measurement in the consensus building process. For example, in some formation control problems, where the formation is moving through space, the information states of each vehicle might be dynamically evolving over time according to some inherent dynamics. Also in most cooperative control problems, the information states of each vehicle might be affected by environmental factors or sensor measurement. As a result, it is essential to incorporate vehicle performance, environmental information, and sensor measurement into the consensus building process as a form of feedback.

The main contribution of this chapter is to study how information feedback can be incorporated into the consensus building process so as to improve the robustness and situational awareness of the whole team. In particular, we study strategies of introducing information feedback to the consensus building process through information flow, external feedback terms, state-dependent weights, and reference states. All of the strategies will be illustrated by application examples. A preliminary version of the chapter was presented at [33].

2 Background and Preliminaries

It is natural to model information exchange between vehicles by directed or undirected graphs. A directed graph consists of a pair $(\mathcal{N}, \mathcal{E})$, where $\mathcal{N}$ is a finite nonempty set of nodes, and $\mathcal{E} \in \mathcal{N} \times \mathcal{N}$ is a set of ordered pairs of nodes, called *edges*. An edge (i, j) in a directed graph denotes that vehicle j can obtain information from vehicle i, but not necessarily vice versa. As a comparison, the pairs of nodes in an undirected graph are unordered, where an edge (i, j) denotes that vehicles i and j can obtain information from one another. Note that an undirected graph can be considered a special case of a directed graph, where an edge (i, j) in the undirected graph corresponds to edges (i, j) and (j, i) in the directed graph. If there is a directed edge from node i to node j, then i is defined as the parent node, and j is defined as the child node.

A directed path is a sequence of ordered edges in a directed graph of the form $(i_1, i_2), (i_2, i_3), \ldots$, where $i_j \in \mathcal{N}$. An undirected path in an undirected graph is defined accordingly. In a directed graph, a cycle is a path that starts and ends at the same node. A directed graph is *strongly connected* if there is a directed path from every node to every other node. An undirected graph is *connected* if there is a path between every distinct pair of nodes. A rooted directed tree is a directed graph, where every node has exactly one parent except for one node, called *root*, which has no parent, and the root has a directed path to every other node. Note that in a rooted directed tree, each edge has a natural orientation away from the root, and no cycle exists. In the case of undirected graphs, a tree is a graph in which every pair of nodes is connected by exactly one path.

A rooted directed spanning tree of a directed graph is a rooted directed tree formed by graph edges that connect all of the nodes of the graph. A graph has or contains a rooted directed spanning tree if a rooted directed spanning tree is a subset of the graph. Note that a directed graph has a rooted directed spanning tree if and only if there exists at least one node having a directed path to all of the other nodes. In the case of undirected graphs, having an undirected spanning tree is equivalent to being connected. However, in the case of directed graphs, having a rooted directed spanning tree is a weaker condition than being strongly connected. The union of a group of graphs is a graph with nodes given by the union of the node sets and edges given by the union of the edge sets of the group of graphs.

Fig. 1 shows a directed graph that contains more than one possible rooted directed spanning tree, but is not strongly connected. Either node 1 or 2 can be the root of a rooted directed spanning tree since they both have a directed path to all of the other nodes. However, the graph is not strongly connected since nodes 3, 4, 5, and 6 do not have a directed path to all of the other nodes.

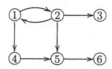

Fig. 1. Information-exchange graph between six vehicles. An arrow from i to j denotes that vehicle j receives information from vehicle i. The directed graph contains more than one possible rooted directed spanning tree with node 1 or 2 being the root, but is not strongly connected.

Suppose that there are n vehicles in the team. The adjacency matrix $A = [a_{ij}] \subset \mathbb{R}^{n \times n}$ of a weighted directed graph is defined as $a_{ii} = 0$ and $a_{ij} > 0$ if $(j, i) \in \mathcal{E}$, where $i \neq j$. The adjacency matrix of a weighted undirected graph is defined accordingly except that $a_{ij} = a_{ji}$, for all $i \neq j$, since $(j, i) \in \mathcal{E}$ implies $(i, j) \in \mathcal{E}$.

Let the matrix $L = [\ell_{ij}] \in \mathbb{R}^{n \times n}$ be defined as $\ell_{ii} = \sum_{j \neq i} a_{ij}$ and $\ell_{ij} = -a_{ij}$, where $i \neq j$. The matrix L satisfies the conditions

$$\ell_{ij} \leq 0, \quad i \neq j,$$

$$\sum_{j=1}^{n} \ell_{ij} = 0, \quad i = 1, \ldots, n. \tag{1}$$

For an undirected graph the *Laplacian matrix* L has the property of symmetric positive semidefiniteness [22]. However, the matrix L for a directed graph does not have this property. For a directed graph, the matrix L is sometimes called the directed graph Laplacian or nonsymmetric Laplacian in the literature. In the case of undirected graphs, all of the eigenvalues of L are nonnegative. In the case of directed graphs, all of the eigenvalues of L have nonnegative real parts. In both cases, 0 is an eigenvalue of L with an associated eigenvector $\mathbf{1}$, where $\mathbf{1} \triangleq [1, \ldots, 1]^T$ is a column vector of all ones. In the case of undirected graphs, 0 is a simple eigenvalue of L if and only if the undirected graph is connected. In the case of directed graphs, 0 is a simple eigenvalue of L if the directed graph is strongly connected [23].

As an example of an adjacency matrix for a weighted directed graph, the matrix

$$A = \begin{bmatrix} 0 & 1.5 & 0 & 0 & 0 & 0 \\ 0.7 & 0 & 0 & 0 & 0 & 0 \\ 0 & 1.1 & 0 & 0 & 0 & 0 \\ 0.8 & 0 & 0 & 0 & 0 & 0 \\ 0 & 0.2 & 0 & 0.3 & 0 & 0 \\ 0 & 0 & 0 & 0 & 1.2 & 0 \end{bmatrix}$$

can be a valid adjacency matrix corresponding to the directed graph in Fig. 1. The matrix L can be defined accordingly from the adjacency matrix A.

Let I_n denote the $n \times n$ identity matrix. Let $M_n(\mathbb{R})$ represent the set of all $n \times n$ real matrices. Given a matrix $S = [s_{ij}] \in M_n(\mathbb{R})$, the directed graph of S, denoted by $\Gamma(S)$, is the directed graph on n nodes i, $i \in \{1, 2, \ldots, n\}$, such that there is a directed edge in $\Gamma(S)$ from j to i if and only if $s_{ij} \neq 0$ [24].

3 Fundamental Consensus Algorithm

For information states with dynamics given by

$$\dot{\xi}_i = u_i, \quad i = 1, \ldots, n, \tag{2}$$

where $\xi_i \in \mathbb{R}^m$ denotes the information state of the ith vehicle and $u_i \in \mathbb{R}^m$ is the control input, a consensus algorithm is proposed in Refs. [1, 3, 4, 25] as

$$u_i = -\sum_{j=1}^{n} g_{ij} k_{ij} (\xi_i - \xi_j), \tag{3}$$

where $k_{ij} > 0$, $g_{ii} \triangleq 0$, and g_{ij} is 1 if information flows from vehicle j to vehicle i and 0 otherwise, $\forall i \neq j$.

By applying the algorithm Eq. 3, Eq. 2 can be written in matrix form as

$$\dot{\xi} = -(L \otimes I_m)\xi,$$

where $\xi = [\xi_1^T, \ldots, \xi_n^T]^T$, $\otimes$ denotes the Kronecker product, and $L = [\ell_{ij}] \in \mathbb{R}^{n \times n}$ is given as $\ell_{ii} = \sum_{j \neq i} g_{ij} k_{ij}$ and $\ell_{ij} = -g_{ij} k_{ij}$, $\forall i \neq j$. Note that L satisfies the property Eq. 1.

Consensus is said to be reached among the n vehicles if $\xi_i(t) \to \xi_j(t)$, $\forall i \neq j$, as $t \to \infty$. With the consensus algorithm Eq. 3, the final consensus value is a weighted average of the vehicles' initial information states. Note that the final consensus value is generally a priori unknown and will depend on the information-exchange topologies as well as weights k_{ij}.

Under a fixed information-exchange topology, the algorithm Eq. 3 achieves consensus asymptotically if and only if the information-exchange topology has a rooted directed spanning tree [25].

In the following, we assume a directed information-exchange topology to take into account the case that sensors might have a limited field of view in the case of information exchange through local sensing. Note that undirected information exchange is a special case of directed information exchange.

4 Consensus Building with Information Feedback through Information Flow

4.1 Basic Result

The most straightforward strategy to introduce feedback to the consensus building process is through information flow between local neighbors.

With the consensus algorithm Eq. 3, the final consensus value is given by $\xi^* = \sum_{i=1}^{n} \alpha_i \xi_i(0)$, where $\alpha = [\alpha_1, \ldots, \alpha_n]^T$ is a left eigenvector of $-L$ associated with eigenvalue 0 with $\alpha_i \geq 0$ and $\sum_{i=1}^{n} \alpha_i = 1$ [25]. Note that $\alpha_i \neq 0$ if vehicle i has a directed path to all the other vehicles in the information-exchange topology and $\alpha_i = 0$ if there does not exist such a directed path [25]. As a result, if a vehicle wants to contribute to the final consensus value, its information needs to flow to all the vehicles in the team directly or indirectly.

Information flow between the vehicles can also be applied to increase the redundancy and robustness of the whole team in the case of failures of certain information-exchange links. For example, if vehicle j only receives data due to its station as strictly a "child" in the directed information-exchange topology or due to unreliable state data transmission, any disturbance to this vehicle will cause inaccuracy in the vehicle team. However, if vehicle j is also a parent of another vehicle, then this disturbance feedback is propagated to the other vehicle.

4.2 Illustrative Example

To illustrate, we consider two information-exchange topologies shown in Fig. 2. Fig. 2a corresponds to a leader-follower topology where vehicle $j + 1$ follows vehicle j, $j = 1, 2, 3$. Fig. 2b corresponds to a topology where feedback is introduced from followers to leaders through information flow. Note that the final consensus value with Fig. 2a is $\xi_1(0)$ while the final consensus value with Fig. 2b is a weighted average of $\xi_1(0)$, $\xi_2(0)$, and $\xi_3(0)$. Also note that in Fig. 2a if vehicle 3 is perturbed by disturbance, vehicles 1 and 2 are unaware of this disturbance and their motions remain unaffected. However, in Fig. 2b if vehicle 3 is perturbed by disturbance, vehicles 1 and 2 are able to adjust their motions according to the motion of vehicle 3 so as to maintain better team performance due to the information flow from vehicle 3 to vehicles 1 and 2 directly or indirectly.

(a)

(b)

Fig. 2. Information-exchange topologies between four vehicles. Fig. 2a denotes a leader-follower topology while Fig. 2b denotes a topology with information flow introduced from followers to leaders.

5 Consensus Building with Information Feedback through External Feedback Terms

5.1 Basic Result

Another strategy to introduce feedback to the consensus building process is through an external feedback term. Consider the following algorithm:

$$u_i = -\sum_{j=1}^{n} g_{ij} k_{ij} (\xi_i - \xi_j) + \rho_i(t, x_i, \{j \in \mathcal{N}_i | x_j\}), \qquad (4)$$

where $\mathcal{N}_i$ denotes the set of vehicles whose information is available to vehicle i, $\rho_i(\cdot, \cdot, \cdot)$ denotes a feedback term introduced to the ith vehicle from its local neighbors, and x_i denotes the state of the ith vehicle. As a result, the consensus building process of each vehicle will be affected by the performance of its local neighbors, which serves as a form of information feedback and therefore improves the robustness of the whole team.

We need the following lemma to analyze Eq. 4.

Lemma 1. *Let $\dot{\xi}_i = -\sum_{j=1}^{n} g_{ij}(t) k_{ij}(t)(\xi_i - \xi_j)$, where $k_{ij}(t) > 0$ is piecewise continuous and uniformly lower and upper bounded. If there exist infinitely many consecutive uniformly bounded time intervals such that the union of the information-exchange graph across each time interval has a rooted directed spanning tree, then $\xi_i \to \xi_j$, $\forall i \neq j$, uniformly exponentially. Furthermore, if $\|\rho_i - \rho_j\|$ is bounded, so is $\|\xi_i - \xi_j\|$, $\forall i \neq j$.*

Proof: See Refs. [26] and [27].   ■

5.2 Illustrative Example

We apply the virtual leader/virtual structure approach [28, 29, 30, 31] to deal with a formation control problem. Let $x_0(s(t))$ denote the parameterized state of the virtual leader of a team of vehicles, where s is a parameter that incorporates error feedback into the whole team through its evolution [31]. Let $x_i^d(s(t))$ represent the desired state of the ith vehicle, which can be defined from $x_0(s(t))$. Suppose that a pointwise control Lyapunov function (CLF) can be found for a smooth parameterized desired trajectory $x_i^d(s)$, that is, $V_i(x_i, x_i^d(s)) = 0$ at $x_i = x_i^d(s)$ for each $s \in [s_s, s_f]$. We also assume that $V_i(x_i, x_i^d(s)) \to \infty$ if $\|x_i - x_i^d(s)\| \to \infty$ for any $s \in [s_s, s_f]$.

In Ref. [32] CLFs are used to define a formation measure function so that a constrained motion control problem of multiple systems is converted into a stabilization problem for one single system. To represent the formation maintenance accuracy, a formation measure function is defined in Ref. [32] as

$$F(x, s) = \sum_{i=1}^{n} \beta_i V_i(x_i(t), x_i^d(s))),$$

where V_i is the pointwise CLF for each vehicle and $\beta_i > 0$. The formation is defined to be preserved if $F(x, s) \leq F_U$, where F_U is an upper bound on the formation measure function $F(x, s)$. Also the evolution speed of s is defined as

$$\dot{s} = \begin{cases} \min \left\{ \frac{v_0}{\delta + \|\frac{\partial x_0(s)}{\partial s}\|}, \frac{-\frac{\partial F}{\partial x} \dot{x}}{\delta + |\frac{\partial F}{\partial s}|} \left(\frac{\sigma(F_U)}{\sigma(F(x,s))} \right) \right\}, & s_s \leq s < s_f \\ 0, & s = s_f \end{cases}, \quad (5)$$

where $\delta > 0$ is a small positive constant, v_0 is the nominal velocity for the formation, and $\sigma(\cdot)$ is a class $\mathcal{K}$ function. Therefore, formation maneuvers are performed in two steps. First, when $s_s \leq s < s_f$, the formation is preserved within some boundary given by F_U. Second, when $s = s_f$, each vehicle is regulated to a constant desired state given by $x_i^d(s_f)$ and reaches (eventually) its final goal.

In Ref. [32], parameter s is implemented at a central location and broadcast to all of the vehicles in the team. The states of each vehicle are also sent to the central location to incorporate error feedback into the evolution of s. Each vehicle then derives its local control law according to the evolution law of s. While this implementation is feasible when a robust central location exists and high bandwidth communication is available, issues such as a single point of failure or stringent intervehicle communication constraints will significantly degrade the overall system performance.

In the following, we study a decentralized scheme, where each vehicle instantiates a local copy of s, denoted as s_i. All the vehicles then exchange their instantiations between local neighbors through intervehicle communication.

The formation measure function for vehicle i is defined as

$$F_i(x_i, s_i) = \beta_i V_i(x_i(t), x_i^d(s_i))),$$

where V_i is the pointwise CLF for each vehicle and $\beta_i > 0$.

The evolution law of s_i is defined as

$$\dot{s}_i = \begin{cases} -\sum_{j=1}^{n} g_{ij}(t) k_{ij}(t)(s_i - s_j) \\ +\min \left\{ \frac{v_0}{\delta + \|\frac{\partial x_0(s_i)}{\partial s_i}\|}, \frac{-\frac{\partial F_i}{\partial x_i} \dot{x}_i}{\delta + |\frac{\partial F_i}{\partial s_i}|} \left(\frac{\sigma(F_{Ui})}{\sigma(F_i(x_i, s_i))} \right) \right\}, \\ \qquad\qquad s_s \leq s_i < s_f \\ -\sum_{j=1}^{n} k_{ij}(t) g_{ij}(t)(s_i - s_j), \quad s_i = s_f \end{cases} \quad (6)$$

where $k_{ij}(t) > 0$ is piecewise continuous and uniformly lower and upper bounded, F_{Ui} is an upper bound on the formation measure function $F_i(x_i, s_i)$. $g_{ii}(t) \triangleq 0$, and $g_{ij}(t)$ is 1 if vehicle i receives s_j from vehicle j at time t and 0 otherwise. In Eq. 6, at $s_s \leq s_i < s_f$ the first term is used to drive $s_i \to s_j$, $\forall i \neq j$, and the second term is used to incorporate feedback from the ith vehicle to the evolution speed of s_i.

Note that the evolution speed of s_i depends on $F_i(x_i, s_i)$ and s_j, $j \in \mathcal{N}_i$, where $\mathcal{N}_i$ represents the ith vehicle's (possibly time-varying) local neighbors. The ith vehicle then derives its local control law according to the evolution law of s_i. The formation is defined to be preserved if $|s_i - s_j| \leq s_U$, $\forall i \neq j$, where s_U is an upper bound on the inconsistency of s_i, and $F_i(x_i, s_i) \leq F_{Ui}$.

Note that

$$\min\left\{ \frac{v_0}{\delta + \left\| \frac{\partial x_0(s_i)}{\partial s_i} \right\|}, \frac{-\frac{\partial F_i}{\partial x_i}\dot{x}_i}{\delta + \left| \frac{\partial F_i}{\partial s_i} \right|} \left(\frac{\sigma(F_{Ui})}{\sigma(F_i(x_i, s_i))} \right) \right\}$$

is bounded at $s_s \leq s < s_f$. From Lemma 1 we know that $|s_i - s_j|$, $\forall i \neq j$, is bounded at $s_s \leq s < s_f$ if there exist infinitely many consecutive uniformly bounded time intervals such that the union of the information-exchange graph across each time interval has a rooted directed spanning tree. Furthermore, we know that $s_i \to s_j$, $\forall i \neq j$, at $s_i = s_f$ under the same condition.

To illustrate, consider fully actuated mobile robot kinematic equations given by

$$\dot{z}_i = u_i, \tag{7}$$

where $z_i = [x_i, y_i]^T$ represents the position of the ith robot, and $u_i = [u_{xi}, u_{yi}]^T$ represents the control input.

We simulate two robots moving in a spiral formation with $s \in [0, 5]$. The desired distance between these two robots is 10 meters. The center of the line connecting the desired positions of the two robots, i.e., the virtual center of the formation, tracks a trajectory $(x_0(s), y_0(s))$ given by $(0, s)$. Also the line connecting the desired positions of the two robots rotates about its center counterclockwise with an angle given by $\omega_0 s$. The desired states for the two robots, denoted by (x_i^d, y_i^d), are given by $(5\cos(\omega_0 s), s + 5\sin(\omega_0 s))$ and $(-5\cos(\omega_0 s), s - 5\sin(\omega_0 s))$, respectively. The two robots start from rest with some initial errors.

Let $V_i = \frac{1}{2}(x_i - x_i^d)^2 + \frac{1}{2}(y_i - y_i^d)^2$, which is a valid (pointwise in s) CLF for the ith robot. Define $F_i(x_i, s_i) = 2V_i(x_i, x_i^d(s))$ as the formation measure function for vehicle i. The local control laws for each robot are derived using the pointwise CLF so that $x_i \to x_i^d(s)$ and $y_i \to y_i^d(s)$ pointwise in s.

In the decentralized scheme, set $F_{U1} = F_{U2} = 0.1$. In Fig. 3, we plot the desired trajectories for robots #1 and #2. The actual trajectories almost coincide with the desired ones. To see the pattern clearly, we let $s \in [0, 15]$. Assume that robots #1 and #2 obtain one another's instantiation of s through communication. Also assume that there exists inconsistency between s_1 and

s_2 at $t = 0$ sec. Fig. 4 shows the inconsistency between s_1 and s_2, denoted by $s_1 - s_2$. Note that $s_1 - s_2$ is bounded and approaches zero as $s_i \rightarrow s_f$, $i = 1, 2$. Fig. 5 shows formation measure function $F_i(x_i, s_i)$, $i = 1, 2$. We can see that $F_i(x_i, s_i)$, $i = 1, 2$, are large at $t = 0$ sec due to the inconsistency between s_1 and s_2 at $t = 0$ sec. Then both $F_i(x_i(t), s_i(t))$, $i = 1, 2$, decrease and stay below F_{Ui} as s_i increases. The tracking errors for robots #1 and #2 are shown in Fig. 6. Note that the tracking errors are bounded and approach zero as s_i approaches s_f.

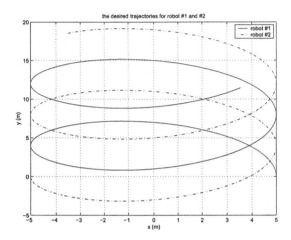

Fig. 3. The desired trajectories for robots #1 and #2.

6 Consensus Building with Information Feedback through State-dependent Weights

6.1 Basic Result

Information feedback can also be introduced to the consensus building process through state-dependent weights. Consider the following algorithm:

$$u_i = -\sum_{j=1}^{n} g_{ij} k_{ij}(t, \xi_i, \{j \in \mathcal{N}_i | \xi_j\})(\xi_i - \xi_j), \qquad (8)$$

where $k_{ij}(\cdot, \cdot, \cdot)$ denotes the state-dependent weights that introduce feedback to the ith vehicle from its local neighbors.

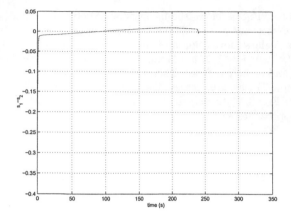

Fig. 4. The inconsistency between s_i using the decentralized scheme.

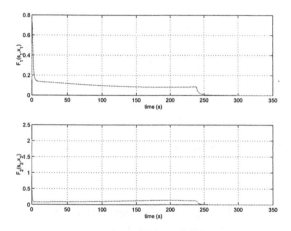

Fig. 5. Formation measure function $F_i(x_i, s_i)$ using the decentralized scheme.

6.2 Illustrative Example

In cooperative control systems, vehicles might move in or out of each other's communication or sensing range. As a result, the information-exchange links between the vehicles might be established or broken randomly. It is relevant to study how a given connectivity pattern between the vehicles can be maintained. The problem of preserving connectivity constraints has been discussed in Refs. [34, 35] recently.

As a preliminary study, we apply the consensus algorithm Eq. 8 to drive multiple mobile agents to reach rendezvous. It is assumed that each agent has

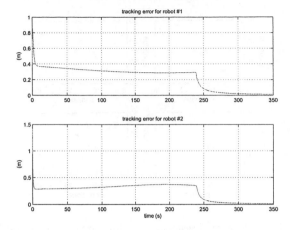

Fig. 6. The tracking errors for robots #1 and #2 using the decentralized scheme.

a limited information-exchange range, and the information-exchange topology is connected initially. Fig. 7 shows the case where the weights k_{ij} are constant. Note that the connectivity of the information-exchange topology cannot be maintained, and the agents form two separated subgroups. As a comparison, Fig. 8 shows the case that the weights k_{ij} are adjusted dynamically such that neighboring vehicles do not move out of their information-exchange range. Note that under the same initial conditions the connectivity between those agents is maintained, and the team reaches rendezvous.

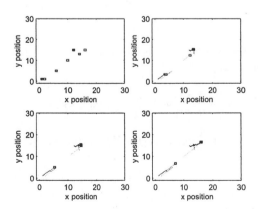

Fig. 7. Rendezvous of seven agents with fixed weights.

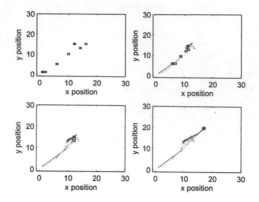

Fig. 8. Rendezvous of seven agents with state-dependent weights.

7 Consensus Building with Information Feedback through Reference States

7.1 Basic Result

Information feedback to the consensus building process can also be introduced through a reference state, which might be a function of vehicle/environmental dynamics or sensor measurement.

Let $\xi^r \in \mathbb{R}^m$ be the reference state and assume that ξ^r satisfies the following dynamics:

$$\dot{\xi}^r = f(t, \xi^r). \tag{9}$$

In the case that only a portion of the vehicles have access to ξ^r, we apply the following consensus algorithm:

$$u_i = \frac{1}{\eta_i} \sum_{j=1}^{n} g_{ij} k_{ij} [u_j - \gamma_i(\xi_i - \xi_j)]$$

$$+ \frac{1}{\eta_i} g_{i(n+1)} \alpha_i [f(t, \xi^r) - \gamma_i(\xi_i - \xi^r)], \tag{10}$$

where $k_{ij} > 0$, $\gamma_i > 0$, $\alpha_i > 0$, $g_{ii} \triangleq 0$, g_{ij} is 1 if information flows from vehicle j to vehicle i and 0 otherwise, $g_{i(n+1)}$ is 1 if vehicle i has access to ξ^r and 0 otherwise, and $\eta_i = g_{i(n+1)} \alpha_i + \sum_{j=1}^{n} g_{ij} k_{ij}$.

Note that k_{ij} and α_{ij} in Eq. 10 can be state-dependent weights. For example, we might choose α_{ij} as $\alpha_i(t, x_i, \{j \in \mathcal{N}_i | x_j\})$ so that the performance of a vehicle and its neighbors can affect how accurate the vehicle wants to track the reference state. Also note that under certain circumstances information feedback can also be introduced directly to the reference model.

7.2 Illustrative Example

We study a formation control problem where four vehicles are required to maintain a certain formation geometry while the formation centroid needs to follow a reference trajectory. We show how feedback can be incorporated into the consensus building process through a reference state, information flow, and state-dependent weights.

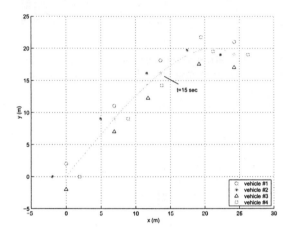

Fig. 9. Formation geometries of the four vehicles in the presence of disturbance, where the information topology is given by Fig. 2a, and the weights α_i are constant.

Letting Eq. 7 denote the vehicle kinematics, the formation control law is given as

$$u_i = \dot{\delta}_i + \frac{1}{\eta_i} \sum_{j=1}^{n} g_{ij} k_{ij} \{u_j - \dot{\delta}_j - \gamma_i[(z_i - z_j) - (\delta_i - \delta_j)]\}$$

$$+ \frac{1}{\eta_i} g_{i(n+1)} \alpha_i(t, z_i, \{j \in \mathcal{N}_i | z_j\}) [\dot{\xi}^r - \gamma_i(z_i - \delta_i - \xi^r)], \qquad (11)$$

where $\delta_i - \delta_j$, $\forall i \neq j$, denotes the desired separation between the vehicle states.

Let $\xi^r = [30\sin(\frac{\pi t}{100}), 20\sin(\frac{\pi t}{50})]^T$, $\delta_1 = [0, 2]^T$, $\delta_2 = [-2, 0]^T$, $\delta_3 = [0, -2]^T$, and $\delta_4 = [2, 0]^T$. Also let $k_{ij} = \alpha_i = \gamma_i = 1$. The desired formation geometry is a diamond shape. Assume that the information-exchange

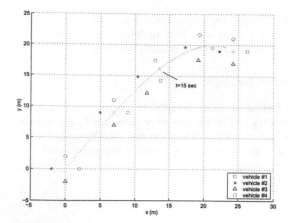

Fig. 10. Formation geometries of the four vehicles in the presence of disturbance, where the information topology is given by Fig. 2b, and the weights α_i are constant.

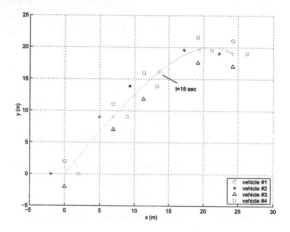

Fig. 11. Formation geometries of the four vehicles in the presence of disturbance, where the information topology is given by Subplot (b) in Fig. 2 and the weights α_i are time-varying and state-dependent.

topologies between the vehicles are given by Fig. 2, and only vehicle 1 has access to ξ^r.

In the following we consider three cases. In Case 1, we assume that the information-exchange topology is given by Fig. 2a, and α_i are constant. In Case 2, we assume that the information-exchange topology is given by Fig. 2b, and α_i are constant. In Case 3, we assume that the information-exchange topology is given by Fig. 2b, and α_i are state-dependent.

Figs. 9, 10, and 11 show the formation geometries of the four vehicles at $t \in \{0, 7.5, 15, 22.5, 30\}$ sec in Cases 1, 2, and 3 respectively, where vehicle 3 is disturbed at $t \in [10, 20]$ sec. The green trajectory represents $\xi^r(t)$, $t \in [0, 30]$ sec and stars represent $\xi^r(t)$ at $t \in \{0, 7.5, 15, 22.5, 30\}$ sec. In Case 1, vehicles 3 and 4 are left behind due to the disturbance to vehicle 3 while vehicles 1 and 2 keep moving forward by following $\xi^r(t)$ as shown in Fig. 9 at $t = 15$ sec. As a result, the desired formation geometry breaks down at $t = 15$ sec. As a comparison, with information feedback introduced from vehicle 3 to vehicle 2 and from vehicle 2 to vehicle 1 in Case 2, vehicles 1 and 2 slow down for vehicles 3 and 4 to catch up as shown in Fig. 10 at $t = 15$ sec. In Case 3, we further introduce state-dependent α_i, which decreases as the formation maintenance accuracy decreases. In other words, if one or more vehicles are left behind due to disturbance, it might be desirable that other vehicles can slow down by sacrificing their performance to track ξ^r. Note that the formation geometry is preserved the best in Case 3 as shown in Fig. 11 at $t = 15$ sec. Note that the motion of ξ^r remains unaffected even if some vehicles are left behind due to distance as shown in Figs. 9, 10, and 11. Although not shown here, it is also possible to introduce feedback directly to the evolution law of ξ^r so that the reference state can adjust its motion according to vehicle performance.

8 Conclusion

We have studied the problem of consensus building in multi-vehicle systems with information feedback. Four strategies of introducing feedback to the consensus building process have been presented including information flow, external feedback terms, state-dependent weights, and reference states. Illustrative examples have also been demonstrated as a proof of concept.

9 ACKNOWLEDGMENTS

The author gratefully acknowledges Nathan Sorensen for his help with the third example in the chapter.

References

1. R. Olfati-Saber and R. M. Murray, *IEEE Transactions on Automatic Control* **49**, 1520, 2004.
2. J. A. Fax and R. M. Murray, *IEEE Transactions on Automatic Control* **49**, 1465, 2004.

3. A. Jadbabaie, J. Lin and A. S. Morse, *IEEE Transactions on Automatic Control* **48**, 988, 2003.
4. Z. Lin, M. Broucke and B. Francis, *IEEE Transactions on Automatic Control* **49**, 622, 2004.
5. L. Xiao and S. Boyd, *Systems and Control Letters* **53**, 65, 2004.
6. L. Moreau, *IEEE Transactions on Automatic Control* **50**, 169, 2005.
7. W. Ren and R. W. Beard, *IEEE Transactions on Automatic Control* **50**, 655, 2005.
8. Y. Hatano and M. Mesbahi, *IEEE Transactions on Automatic Control* **50**, 1867, 2005.
9. D. Bauso, L. Giarre and R. Pesenti, Distributed consensus in networks of dynamic agents, in *Proceedings of the IEEE Conference on Decision and Control*, Seville, 2005.
10. L. Fang and P. J. Antsaklis, Information consensus of asynchronous discrete-time multi-agent systems, in *Proceedings of the American Control Conference*, (Portland, OR, 2005).
11. J. Lin, A. S. Morse and B. D. O. Anderson, The multi-agent rendezvous problem - the asynchronous case, in *Proceedings of the IEEE Conference on Decision and Control*, Paradise Island, 2004.
12. S. Martinez, J. Cortes and F. Bullo, On robust rendezvous for mobile autonomous agents, in *IFAC World Congress*, Prague, 2005.
13. J. S. Caughman, G. Lafferriere, J. J. P. Veerman and A. Williams, *Systems and Control Letters* **54**, 899, 2005.
14. R. Olfati-Saber, *IEEE Transactions on Automatic Control* **51**, 401, 2006.
15. H. G. Tanner, A. Jadbabaie and G. J. Pappas, Stable flocking of mobile agents, part ii: Dynamic topology, in *Proceedings of the IEEE Conference on Decision and Control*, Maui, 2003.
16. J. R. Lawton and R. W. Beard, *Automatica* **38**, 1359, 2002.
17. W. Ren and R. W. Beard, *AIAA Journal of Guidance, Control, and Dynamics* **27**, 73, 2004.
18. M. Alighanbari and J. How, Decentralized task assignment for unmanned air vehicles, in *Proceedings of the IEEE Conference on Decision and Control*, Seville, 2005.
19. L. Xiao, S. Boyd and S. Lall, A scheme for robust distributed sensor fusion based on average consensus, in *Proceedings of the International Conference on Information Processing in Sensor Networks*, Los Angeles, CA, 2005.
20. D. P. Spanos and R. M. Murray, Distributed sensor fusion using dynamic consensus, in *IFAC World Congress*, Prague, 2005.
21. R. Olfati-Saber and J. S. Shamma, Consensus filters for sensor networks and distributed sensor fusion, in *Proceedings of the IEEE Conference on Decision and Control*, Seville, 2005.
22. G. Royle and C. Godsil, *Algebraic Graph Theory*, Springer Graduate Texts in Mathematics #207, New York, 2001.
23. F. R. K. Chung, Spectral graph theory, in *Regional Conference Series in Mathematics*, American Mathematical Society, 1997.
24. R. A. Horn and C. R. Johnson, *Matrix Analysis* Cambridge University Press, 1985.
25. W. Ren, R. W. Beard and T. W. McLain, Coordination variables and consensus building in multiple vehicle systems, in *Cooperative Control: A Post-Workshop*

Volume 2003 Block Island Workshop on Cooperative Control, V. Kumar, N. E. Leonard and A. S. Morse (eds), Springer-Verlag Series: Lecture Notes in Control and Information Sciences, 2005.

26. W. Ren, R. W. Beard and D. B. Kingston, Multi-agent Kalman consensus with relative uncertainty, in *Proceedings of the American Control Conference*, Portland, OR, 2005.

27. D. B. Kingston, W. Ren and R. W. Beard, Consensus algorithms are input-to-state stable, in *Proceedings of the American Control Conference*, Portland, OR, 2005.

28. M. A. Lewis and K.-H. Tan, *Autonomous Robots* 4, 387, 1997.

29. R. W. Beard, J. R. Lawton and F. Y. Hadaegh, *IEEE Transactions on Control Systems Technology* 9, 777, 2001.

30. N. E. Leonard and E. Fiorelli, Virtual leaders, artificial potentials and coordinated control of groups, in *Proceedings of the IEEE Conference on Decision and Control*, Orlando, Florida, 2001.

31. M. Egerstedt, X. Hu and A. Stotsky, *IEEE Transactions on Automatic Control* 46, 1777, 2001.

32. P. Ogren, M. Egerstedt and X. Hu, *IEEE Transactions on Robotics and Automation* 18, 847, 2002.

33. W. Ren, Consensus Building in Multi-vehicle Systems with Information Feedback, IEEE International Conference on Mechatronics and Automation, Luoyang, China, June 2006, pp. 37–42.

34. M. M. Zavlanos and G. J. Pappas, Controlling connectivity of dynamic graphs, in *Proceedings of the IEEE Conference on Decision and Control*, Seville, 2005.

35. D. P. Spanos and R. M. Murray, Robust connectivity of networked vehicles, in *Proceedings of the IEEE Conference on Decision and Control*, Paradise Island, Bahamas, 2004.

5 Levy Flights in Robot Swarm Control and Optimization

Yechiel J. Crispin

Summary. A stochastic method for the cooperative control of a swarm of mobile robots is presented. The network of mobile robots is modeled by a swarm performing a directed random walk. The swarm dynamics are governed by a system of stochastic difference equations. The motion is controlled by a robot leader, which transmits the coordinates of the best solution to the swarm as the network cooperative control signal. We first treat the problem where the swarm searches for the minimum of a noisy function in a given domain. We then consider the problem of gathering a swarm of robots which is initially randomly dispersed over a domain in the plane. We study the case where the control signal is corrupted by noise and find that the gathering process is robust to noise and efficient. The swarm dynamics display anomalous diffusion and Levy flights, where the robots move along straight lines over many time steps, followed by short random walks in the vicinity of the minimum point or the gathering point.

1 Introduction

The term Levy flight was introduced by Mandelbrot and is described in his book on fractals [8] as a sequence of jumps separated by stopovers. In plates 296 and 297 in his book, he gives the example where each stopover is a star, a galaxy or a cluster of stars or galaxies, thus showing that the global structure of matter distribution in the universe is composed of clusters separated by Levy flights. The clusters themselves can be decomposed into self-similar miniclusters, resulting in a fractal structure. Since then, other phenomena have been described as displaying Levy flights and anomalous diffusion. Levy flights and anomalous diffusion of tracer particles have been observed in a two-dimensional rotating fluid flow, see [12, 13, 15, 16]. Levy flights in physics have been described in [10]. In the present work, we show that controlled swarm robotic motion displays anomalous diffusion and Levy flights.

Robots are used in many practical applications such as industrial robots in manufacturing, spacecraft and rover robots for space exploration and unmanned aerial vehicles (UAVs) for reconnaissance, surveillance and tactical military missions. Other applications include underwater missions by autonomous underwater vehicles (AUVs) such as formation control and rendezvous, search and rescue missions and exploration and mapping of unknown environments. In many applications, single robots are employed in the performance of a given task. It has been recognized for some time, however, that the use of collaborating multiple mobile robots can have significant advantages in achieving complex tasks and missions, which otherwise might not be achievable through the use of single robots. Consequently, in recent years, there has been an interest in the cooperative control of networked collaborating mobile robots with distributed resources such as sensors, computing power and communications [4, 11, 14].

We first treat the problem where a group of mobile robots such as a group of autonomous underwater vehicles is assigned with the collective task of exploring a limited domain of a body of water such as a lake, sea or a limited part of the ocean. Let's consider a specific task where the robots have to collectively find the extremum of an otherwise unknown function using a given experimental method where they can measure a property of the environment such as water depth or temperature at any required point, see for example [1, 3]. The simplest such test problem is for the group of robots to collectively search the maximum depth in a limited area; for example a group of 20 robots searching an area on the order of 2 km by 2 km and measuring the depths using a depth finder device. Another problem might be to find the minimum temperature or maximum concentration of some chemical compound or density of plankton in a three dimensional domain.

We then consider the problem where of a group of mobile robots has been dispersed in a given area, and that it is required to gather the robots in the vicinity of a given point. For example, consider the case where a group of robots has to perform a mission in a remote area, where they have landed by parachutes. The robots are now randomly scattered in a wide area and need to be gathered into a much smaller area in the vicinity of a designated location before starting their mission. The specific task now is for the robots to collectively move towards the gathering point. We consider swarms on the order of 200 robots dispersed in a two-dimensional domain on the order of 1 km by 1 km.

Each autonomous robot is equipped with a compass and is capable of moving in a given azimuthal direction for a given distance. Each robot has a low level control and navigation system that can detect its location at all times and guide it from one point in the domain to the next at the right speed and orientation. It is also assumed that each autonomous robot is equipped with a collision and obstacle avoidance control system for preventing collisions with other robots and obstacles. The robots' network architecture consists of a leader robot acting as a server and communicating with the other robots as clients.

Details of the proposed robot swarm cooperative control method will be described in the next section. Each robot has a microprocessor computing device on board capable of running the robot swarm algorithm. We propose to use this algorithm as a top level discrete event controller for the cooperative control of the swarm. Each robot sends the best solution found at any given time to the leader or other central processing station through its communication channel. The leader in turn computes the global best solution and transmits the result as a control signal to the network. The Robot Swarm Optimization (RSO) is a stochastic population based method that belongs to the class of biologically inspired algorithms. It is based on the paradigm of a swarm of insects performing a collaborative task such as ants or bees foraging for food using chemical or some other type of communication, see for example [2, 5]. The swarm intelligence method was originally developed by Eberhart [6] and later described in great detail in [7]. An overview of the method as extensively applied to various function optimization problems of increasing difficulty has recently been presented by Parsopoulos [9].

In the next section we develop the robot swarm algorithm with communication noise and we explain how it can be applied to solve the problems described above, namely, the collaborative search for a minimum of a function and the swarm gathering problem. The search for a function's minimum is described in section 3 for a two-dimensional domain and in section 4 for a three-dimensional domain. In section 5, results of simulations are described for a swarm of 200 robots, gathering in a noisy environment. We show that the robots trajectories follow Levy flights and compute the probability distribution for the flights' lengths.

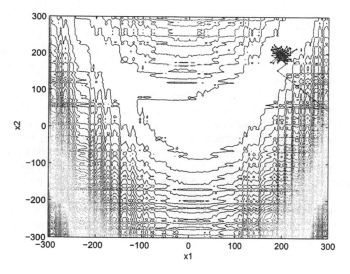

Fig. 1. Contour plot of the noisy banana function. Unlike gradient methods, the robot swarm algorithm can deal with rugged topography.

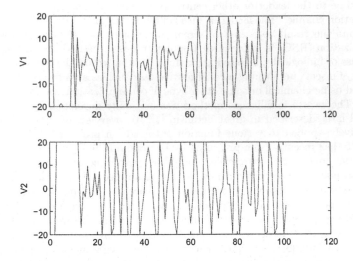

Fig. 2. Velocity components V_1 and V_2 in the X_1 and X_2 directions as a function of the time counter k. The constraint $|V| < \Delta X_{\max}/\Delta t$ is apparent.

2 Cooperative Control of the Robot Swarm

In developing the robot swarm cooperative control method, we incorporate physical effects or constraints in order to implement the search method by actual mobile robots such as land vehicles, autonomous underwater vehicles or autonomous unmanned aerial vehicles. The first effect imposes a limitation on the speed of the vehicle, or equivalently, a limit on the distance ΔX_{max} it can move in a given typical time step Δt. Another effect taken into account is imperfect and noisy communication between the robots. At any given time, communication with one or more robots can be attenuated or corrupted by noise. Therefore, rather than assuming that the global minimum is available to the swarm at all times as in the case of perfect communication, we introduce noise in the control signal transmitted to all members of the swarm.

The robot swarm cooperative control algorithm without any robot speed constraints and with perfect communication consists of minimizing a function of several variables:

$$\text{minimize} f(X), \text{where } X \in \Omega \subset \mathbb{R}^n \text{ and} f : \Omega \mapsto \mathbb{R}$$

subject to the side constraints

$$X_{\min} \leq X \leq X_{\max}$$

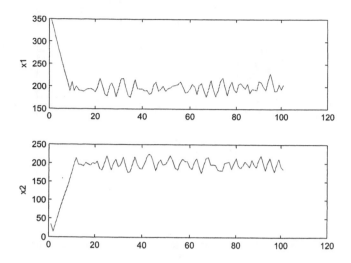

Fig. 3. Coordinates of the best trajectory X_1 and X_2 as a function of the time counter k. A Levy flight is followed by sticking random walks.

using a directed random walk process described by the following system of stochastic difference equations:

$$X^i(k+1) = X^i(k) + \Delta X^i(k+1) \tag{1}$$

$$\Delta X^i(k+1) = w(k)X^i(k) + c_1 r_1^i(P^i(k) - X^i(k)) + c_2 r_2^i(P^g(k) - X^i(k)) \tag{2}$$

Here k is the discrete time counter, c_1 and c_2 are real constants, r_1^i and r_2^i are random variables uniformly distributed between 0 and 1. The superscript index i denotes robot number $i \in [1, N_R]$ where N_R is the number of robots in the swarm. The location $P^i(k)$ is the best solution found by robot i at time $t = k$ and $P^g(k)$ is the global minimum at time $t = k$. The factor $w(k)$ can be either constant or time dependent. If it decreases with time, the search process can usually be improved as the search approaches the global minimum and smaller steps are needed for better resolution. For example, the parameter $w(k)$ can be set to decrease from an initial value of $w_0 = 0.8$ to a final value of $w_f = 0.2$ after N time steps:

$$w(k) = w_f + (w_0 - w_f)(N - k)/N \tag{3}$$

The system of equations (1)-(2) describes a directed random walk for each robot i in the swarm, similar to a Brownian motion of a tracer particle in a fluid. Whereas Brownian motion is an undirected random motion, the motion of a robot in the swarm will have a velocity that will start as a random

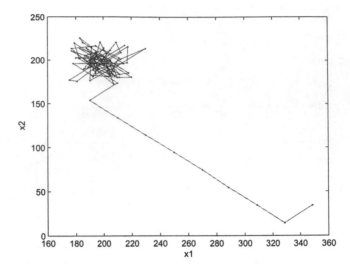

Fig. 4. Same trajectory as in Fig. 1 with the end point at the minimum (200,200). Notice the Levy flight along a straight line segment and random short walks around the minimum.

motion, but will eventually decay as the particle approaches a point $P^i(k)$ in the domain where the function reaches a local minimum and as the swarm as a whole approaches a point $P^g(k)$ of the domain where the function reaches a global minimum, that is,

$$P^i(k) = \mathrm{argmin}\{f(X^i(k))\},$$
$$P^g(k) = \mathrm{argmin}\{f(P^i(k))\}, \ i \in [1, N_R] \tag{4}$$

The following initial conditions are needed in order to start the solution of the system of difference equations

$$X^i(0) = X_{\min} + r^i \Delta X_{\max} \tag{5}$$

$$\Delta X_{\max} = (X_{\max} - X_{\min})/N_x \tag{6}$$

N_x is a typical number of grid segments along each component of the position vector X. For example, if the domain consists of a two dimensional square domain of $a = 2000$ m by $a = 2000$ m, then with $N_x = 100$, we can use a typical distance segment of $\Delta X_{\max} = a/N_x = 20$ m. If we take a typical speed of an autonomous robot as $V_c = 1$ m/s, then the characteristic time will be $t_c = \Delta X_{\max}/V_c = 20$ s. We can now measure X in units of $\Delta X_{\max}$, V in units of V_c and Δt in units of t_c. The equations will then have exactly the same form in non-dimensional variables.

Placing a limit on the magnitude of the velocity component of each robot in any given direction for a given time step, we can impose a constraint on the magnitude of the distance traveled in any time step as:

$$|\Delta X^i(k+1)| < \Delta X_{\max} \tag{7}$$

Under these assumptions, the equations of motion of the swarm become:

$$X^i(k+1) = X^i(k) + \text{sign}(\Delta X^i(k+1))(\min[|\Delta X^i(k+1)|, \Delta X_{\max}]) \tag{8}$$

$$\Delta X^i(k+1) = w(k)X^i(k) + c_1 r_1^i(P^i(k) - X^i(k)) + c_2 r_2^i(P^g(k) - X^i(k)) \tag{9}$$

subject to the side constraint

$$X_{\min} \leq X^i(k+1) \leq X_{\max} \tag{10}$$

The signum function term $\text{sign}(\Delta X^i(k+1))$ is added in order to keep the original direction of the motion while reducing the length of the step.

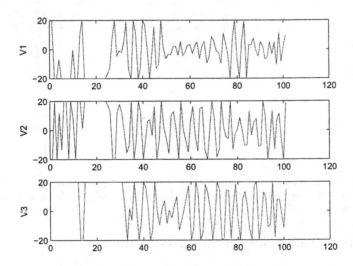

Fig. 5. Velocity components V_1, V_2 and V_3 in the X_1, X_2 and X_3 directions as a function of the time counter k. The speed constraint $|V|_{\max} < \Delta X_{\max}/\Delta t$ is apparent.

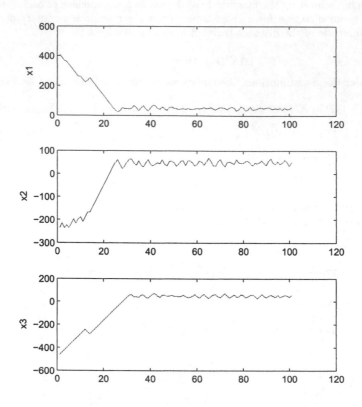

Fig. 6. Coordinates of a typical trajectory X_1, X_2 and X_3 as a function of the time counter k. Levy flights are followed by sticking random walks in the vicinity of the minimum.

3 Collaborative Search in a 2-D Domain

The cooperative control method developed in the previous section is applied to the problem of experimentally finding the minimum of a noisy scalar function of two real variables. For instance, in the context of a group of underwater vehicles, the problem consists of finding the minimum of a scalar quantity such as depth, temperature, or the concentration of a chemical or biological species, through the measurement of the scalar quantity by the autonomous robots as they perform a search process in the domain. We would like to keep the robots' resources to a minimum, so we limit the number of robots to 20, although we were able to minimize 2-D functions with as little as 6 robots.

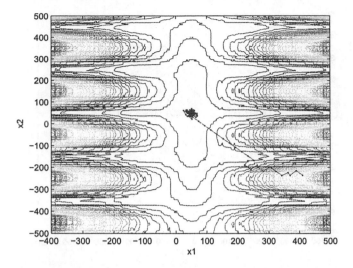

Fig. 7. Contour plot of the projection of $f(X_1, X_2, X_3)$ on the plane $X_3=50$, with the trajectory of one robot. The minimum is located at (50,50,50).

We treat the case where the function is not smooth. The function and its derivatives can have many discontinuities in the domain. For example, the measurements of the robots' coordinates and the measurements of the function can be corrupted by noise, so that we cannot rely on methods that are based on gradient techniques. As an example, let's select a two-dimensional test function for which it is not easy to find the minimum, such as the banana function that has a deep, narrow and curved valley. We then introduce noise in the measurements of the coordinates X_1 and X_2 such as to obtain a noisy and rugged type of topography for the contours of the function of two variables. The banana function is defined by:

$$f(X_1, X_2) = 10(X_1/d)^4 - 20(X_1/d)^2(X_2/d)+$$
$$+10(X_2/d)^2 + (X_1/d)^2 - 2(X_1/d) + 5$$

where $d = 200$ m is a scaling length. Consider a square two dimensional domain of 2000 m by 2000 m, defined by the coordinates:

$$X_1 \in [X_{\min}, X_{\max}] = [-1000, 1000], \quad X_2 \in [X_{\min}, X_{\max}] = [-1000, 1000],$$

where we choose the origin in the center of the square and the number of grid segments as $N_x = 100$, so that the maximum distance traveled by any robot in any direction X_1 or X_2 in one time step is 20 m, which is one distance unit

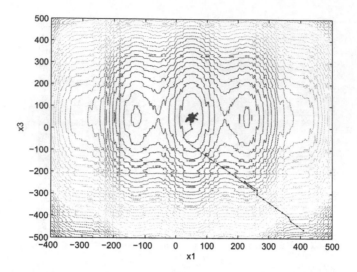

Fig. 8. Contour plot of the projection of $f(X_1,X_2,X_3)$ on the plane $X_2=50$, with the trajectory of one robot. The minimum is located at (50,50,50).

or 1 DU. The equivalent time unit TU is the time to travel along 1 DU at a typical speed of 1 m/s, i.e. 20 s.

$$\Delta t = 1\text{TU} = 20\text{s}$$

$$\Delta X_{\max} = (X_{\max} - X_{\min})/N_x = 20\text{m} = 1\text{DU}$$

$$|V|_{\max} \leq \Delta X_{\max}/\Delta t = 1\text{DU}/1\text{TU} = 20\text{m}/20\text{s} = 1\text{m/s}$$

The measured coordinates are corrupted by noise, so that every time the function is evaluated, there is a random error given by:

$$\Delta f(X_1, X_2) = f(X_1 + \eta, X_2 + \eta) - f(X_1, X_2)$$
$$\eta = \mathcal{N}(0, \sigma)$$

where $\mathcal{N}(0, \sigma)$ denotes random numbers having a normal distribution with zero mean and standard deviation σ. For the Gaussian noise η the standard deviation is

$$\sigma = \delta \Delta X_{\max}$$

We show results for $\delta = 0.4$. The number of autonomous robots is $N_R = 20$. The other parameters appearing in the equations of motion are $c_1 = c_2 = 2$. We chose a constant value of $w(k)$, i.e. an initial value of $w_0 = 0.8$ and a final value of $w_f = 0.8$.

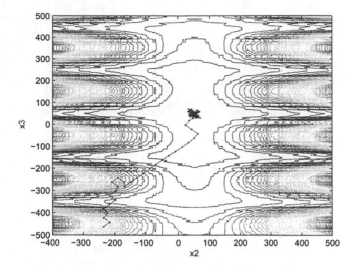

Fig. 9. Contour plot of the projection of $f(X_1,X_2,X_3)$ on the plane $X_1=50$, with the trajectory of one robot. The minimum is located at $(50,50,50)$.

$$w(k) = w_0 = \text{const}$$

The results of a simulation of the 20 robots as they search for the minimum of the noisy and rugged banana function are given in Figs. 1-4. The 20 robots were spread randomly over the domain at the start of the simulation, which was run over $N = 100$ steps. At the end of the simulation the trajectory of the robot that came closest to the location of the minimum at the point $(X_1, X_2) = (200, 200)$ was chosen for display.

A contour plot of the noisy banana function is given in Fig. 1, which also displays a typical trajectory of one robot in the swarm. The velocity components are shown in Fig. 2, with the limitation on the absolute values of the velocity components shown along some segments of the motion. The trajectory in parametric form, i.e., with the event counter k as a parameter, is given in Fig. 3. Notice the long segment of motion, known as a "Levy flight" with maximum speed along a straight line and segments where the robot performs a random walk about the same location. For clarity, the same trajectory is shown in Fig. 4 without the contour lines. The long Levy flights along straight lines are noticeable in the figures.

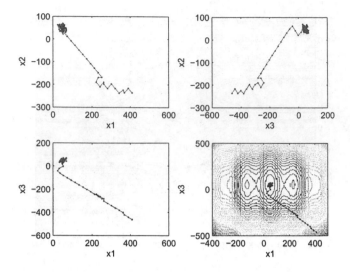

Fig. 10. Projections of a typical trajectory on the plane (X_1, X_2), the plane (X_2, X_3) and the plane (X_1, X_3). Levy flights are followed by sticking random walks in the vicinity of the minimum.

4 Collaborative Search in a 3-D Domain

Here we consider a more difficult search problem, in which the swarm of robots is performing a search for the minimum of a scalar function of three real variables. In the context of autonomous underwater vehicles, the task here is to find a minimum temperature or maximum concentration of a chemical or biological species in a three-dimensional domain. The number of robots is limited to 20. We select a 3-D test function taken from the literature, for example the Levy No. 8 function [9]. Consider a three-dimensional body of water with sides $a = 1000$ m by $a = 1000$ m by $a = 1000$ m deep. Select the origin of a cartesian system of coordinates in the center of this cube, such that the domain is defined by:

$$X_1 \in \mathcal{D}, X_2 \in \mathcal{D}, X_3 \in \mathcal{D}$$

$$\mathcal{D} = [X_{\min}, X_{\max}] = [-500, 500]$$

The Levy No. 8 function is defined by

$$f(X_1, X_2, X_3) = sin^2(\pi y_1) + f_1(y_1, y_2) + f_2(y_2, y_3) + (y_3 - 1)^2$$

$$f_1(y_1, y_2) = (y_1 - 1)^2(1 + 10sin^2(\pi y_2))$$

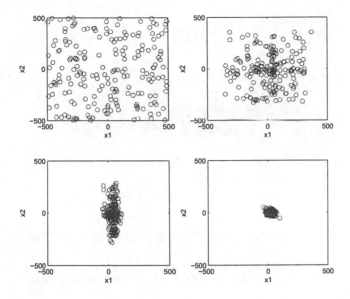

Fig. 11. Top left: Initial random swarm distribution in the domain. Top right: Distribution after 10 time steps. Bottom left: after 40 time steps. Bottom right: after 80 time steps. $N_R = 200$ robots.

$$f_2(y_2, y_3) = (y_2 - 1)^2 (1 + 10 sin^2(\pi y_3))$$

$$y_1 = 1 + (x_1 - 1)/4, \; y_2 = 1 + (x_2 - 1)/4, \; y_3 = 1 + (x_3 - 1)/4$$

where the coordinates x_1, x_2, x_3 are scaled by a length $d = 50$ m:

$$x_1 = X_1/d, \; x_2 = X_2/d, \; x_3 = X_3/d$$

The number of autonomous robot vehicles is $N_R = 20$. The other parameters appearing in the equations of motion are the same as in the previous case of a 2-D function. The results of a simulation of the group of robots collaboratively searching for the minimum of the function of three variables are given in Figs. 5-10. The three velocity components are shown in Fig. 5. The constraint on the absolute values of the velocity components is active along some segments of the trajectory, especially at the beginning of the motion. The three coordinates as a function of the event counter k are given in Fig. 6. A typical trajectory of a robot in the swarm is also shown in Figs.7-10 together with contour plots of the projections of the f(X_1,X_2,X_3) function on the plane $X_3 = 50, X_2 = 50$ and $X_1 = 50$, respectively. The minimum is located at the point (50, 50, 50). Again there are segments of relatively long Levy flights at

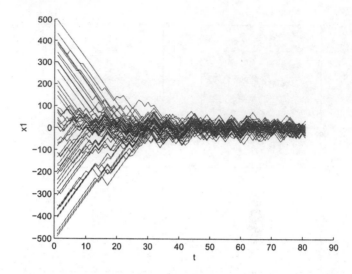

Fig. 12. Trajectories $X_1(t)$ of the first 50 robots in the swarm.

maximum speed along straight lines and segments where the robot performs a random walk in a limited area.

5 Swarm Gathering in a 2-D Domain

The cooperative control method is applied to the problem of gathering a swarm of robots at a given point in the plane. We consider a two-dimensional domain $\Omega \subset \mathbb{R}^2$, defined by the coordinates:

$$X_1 \in [X_{\min}, X_{\max}] = [-500, 500], \quad X_2 \in [X_{\min}, X_{\max}] = [-500, 500]$$

which forms a square of 1000 m by 1000 m, with the origin at the center of the square. We choose the number of grid segments as $N_x = 50$, so that the maximum distance traveled by any robot in any direction X_1 or X_2 in one time step is 20 m, which is defined as one distance unit or 1 DU. The equivalent time unit $\Delta t =$TU$= 20$ s is the time it takes a robot to travel along 1 DU at a typical speed of 1 m/s.

$$\Delta X_{\max} = (X_{\max} - X_{\min})/N_x = 20\text{m} = 1\text{DU}$$

$$|V|_{\max} \leq \Delta X_{\max}/\Delta t = 1\text{DU}/1\text{TU} = 1\text{m/s}$$

Initially, the swarm is randomly distributed in the domain Ω or in a subset domain of Ω. At time $k = 0$, the control is started and the swarm is set in

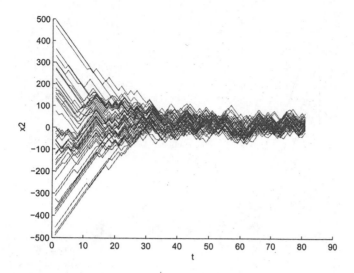

Fig. 13. Trajectories $X_2(t)$ of the first 50 robots in the swarm.

motion. Each robot in the swarm is programmed to minimize its distance from the gathering point, by minimizing the function:

$$f(X_1^i, X_2^i) = (X_1^i - P_1^g)^2 + (X_2^i - P_2^g)^2$$

Here the control signal $P^g = (P_1^g, P_2^g)$ transmitted to each member of the swarm specifies the gathering point and (X_1^i, X_2^i) is the location of the ith robot in the swarm, where $i \in [1, N_R]$. The communication signal is corrupted by additive noise η:

$$P^g = (P_1^g + \eta, P_2^g + \eta)$$

$$\eta = \mathcal{N}(0, \sigma)$$

$$\sigma = \delta \Delta X_{\max}$$

where $\mathcal{N}(0, \sigma)$ denotes random numbers having a normal distribution with zero mean and standard deviation σ. Without loss of generality, we choose the gathering point at the origin, i.e., $(P_1^g, P_2^g) = (0, 0)$.

For the Gaussian noise η we chose a standard deviation $\sigma = \delta \Delta X_{\max}$ with $\delta = 5$. As the noise level is increased, say with values of $\delta = 10, 20, 30$, it becomes more difficult to gather all the swarm in the vicinity of the origin. The other parameters appearing in the equations of motion are $c_1 = c_2 = 2$ and $w_0 = w_f = 0.8$. The results of a simulation of the gathering of the 200 robots are given in Figs. 11-16. The simulation was run for N= 80 time steps.

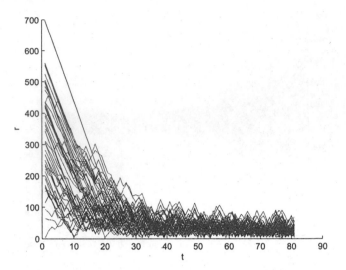

Fig. 14. Radial distances r(t) of the first 50 robots in the swarm.

Fig. 11 shows the locations of the robots as they were spread randomly over the domain at the start of the simulation, the robots locations after 10 time steps, the locations after 40 time steps and the locations of the swarm as the robots gathered in the vicinity of the origin after 80 time steps.

The trajectories of the first 50 robots in the swarm are shown in Figs. 12-14. Fig. 12 displays the coordinates $X_1(t)$ as a function of time. The coordinates $X_2(t)$ as a function of time are shown in Fig. 13. Most Levy flights occur at the beginning of the gathering motion, up to about 30 time steps. After that the swarm aggregates in the vicinity of the gathering point. This can also be seen in Fig. 14, which displays the radial distances $r(t)$ from the origin, where $r^2(t) = X_1^2(t) + X_2^2(t)$ for the first 50 robots in the swarm.

In order to obtain the probability distribution of the Levy flights, the lengths of flights along straight lines are followed for each robot in the swarm. Then the number of flights for each given length are counted for all the robots in the swarm and put in 25 bins ordered from the shortest to the longest flights. Then a histogram is plotted showing the frequency of occurrence of the various flight lengths. Such a histogram is shown in Fig. 15. The histogram does not follow a Gaussian distribution, but rather a Levy distribution, which has a very long tail and an infinite variance.

Levy flights follow power laws of the form

$$N = (L/L_0)^\alpha$$

which appear as straight lines with slope α when displayed on a log-log scale

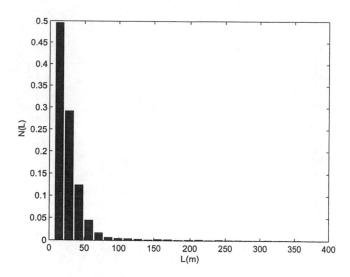

Fig. 15. Probability distribution of the Levy flights for the 200 robots.

$$logN = \alpha log(L/L_0) = \alpha logL - \alpha logL_0$$

where N is the number of flights of length L and L_0 is a characteristic length. For the noise level described above, a value of $\alpha = -2.446$ and a characteristic length $L_0 = 11.62$ were obtained. Such a plot is shown in Fig. 16.

6 Conclusions

A method for the cooperative control of a group of robots based on a stochastic model of swarm motion has been developed. The network of mobile robots is modeled by a swarm moving randomly in the search domain with the global motion of the swarm directed and controlled by a central unit which can be a leader robot or a central server. The motion of each robot in the swarm is governed by a system of two stochastic difference equations.

We first treated the problem where the robots swarm searches for the minimum of a noisy function in a given two-dimensional domain as well as in a three-dimensional domain. In both cases the swarm detects the location of the minimum point of the function despite the fact that the function is not smooth and its arguments can be corrupted by noise originating in the measuring device or environmental noise such as fluid turbulence.

Usually, in the robot swarm method developed in this work, the best solution found collectively by the swarm serves as the control signal for the

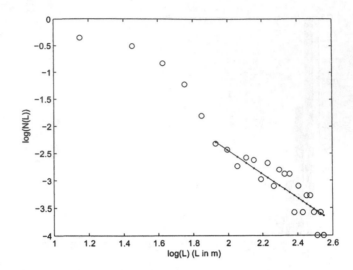

Fig. 16. Power law of the Levy flights for the tail of the distribution.

network of robots. However, in the swarm gathering problem, the problem is simpler, since the coordinates of the gathering point, which serve as the control signal, are fixed, except for additive noise that is present in the communication system.

The method was then used to solve the basic problem of collaborative gathering of a swarm of robots in a two-dimensional domain. It was found that the swarm can gather successfully in the vicinity of a designated point in the plane despite significant noise in the network communications. Moreover, it was found that the gathering process is efficient, in the sense that the robots' trajectories exhibit anomalous diffusion, performing long distance Levy flights along straight lines, followed by local sticking random walks in a limited area of the domain in the vicinity of the gathering point.

References

1. Bachmayer, R. and Leonard, N.E., "Vehicle Networks for Gradient Descent in a Sampled Environment", in Proceedings of the 41st IEEE Conference on Decision and Control, 2002.
2. Bonabeau, E., Dorigo, M. and Theraulaz, G., Swarm Intelligence, From Natural to Artificial Systems, Santa Fe Institute in the Sciences of Complexity, Oxford University Press, New York, 1999.

3. Burian, E., Yoerger, D., Bradley, A. and Singh, H., "Gradient Search with Autonomous Under-water Vehicles Using Scalar Measurements", in Proceedings of the IEEE Symposium on Autonomous Underwater Vehicle Technology, 1996.

4. De Sousa, J.B. and Pereira, F.L., "On Coordinated Control Strategies for Networked Dynamic Control Systems: an Application to AUVs", in Proceedings of the 15th International Symposium of Mathematical Theory of Networks and Systems, 2002.

5. Dorigo, M., Di Caro, G. and Gambardella, L.M., "Ant Algorithms for Discrete Optimization", Artificial Life, Vol.5, pp.137-172, 1999.

6. Eberhart, R.C. and Kennedy, J., "A New Optimizer Using Particle Swarm Theory", Proceedings of the 6th Symposium on Micro Machine and Human Science, IEEE Service Center, Piscataway, NJ, 1995.

7. Kennedy, J. and Eberhart, R.C., Shi, Y., Swarm Intelligence, Morgan Kaufman Publishers, Academic Press, San Francisco, 2001.

8. Mandelbrot, B. B., The Fractal Geometry of Nature, W.H. Freeman and Company, San Francisco, CA, Revised Edition, 1983.

9. Parsopoulos, K.E. and Vrahatis, M.N., "Recent Approaches to Global Optimization Problems Through Particle Swarm Optimization", Natural Computing, Vol.1, pp.235-306, 2002.

10. Schlesinger, M.F., Zaslavsky, G.M. and Frisch, U. (eds), Levy Flights and Related Topics in Physics, Springer-Verlag, Berlin, 1995.

11. Silva, J., Speranzon, A., de Sousa, J.B. and Johansson, K.H., "Hierarchical Search Strategy for a Team of Autonomous Vehicles", IFAC 2004.

12. Solomon, T.H., Weeks, E.R. and Swinney, H.L., "Observation of Anomalous Diffusion and Levy Flights in a Two-Dimensional Rotating Flow", Physical Review Letters, Vol.71, p.3975, 1993.

13. Solomon, T.H., Weeks, E.R. and Swinney, H.L., "Chaotic Advection in a Two-dimensional Flow: Levy Flights and Anomalous Diffusion", Physica D, 76, pp.70-84, 1994.

14. Speranzon, A., "On Control Under Communication Constraints in Autonomous Multi-Robot Systems", Licentiate Thesis, KTH Signals, Sensors and Systems, Royal Institute of Technology, Stockholm, Sweden, 2004.

15. Weeks, E.R., Urbach, J.S. and Swinney, H.L., "Anomalous Diffusion in Asymmetric Random Walks with a Quasi-geostrophic Flow Example", Physica D, 97, pp.291-310, 1996.

16. Weeks, E.R. and Swinney, H.L., "Anomalous Diffusion Resulting From Strongly Asymmetric Random Walks", Physical Review E, 57, 5, pp.4915-4920, 1998.

6 Eavesdropping and Jamming Communication Networks

Clayton W. Commander, Panos M. Pardalos, Valeriy Ryabchenko, Oleg Shylo, Stan Uryasev and Grigoriy Zrazhevsky

Summary. Eavesdropping and jamming communication networks is an important part of any military engagement. However, until recently there has been very limited research in this area. Two recent papers by Commander et al. [2, 3] addressing this problem represent the current state-of-the-art in the study of the WIRELESS NETWORK JAMMING PROBLEM. In this chapter, we survey the problem and highlight the results from the aforementioned papers. Finally, we indicate future areas of research and mention several extensions which are currently under investigation.

1 Introduction

In any military engagement, disrupting the communication mechanism of one's enemy is an important strategic maneuver. Depending on the circumstances the goal may be to intercept the communication, neutralize the communication network, or both. These problems can be modeled in the same manner. Without the loss of generality, we will refer the problem of determining the optimal placement and quantity of eavesdropping or jamming devices to either intercept or destroy the communication as the WIRELESS NETWORK JAMMING PROBLEM (WNJP) and our descriptions will be in this context. Hence jamming devices can also be thought of as eavesdropping devices.

Research on optimization problems in telecommunications is abundant [30]. However, until recently there has been very limited research on the use of optimization techniques for jamming wireless networks. Two recent papers by Commander et al. [2, 3] addressing this problem represent the current state-of-the-art in the study of the WIRELESS NETWORK JAMMING PROBLEM. In this chapter, we survey the problem and highlight the results from the aforementioned papers. Finally, we indicate future areas of research and mention several extensions which are currently under investigation.

The organization of the chapter is as follows. Section 2 contains several formulations of the WNJP. The models considered are formulated with the assumption that the locations of the set of communication devices to be jammed is known. In Section 3, we will review several recent results on jamming wireless networks when no information about the topology of underlying network is assumed. In this case, bounds are derived for the optimal number of jamming devices required to cover an area in the plane. Finally in Section 4, conclusions and future directions of research will be addressed.

2 Deterministic Formulations

Before continuing, we will introduce some of the idiosyncrasies, symbols, and notations we will employ throughout this chapter. Denote a graph $G = (V, E)$ as a pair consisting of a set of vertices V, and a set of edges E. All graphs in this chapter are assumed to be undirected and unweighted. We use the symbol "$b := a$" to mean "the expression a defines the (new) symbol b" in the sense of King [7]. Of course, this could be conveniently extended so that a statement like "$(1 - \epsilon)/2 := 7$" means "define the symbol ϵ so that $(1 - \epsilon)/2 = 7$ holds." Any other locally used terms and symbols will be defined in the sections in which they appear.

In [3], the authors provide several formulations for the WNJP. Some assumptions made about the network nodes and the jamming devices are as follows. The jamming devices and network nodes are assumed to have omnidirectional antennas and operate as both transmitters and receivers. In other words, given a graph $G = (V, E)$, we can represent the enemy communication devices as the vertices of the graph. An undirected edge would connect two nodes if they are within a certain communication threshold.

2.1 Coverage Based Formulations

Given a set $\mathcal{M} = \{1, 2, \ldots, m\}$ of communication nodes to be jammed, our goal is to find a set of locations for placing jamming devices in order to suppress the functionality of the network. We assume that the *jamming effectiveness* of device j is a function $d : \mathbb{R} \to \mathbb{R}$, where d is a decreasing function of the distance from the device to the node being jammed. For the problem at hand, we are considering radio transmitting nodes, and correspondingly, jamming devices which emit electromagnetic waves. Thus the jamming effectiveness of a device depends on the power of its electromagnetic emission, which is assumed to be inversely proportional to the squared distance from the device to the node begin jammed. That is,

$$d_{ij} := \frac{\lambda}{r^2(i, j)},$$

where $\lambda \in \mathbb{R}$ is a constant, and $r(i, j)$ represent the distance between node i and jammer location j. Without the loss of generality, we can set $\lambda = 1$.

Define the cumulative amount of jamming energy received at node $i \in \mathcal{M}$ as

$$Q_i := \sum_{j=1}^{n} d_{ij} = \sum_{j=1}^{n} \frac{1}{r^2(i, j)},$$

where n is the number of jamming devices placed. Then, we can formulate the WIRELESS NETWORK JAMMING PROBLEM (WNJP) as the minimization of the number of jamming devices placed, subject to a set of covering constraints:

$$\textbf{(WNJP)} \quad \text{Minimize} \quad n \tag{1}$$

$$\text{s.t.} \quad Q_i \geq C_i, \quad i = 1, 2, \dots, m. \tag{2}$$

The solution to this problem provides the optimal number of jammers needed to ensure a certain jamming threshold at every node. As mentioned in [3], a continuous optimization approach where one is seeking the optimal placement coordinates $(x_j, y_j), j = 1, 2, \dots, n$ for jamming devices given the coordinates $(X_i, Y_i), i = 1, 2, \dots, m$, of network nodes, leads to highly non-convex formulations. For example, consider the covering constraint for some node $i \in \mathcal{M}$,

$$\sum_{j=1}^{n} \frac{1}{(x_j - X_i)^2 + (y_j - Y_i)^2} \geq C_i.$$

It is easy to verify that this constraint is non-convex.

To overcome the non-convexity of the above formulation, we now develop some integer programming models for the problem. Suppose now that along with the set of communication nodes $\mathcal{M} = \{1, 2, \dots, m\}$, there is a fixed set $\mathcal{N} = \{1, 2, \dots, n\}$ of possible locations to place the jamming devices. This assumption is valid, because in reality the set of possible placement locations may be limited. Define the decision variable x_j as

$$x_j := \begin{cases} 1, & \text{if device installed at location } j \\ 0, & \text{otherwise.} \end{cases} \tag{3}$$

Then if C_i and d_{ij} are as defined above, we can formulate the OPTIMAL NETWORK COVERING (ONC) formulation of the WNJP as

$$\textbf{(ONC)} \quad \text{Minimize} \quad \sum_{j=1}^{n} c_j x_j \tag{4}$$

$$\text{s.t.}$$

$$\sum_{j=1}^{n} d_{ij} x_j \geq C_i, \quad i = 1, 2, \dots, m, \tag{5}$$

$$x_j \in \{0, 1\}, \quad j = 1, 2, \dots, n. \tag{6}$$

Here our objective is to minimize the number of jamming devices while achieving some minimum level of jamming at each node. The coefficients c_j in (4) represent the costs of installing a jamming device at location j. In a battlefield scenario, placing a jamming device in a direct proximity of a network node may be theoretically possible; however, such a placement might be undesirable to do security considerations. In this case, the location considered would have a higher placement cost than a safer location [3]. If there are no preferences for device locations, then

$$c_j = 1, \quad j = 1, 2, \ldots, n.$$

Notice that ONC is formulated as a MULTIDIMENSIONAL KNAPSACK PROBLEM which is well known to be $\mathcal{NP}$-hard [18].

In a philosophical sense, the objective of the WNJP is not to simply jam the communication nodes, but rather to destroy the *functionality* of the underlying communication network. Therefore, we need to develop a method for characterizing the connectivity of the network nodes. In the next section, we provide an alternate formulation based on constraining the number of nodes any given vertex can communicate with. The intuition is that by limiting the *connectivity index* of the nodes, we can place the jamming devices in such a way that disconnects the network into several small components. This will prohibit global information sharing and will effectively dismantle the network.

2.2 Connectivity Based Formulation

Given a graph $G = (V, E)$, the *connectivity index* of a node is defined as the number of nodes reachable from that vertex (see Figure 1 for examples). In this section, we develop a formulation for the WNJP based on constraining connectivity indices of the network nodes. We will assume that the topology of the network is known. That is, we know that there is a set of communication nodes $M = \{1, 2, \ldots, m\}$ to be jammed and a set of possible jammer locations $N = \{1, 2, \ldots, n\}$. Let $S_i := \sum_{j=1}^{n} d_{ij} x_j$ denote the cumulative level of jamming at node i. Then as before, node $i \in \mathcal{M}$ is said to be jammed if S_i exceeds some minimum threshold value C_i. We say that a communication link

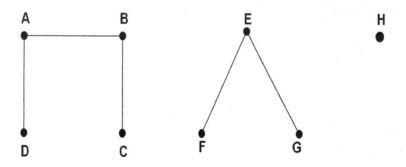

Fig. 1. Connectivity index of nodes A,B,C,D is 3. Connectivity index of E,F,G is 2. Connectivity index of H is 0.

between nodes i and j is severed if at least one of the nodes is jammed. Further, let $y : V \times V \to \{0,1\}$ be a surjection where $y_{ij} := 1$ if there exists a path from node i to node j in the jammed network. Lastly, define $z : V \to \{0,1\}$ to be a surjective mapping where z_i returns 1 if node i is not jammed.

The objective of the CONNECTIVITY INDEX PROBLEM (CIP) formulation of the WNJP is to minimize total jamming cost subject to a constraint that the connectivity index of each node does not exceed some pre-described level L. The corresponding optimization problem is the following:

$$\textbf{(CIP)}\quad \text{Minimize} \sum_{j=1}^{n} c_j x_j \tag{7}$$

$$\text{s.t.}$$

$$\sum_{j \neq i} y_{ij} \leq L,\ \forall\ i,j \in \mathcal{M} \tag{8}$$

$$M(1 - z_i) \geq S_i - C_i \geq -M z_i,\ \forall\ i \in \mathcal{M} \tag{9}$$

$$x_j \in \{0,1\},\ \forall\ j \in \mathcal{N} \tag{10}$$

$$z_i \in \{0,1\}\ \forall\ i \in \mathcal{M}, \tag{11}$$

$$y_{ij} \in \{0,1\}\ \forall\ i,j \in \mathcal{M}, \tag{12}$$

where M is some large constant. Constraint (8) is the connectivity constraint, which ensures that there are less than L nodes reachable from any vertex. Further, (9) implies that those nodes $i \in \mathcal{M}$ which receive a cumulative level of energy in excess of the threshold level C_i are considered jammed. Finally, (10)-(12) define the domains of the decision variables used.

Now, let $v : V \times V \to \{0,1\}$ and $v' : V \times V \to \{0,1\}$ be defined as follows:

$$v_{ij} = \begin{cases} 1, & \text{if } (i,j) \in E, \\ 0, & \text{otherwise}, \end{cases} \tag{13}$$

and

$$v'_{ij} = \begin{cases} 1, & \text{if } (i,j) \text{ exists in the jammed network}, \\ 0, & \text{otherwise}. \end{cases} \tag{14}$$

Then, as seen in [3], an equivalent integer programming formulation is given by

$$\textbf{(CIP-1)} \quad \text{Minimize} \quad \sum_{j=1}^{n} c_j x_j, \tag{15}$$

s.t.

$$y_{ij} \geq v'_{ij}, \ \forall \, i,j \in \mathcal{M}, \tag{16}$$

$$y_{ij} \geq y_{ik} y_{kj}, \ k \neq i,j; \ \forall \, i,j \in \mathcal{M}, \tag{17}$$

$$v'_{ij} \geq v_{ij} z_j z_i, \ i \neq j; \ \forall \, i,j \in \mathcal{M}, \tag{18}$$

$$\sum_{j=1}^{m} y_{ij} \leq L, \ j \neq i, \ \forall \, i \in \mathcal{M}, \tag{19}$$

$$M(1-z_i) \geq S_i - C_i \geq -M z_i, \ \forall \, i \in \mathcal{M}, \tag{20}$$

$$z_i \in \{0,1\}, \ \forall \, i \in \mathcal{M}, \tag{21}$$

$$x_j \in \{0,1\}, \ \forall \, j \in \mathcal{N}, \quad y_{ij} \in \{0,1\} \ \forall \, i,j \in \mathcal{M}, \tag{22}$$

$$v_{ij} \in \{0,1\}, \ \forall \, i,j \in \mathcal{M}, \quad v'_{ij} \in \{0,1\}, \ \forall \, i,j \in \mathcal{M}. \tag{23}$$

We present the following lemma from [3] without proof, which establishes the equivalency between formulations CIP and CIP-1.

Lemma 1 *If* CIP *has an optimal solution then,* CIP-1 *has an optimal solution. Further, any optimal solution* x^* *of the optimization problem* CIP-1 *is an optimal solution of* CIP.

Now by applying some basic linearization techniques, we can formulate the linear 0–1 program, CIP-2 as

$$\textbf{(CIP-2)} \quad \text{Minimize} \quad \sum_{j=1}^{n} c_j x_j \tag{24}$$

s.t.

$$y_{ij} \geq v'_{ij}, \ \forall \, i,j = 1,\ldots,\mathcal{M}, \tag{25}$$

$$y_{ij} \geq y_{ik} + y_{kj} - 1, \ k \neq i,j; \ \forall \, i,j \in \mathcal{M}, \tag{26}$$

$$v'_{ij} \geq v_{ij} + z_j + z_i - 2, \ i \neq j; \ \forall \, i,j \in \mathcal{M}, \tag{27}$$

$$\sum_{j=1}^{I} y_{ij} \leq L, \ j \neq i, \ \forall \, i \in \mathcal{M}, \tag{28}$$

$$M(1-z_i) \geq S_i - C_i \geq -M z_i, \ \forall \, i \in \mathcal{M}, \tag{29}$$

$$z_i \in \{0,1\}, \ \forall \, i \in \mathcal{M}, \tag{30}$$

$$x_j \in \{0,1\}, \ \forall \, j \in \mathcal{N}, \quad y_{ij} \in \{0,1\} \ \forall \, i,j \in \mathcal{M}, \tag{31}$$

$$v_{ij} \in \{0,1\}, \ \forall \, i,j \in \mathcal{M}, \quad v'_{ij} \in \{0,1\}, \ \forall \, i,j \in \mathcal{M}. \tag{32}$$

Similar to the previous lemma, the following establishes the equivalency between CIP-1 and CIP-2.

Lemma 2 *If* CIP-1 *has an optimal solution then* CIP-2 *has an optimal solution. Any optimal solution* x^* *of* CIP-2 *is an optimal solution of* CIP-1.

We have as a result of the above lemmata the following theorem which states that the optimal solution to the linearized integer program CIP-2 is an optimal solution to the original connectivity index problem CIP. The proof of which follows directly from **Lemma 1** and **Lemma 2** above.

Theorem 1 *If* CIP *has an optimal solution then* CIP-2 *has an optimal solution. Any optimal solution of* CIP-2 *is an optimal solution of* CIP.

In addition to the above formulations, in [3] the authors present coverage and connectivity formulations with the addition of percentile risk constraints. Prevalent in financial engineering applications and stochastic optimization problems, risk measures can also be applied in deterministic setups in order to control the loss associated with the objective function or set of constraints. In particular, suppose it is determined that in order to neutralize a network in a given instance of the WNJP, that only a percentage of the nodes must to be jammed. For instances such as these, percentile constraints can be applied which can result in near optimal solutions while providing a significant reduction in cost.

In [3] the authors consider the Value-at-Risk (VaR) [5] and Conditional Value-at-Risk (CVaR) [10] measures and present formulations of the ONC and CIP with each. Case studies indicate that models incorporating the CVaR constraints provide solutions which are comparable to the models using VaR type constraints and can be solved up to two orders of magnitude faster.

3 Jamming Under Complete Uncertainty

In this section, we highlight the contributions of Commander et al. from [2]. Considered in this chapter was the case of jamming a communication network when no a priori information was assumed other than a general region (i.e. a map grid) known to contain the network. The authors provide upper and lower bounds on the optimal number of jamming devices required to cover the region when the jammers are placed at the nodes of a uniform grid covering the area of interest.

If we ignore the cumulative effect of the jamming devices, then the problem reduces to determining the optimal covering of an area in the plane with uniform circles. This covering problem was solved in 1936 by Kershner [6]. In this section, we will review the recent work from [2] which shows that accounting for the cumulative effect of all the devices can lead to significant decreases in costs, i.e. the required number of jamming devices needed to destroy the network [2].

Again, we assume that energy received at a point X from a jamming device decreases reciprocally with the squared distance from a device. In particular we have the following definitions about the considered scenario.

Definition 1 *A point (communication node) X is said to be jammed or covered if the cumulative energy received from all jamming devices exceeds some threshold value E:*

$$\sum_i \frac{\lambda}{r^2(X, i)} \geq E, \tag{33}$$

where $\lambda \in \mathbb{R}$ and $r(X, i)$ represents the distance from X to jamming device i. This condition can be rewritten as:

$$\sum_i \frac{1}{r^2(X, i)} \geq \frac{1}{L^2}, \tag{34}$$

where $L = \sqrt{\frac{\lambda}{E}}$.

The latter inequality implies that a jamming device covers any point inside a circle of radius L.

Definition 2 *A connection (arc) between two communication nodes is considered blocked if any of the two nodes is covered.*

The intuition behind the approach is as follows. Consider a square region in the plane with side length $a \in \mathbb{R}$ which is known to contain the network to be jammed. Then, in order to place the minimum number of jamming devices on a uniform grid over the region, the grid step size, R, should be as large as possible and still jam every point in the region. Without the loss of generality, it is assumed that if the length of a square side a is not a multiple of R, then we cover a larger square with a side of length $R(\lceil \frac{a}{R} \rceil + 1)$. Figure 2 provides a graphical representation of this scenario.

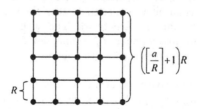

Fig. 2. Uniform grid with jamming devices.

At the very basis of the theorems in which the bounds are derived is the result of a lemma which establishes that a point located in a corner square of the bounding area will receive the least amount of cumulative energy from all the jamming devices.

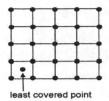

least covered point

Fig. 3. The least covered point is shown in the lower left grid cell.

Lemma 3 *For any covering of a square with a uniform grid, a point which receives the least amount of jamming energy lies inside a corner grid cell (see Figure 3).*

This is intuitively clear as with the nature of wireless broadcasting. In the same manner, we would expect that the points in the center of the region would receive the most energy since the maximum distance to any other point is minimum. With this lemma it is clear that if a grid step size guarantees coverage of the corner cells of the grid, then all points in the area will be jammed. With this established, a lower bound on the optimal grid step size can be established as follows. In all formulated theorems, we consider covering a square with side length a.

Theorem 2 *The unique solution of the equation*

$$\frac{1}{2R^2}\left(\pi\ln(\frac{a}{R}+1)+\pi-3\right)=\frac{1}{L^2} \tag{35}$$

is a lower bound $\underline{R}$ for the optimal grid step size R^*.

Thus, a uniform grid with step size $\underline{R}$ jams any point P inside a corner cell. According to Lemma 3, the grid jams the least covered point in the square implying that the whole square is jammed. Furthermore, since the function $f(R)=\frac{1}{2R^2}(\pi\ln(\frac{a}{R}+1)+\pi-3)$ is monotonic, the solution to the equation $f(R)=\frac{1}{L^2}$ is unique, and can be easily determined by a numerical procedure such as a binary search. Therefore, using (35), we can obtain a step size $\underline{R}$ such that the corresponding uniform grid covers the entire square. Further, the number of jamming devices in the grid does not exceed

$$N_1=\left(\frac{a}{\underline{R}}+2\right)^2. \tag{36}$$

The result from Kershner [6] using only circles, i.e not accounting for the cumulative effect of the jamming devices is that in the limit, the minimum number of circles required to cover the considered square region having area a^2 is

$$N_2=\frac{2a^2}{3\sqrt{3L^2}}. \tag{37}$$

In [2] it was shown that $\lim_{a \to \infty} \frac{N_2}{N_1} = \infty$, which further makes clear the benefit of the wireless broadcast advantage.

Similarly to the derivation of the lower bound on the optimal grid step size, an upper bound $\bar{R}$ can be computed and is given by the following theorem.

Theorem 3 *The unique solution of the equation*

$$\frac{1}{R^2} \left(\frac{\pi}{2} \ln \left(\frac{2a}{R} + 1 \right) - \frac{1}{6(\frac{a}{R} + 1)} + \frac{\pi}{2} + \frac{19}{3} \right) = \frac{1}{L^2} \tag{38}$$

is an upper bound $\overline{R}$ of the optimal grid step size R^*.

The result of the grid step $\overline{R}$ is that all points except those located in the corner cells are jammed.

A numerical example demonstrating the effectiveness of the bounds was given in [2]. In Figure 4 we are covering a 40×40 square and the required jamming level at each point is 3.0 units. In part (a), we see the coverage associated with the required number of devices from the lower bound of Theorem 2. In this case, $20^2 = 400$ jamming devices are used to cover the area. The scallop shell outside the bounding box indicates the boundary of the coverage from the jamming devices. In part (b), we see the coverage corresponding to the placement of the jamming devices on a uniform grid according to the upper bound of Theorem 3. Here, the required number of devices is $19^2 = 361$. Notice the holes located at the four corners of the region indicating that these points are uncovered. This validates the theoretical results obtained in Theorem 2 and Theorem 3 [2].

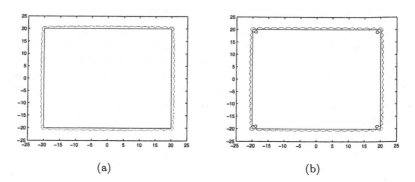

(a)   (b)

Fig. 4. (a) The coverage of when jamming devices are placed according to the lower bound from Theorem 2. The total number of jamming devices required is $20^2 = 400$. (b) We see the coverage associated with the result obtained from Theorem 3. In this case, $19^2 = 361$ devices are placed. Notice the corner points are not jammed.

One final result from [2] was the following convergence result indicating that the derived upper and lower bounds are indeed tight within a constant.

Theorem 4

$$\lim_{a \to \infty} \frac{\overline{R}}{\underline{R}} = 1,$$

(39)

where $\overline{R}$ and $\underline{R}$ are bounds obtained from equations (35) and (38), correspondingly. Moreover, the following inequality holds:

$$1 \leq \frac{\overline{R}}{\underline{R}} \leq \sqrt{1 + \frac{c}{\ln(a)}},$$

(40)

for constants $M \in \mathbb{R}, c \in \mathbb{R}$.

4 Concluding Remarks and Future Research

Using optimization approaches to jam wireless communication networks is a novel approach which was only recently introduced [2, 3]. There is still a great deal of work to be done on this problem which will help military strategists ensure the best level of performance against a hostile force. Below we briefly outline several extensions and areas in which our current efforts are focused.

4.1 Alternative Formulations

A generalization of the node coverage formulation including uncertainties in the number of communication nodes and their coordinates might be considered. For the connectivity index problem, there might exist uncertainties in the number of network nodes, their locations, and the probability that a node will recover a jammed link. Furthermore, there are several alternative formulations which can be formulated and considered. Some possibilities include a formulation problem based on maximum network flows [1]. That is, consider the communication graph $G = (V, E)$ with a set U of arc capacities, a source node $s \in V$, and a sink node $t \in V$. Each node in the graph acts as both a server and client with the exception of s which only transmits and t which only receives data. Suppose further that each receiver is equipped with a k-sectored antenna. A sectored antenna is a set of directional antennas that can cover all directions but can isolate the sectors. Thus, an arc can be jammed only by those jamming devices that are located in the same sector as the transmitter. Then the objective of the MAXIMUM NETWORK FLOWS formulation is to find locations of jamming devices such that the expected maximum flow on the network is minimized. We are considering this formulation and with the incorporation of various uncertainties in the arc capacities, the number and coordinates of the communication nodes, type of sectored antennas, and the probability of nodes recovering from a jammer.

4.2 Heuristics

The inherent complexity of the aforementioned formulations motivates the need for efficient heuristics to solve real-world instances within reasonable computing times. Currently, we are working on a randomized local search heuristic for the case of jamming under complete uncertainty. Preliminary results are promising. Another heuristic currently being tested is a Greedy Randomized Adaptive Search Procedure (GRASP) [9] for the OPTIMAL NETWORK COVERING formulation. Finally, we are designing a combinatorial algorithm for the CONNECTIVITY INDEX PROBLEM formulation. Other endeavors involving heuristic development and design along with the implementation of advanced cutting plane techniques would indeed be helpful.

Acknowledgements

The authors are grateful to the Air Force Office of Scientific Research for providing funding under project number FA9550-05-1-0137.

References

1. R.K. Ahuja, T.L. Magnanti and J.B. Orlin, *Network Flows: Theory, Algorithms, and Applications*, Prentice-Hall, 1993.
2. C.W. Commander, P.M. Pardalos, V. Ryabchenko, O. Shylo, S. Uryasev, and G. Zrazhevsky, Jamming communication networks under complete uncertainty, *Optimization Letters*, published online, DOI 10.1007/s11590-006-0043-0, 2007.
3. C.W. Commander, P.M. Pardalos, V. Ryabchenko, S. Uryasev, and G. Zrazhevsky, The wireless network jamming problem, to appear in *Journal of Combinatorial Optimization*, 2008.
4. M.R. Garey and D.S. Johnson, *Computers and Intractability: A Guide to the Theory of NP-Completeness*, W.H. Freeman and Company, 1979.
5. G. Holton, *Value-at-Risk: Theory and Practice*, Academic Press, 2003.
6. R. Kershner, The number of circles covering a set, *American Journal of Mathematics*, 61(3), pp. 665–671, 1939.
7. J. King, Three problems in search of a measure, *American Mathematical Monthly*, 101, pp. 609–628, 1994.
8. M.G.C. Resende and P.M. Pardalos, *Handbook of Optimization in Telecommunications*, Springer, 2006.
9. M.G.C. Resende and C.C. Ribeiro, Greedy randomized adaptive search procedures, in F. Glover and G. Kochenberger (eds), *Handbook of Metaheuristics*, pp. 219–249, Kluwer Academic Publishers, 2003.
10. R.T. Rockafellar and S. Uryasev, Optimization of conditional value-at-risk, *The Journal of Risk*, 2(3), pp. 21–41, 2000.

7 Network Algorithms for the Dual of the Constrained Shortest Path Problem

Bogdan Grechuk, Anton Molyboha and Michael Zabarankin

Summary. A general iterative procedure for finding an exact solution to the dual for the Constrained Shortest Path Problem (CSPP) has been developed. The procedure couples an extended shortest path algorithm with one of the contractive algorithms. Two contractive algorithms minimizing the region of feasible solutions in the domain-range plane of the Lagrange function for the CSPP: the "domain-optimal algorithm" and the "range-optimal algorithm" have been suggested. Correctness, finiteness and complexity of the extended shortest path algorithm and both contractive algorithms have been proved.

1 Introduction

This chapter develops fast optimization algorithms for solving the dual of the constrained shortest path problem (CSPP).

The CSPP is also known as a restricted shortest path problem and formulated as follows. In a directed network (graph), each arc (edge) is assigned two positive values: cost and weight. The CSPP is finding a path from a source node s to a sink node t with minimal cost subject to a constraint on the path weight.

The CSPP arises in various applications. These include minimum-risk routing of military vehicles (e.g. Latourell et al. [17], Zabarankin et al. [26, 25]), signal routing in communications networks with quality-of-service guarantees (e.g. Korkmaz et al. [16]). Dooms et al. [6] mention the CSPP problem in the context of biochemical networks analysis.

Joksch [14] was probably the first who encountered the CSPP and presented an algorithm based on dynamic programming. The CSPP is a special case of the Resource Constrained Shortest Path Problem (RCSPP) [2] and a special case of the Shortest Path Problem with Time Windows (SPPTW) [5]. The CSPP is listed as problem ND30 (shortest weight-constrained path) in Garey and Johnson [10]. In contrast to a shortest path problem (SPP), the CSPP is *weakly NP complete* [10]. Hassin [13] applied a standard technique of rounding and scaling to obtain a fully polynomial ϵ-approximation scheme (PTAS) for the CSPP. Based on Hassin's result, Lorenz and Raz [19] suggested an improved approximation scheme with complexity $O(mn(\log \log n + 1/\epsilon))$, where m and n are the number of arcs and nodes in the network, respectively. This scheme was applied to Quality of Service (QoS) routing in the Internet. Recently, Dumitrescu and Boland [7, 8, 9], and Ziegelmann [27] surveyed and compared different approximation schemes and exact algorithms for solving the CSPP and RCSPP.

One of the most efficient approaches for solving the one-resource constrained shortest path problem, which the CSPP is, utilizes upper and lower bounds for optimal cost obtained from Lagrangean relaxation [2, 4, 9, 12, 18, 27]. In this case, the dual of the CSPP (also known as the Lagrangean dual) reduces to maximizing a concave piecewise linear function of a single variable. Since in general the CSPP is a non-convex optimization problem, optimal value for the dual function provides only a lower bound for the optimal cost in the CSPP. This fact is known as duality gap. To find maximum for the dual function, Handler and Zang [12] suggested an exact algorithm based on Kelley's cutting plane [15]. Kelley's cutting plane, also known as Benders decomposition [3], is closely related to Gomory's integer linear programming algorithm [11] and the Analytic Center Cutting Plane Method (ACCPM) [23]. Using parametric shortest path computations, Ziegelmann [27] showed that in the case of integer costs and weights, a solution to the Lagrangean dual can be found in $O(\log(nRC))$ time, where C and R are the maximal arc cost and maximal arc weight in the graph, respectively. Some issues on complexity of solving the dual problem are discussed in [21, 22]. As the next step, information obtained from Lagrangean relaxation is used in a preprocessing procedure [1, 2, 9, 20]. The idea of preprocessing (graph pruning) is eliminating such nodes and arcs in the graph that, if used in completing a path from the source node to the sink node, violate either the constraint on path weight or an upper bound for the optimal cost. Dumitrescu and Boland [9] discussed and compared several modifications for the preprocessing procedure. The last step in solving the CSPP is closing the duality gap. Various techniques are available for closing the duality gap: enumerating near-shortest paths [4]; k-shortest path algorithm (path ranking) [24]; branch-and-bound procedure [2, 21]; dynamic programming algorithm of a label-correcting type [27]; and label-setting algorithm [9, 5], which has pseudopolynomial time complexity [5]. The reader interested in details and further discussion of the CSPP and RCSPP may refer to [7, 8, 9, 27].

Lagrange relaxation can be used alone to obtain an approximate solution to the CSPP. The advantage of this approach is that polynomial-time algorithms are available for solving the dual problem, while all known exact algorithms have exponential complexity. Consequently, if the graph in consideration is relatively large, finding an exact solution becomes impractical because of the amount of computational resources required. In this case even solving the dual problem can take significant running time, thereby necessitating the development of efficient algorithms for solving the dual problem.

This chapter challenges the aforementioned task. We develop the Extended Shortest Path (ESP) algorithm as an extension for the Dijkstra algorithm and use the ESP algorithm in a general iterative procedure for solving the dual problem. We prove that the procedure with the ESP algorithm obtains an exact solution to the dual problem in finitely many iterations.

We further propose two measures of closeness between approximate solution, obtained at a particular iteration, and an optimal solution and develop two realizations of the general iterative procedure, each of which minimizes a corresponding measure of closeness. We evaluate the complexity of these two realizations of the procedure.

The chapter is organized as follows. Section 2 formulates the CSPP and Lagrangian dual to the CSPP. Section 3 develops the ESP algorithm. Section 4 suggests a general iterative procedure for solving the dual, based on the ESP algorithm. Section 5 estimates complexity of this procedure, satisfying certain criteria. Section 6 introduces two criteria of optimality for choosing next approximation to an optimal solution, and develops algorithms satisfying those criteria. Section 7 concludes the chapter.

2 Problem Formulation

Let $\mathcal{G} = (\mathcal{N}, \mathcal{A})$ be a graph with the set of nodes $\mathcal{N} = \{1, 2, \ldots, n\}$ and the set of arcs $\mathcal{A} \subset \{(i,j) | i, j \in \mathcal{N}\}$. Let m be a total number of arcs. With each arc (i,j), we associate two values: positive cost $c_{ij} > 0$ and positive weight $w_{ij} > 0$. A path $\mathcal{P}$ in the graph is defined as a sequence of nodes $\langle j_0, j_1, \ldots, j_p \rangle$ such that $j_0 = s$, $j_p = t$ and $(j_{k-1}, j_k) \in \mathcal{A}$ for all k from 1 to p, where $s \in \mathcal{N}$ and $t \in \mathcal{N}$ are source and sink nodes, respectively. Let functions $f(\mathcal{P})$ and $\bar{g}(\mathcal{P})$ define the total cost and weight accumulated along the path $\mathcal{P}$. Let $\bar{g}(\mathcal{P}) \leq w$ be a constraint on the path weight, where $w > 0$ is given. Then we say that a path $\mathcal{P}$ is weight feasible if it satisfies the constraint. The constrained shortest path problem (CSPP) is finding such a feasible path $\mathcal{P}$ from node s to node t that minimizes total cost $f(\mathcal{P})$, i.e.,

$$\min_{\mathcal{P}} f(\mathcal{P})$$
$$\text{s.t. } \bar{g}(\mathcal{P}) \leq w. \tag{1}$$

The dual problem is formulated as follows. The Lagrangian for (1) is defined by

$$L(\mathcal{P}, \lambda) = f(\mathcal{P}) + \lambda g(\mathcal{P}), \tag{2}$$

where

$$g(\mathcal{P}) = \bar{g}(\mathcal{P}) - w. \tag{3}$$

Denote

$$\Phi(\lambda) = \min_{\mathcal{P}} L(\mathcal{P}, \lambda) = \min_{\mathcal{P}}(f(\mathcal{P}) + \lambda g(\mathcal{P})). \tag{4}$$

The dual problem is then

$$\max_{\lambda \geq 0} \Phi(\lambda). \tag{5}$$

This chapter develops efficient algorithms for solving problem (5).

The function $\Phi(\lambda)$ is continuous, concave, and piecewise linear. The problem of finding $\Phi(\lambda)$ is the shortest-path problem and can be solved efficiently by the Dijkstra algorithm. Let S_λ be the set of optimal paths that solve (4)

$$S_\lambda = \{\mathcal{P}|\mathcal{P} \in \arg\min_{\mathcal{P}} L(\mathcal{P}, \lambda)\} \tag{6}$$

The dual function $\Phi(\lambda)$ possesses the following properties:

1. If path $\mathcal{P}_\lambda$ is a solution to (4) for some λ then $g(\mathcal{P}_\lambda) \in \partial(\Phi(\lambda))$, where $g(.)$ is defined in (3), and $\partial(\Phi(\lambda))$ denotes the set of supergradients for $\Phi(.)$ at the point λ.
2. Suppose that $\lambda_2 > \lambda_1 \geq 0$, and let $\mathcal{P}_i$ be a solution to (4) for λ_i, $i = 1, 2$. In this case, if $g(\mathcal{P}_1) > 0$ and $g(\mathcal{P}_2) < 0$ then an optimal solution, λ^*, to problem (5) belongs to the closed interval $[\lambda_1, \lambda_2]$.

From the convex analysis it is known, that if the primal problem is convex, then the optimal values of primal and dual problems coincide. If the primal problem is not convex, but the dual function is smooth, then the optimal values are equal as well. In our case, the primal problem is discrete, and hence, is non-convex; the dual function $\Phi(\lambda)$ is piecewise linear and, hence, is non-smooth. Consequently, the optimal values of the primal and dual problems may differ. The difference between these values is called duality gap.

Because of the duality gap, a solution to dual (5) does not provide a solution to the primal problem (1). However, a solution to the dual problem provides the following upper and lower bounds for the optimal cost for the CSPP

1. Let $\mathcal{P}_0$ be a solution to (1). Then

$$\max_{\lambda \geq 0} \Phi(\lambda) \leq f(\mathcal{P}_0).$$

This means that the maximum for the Lagrange function provides the lower bound to the optimal cost in the CSPP.
2. If λ^* is a solution to dual problem (5), then there exists a path $\mathcal{P}_+ \in S_{\lambda^*}$ such that $g(\mathcal{P}_+) \leq 0$. Then $\mathcal{P}_+$ is a feasible solution to problem (1) and $f(\mathcal{P}_+) \geq f(\mathcal{P}_0)$. That is, $f(\mathcal{P}_+)$ is an upper bound for the solution to the primal problem.

   Notice, that such $\mathcal{P}_+$ can be obtained only if we know the exact solution λ^* to the dual problem. Given an approximate solution $\lambda \in (\lambda^* - \epsilon, \lambda^* + \epsilon)$, one cannot guarantee that there exists a feasible path $\mathcal{P} \in S_\lambda$ regardless of how small $\epsilon > 0$ is.

Because of the duality gap, these upper and lower bounds do not necessarily coincide. But in many practical applications they are often quite close one to the other. In those cases, a solution to the dual problem closely approximates a solution to the CSPP.

In this chapter we propose a new algorithm to solve the dual for the CSPP.

3 Extended Shortest Path Algorithm

Dual (5) is a problem of finding the maximum for the piecewise linear concave function $\Phi(\lambda)$. Several well-known iterative algorithms are available for solving this problem. However, most of them find only an approximate solution.

One of the iterative algorithms for solving the dual problem is Kelley's cutting plane algorithm [12]. To our knowledge, this algorithm is the only one that finds an exact solution. It takes advantage of the piecewise linear structure of $\Phi(\lambda)$ by choosing the next point at the intersection of supergradients at λ_- and λ_+, see Figure 1. At the last iteration, the point of intersection coincides with the maximum for $\Phi(\lambda)$. For the detailed discussion of the algorithm, we refer the reader to [12].

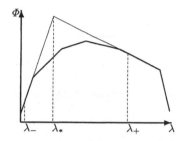

Figure 1 Kelley's Cutting Plane algorithm

At each iteration the problem of calculating $\Phi(\lambda)$ and its supergradient at any λ_0 reduces to the shortest path problem, which can be solved using the Dijkstra algorithm. This algorithm finds a path $\mathcal{P} \in S_{\lambda_0}$, value $\Phi(\lambda_0) = f(\mathcal{P}) + \lambda_0 g(\mathcal{P})$ and supergradient $g(\mathcal{P})$. We will extend this algorithm to calculate the value of function $\Phi(.)$ and its supergradient in some neighborhood of λ_0. This extension of the Dijkstra algorithm will be called the Extended Shortest Path (ESP) algorithm. We will use it in the next section in a general iterative procedure to solve the dual problem. We will show that the procedure obtains an exact solution in finite number of iterations regardless of the method for choosing the next point $\lambda_* \in [\lambda_-, \lambda_+]$. The ESP algorithm is the subject of this section.

Behavior of function $\Phi(.)$ in the neighborhood of a point λ_0 can be characterized by the following proposition.

Proposition 1. *For each* $\lambda_0 > 0$, *there exist* $\Delta\lambda_- > 0$, $\Delta\lambda_+ > 0$ *and paths* $\mathcal{P}_- \in S_\lambda$ *and* $\mathcal{P}_+ \in S_\lambda$, *which may coincide, such that*

$$\Phi(\lambda) = \begin{cases} \Phi(\lambda_0) + (\lambda - \lambda_0)g(\mathcal{P}_-) & \lambda \in [\lambda_0 - \Delta\lambda_-, \lambda_0], \\ \Phi(\lambda_0) + (\lambda - \lambda_0)g(\mathcal{P}_+) & \lambda \in [\lambda_0, \lambda_0 + \Delta\lambda_+]. \end{cases} \tag{7}$$

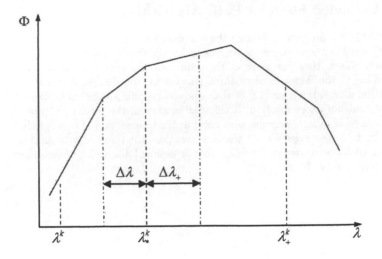

Figure 2 Illustration of $\Delta\lambda_-$ and $\Delta\lambda_+$

This fact is an immediate consequence of piecewise linearity of $\Phi(.)$, see Figure 2. To calculate $\Phi(\lambda)$ at $\lambda = \lambda_0$, we should find a shortest path $\mathcal{P}$ in the graph with arc costs $l_{ij} = c_{ij} + \lambda_0 w_{ij}$. Then $\mathcal{P} \in S_{\lambda_0}$, $\Phi(\lambda_0) = f(\mathcal{P}) + \lambda_0 g(\mathcal{P})$ and $g(\mathcal{P})$ is a supergradient of $\Phi(.)$ at the point λ_0. While the Dijkstra algorithm finds an arbitrary path $\mathcal{P} \in S_{\lambda_0}$, we will develop an algorithm which finds two paths $\mathcal{P}_-$ and $\mathcal{P}_+$ that are shortest for any $\lambda \in [\lambda_0 - \Delta\lambda_-, \lambda_0]$ and $\lambda \in [\lambda_0, \lambda_0 + \Delta\lambda_+]$ respectively. Then in the interval $[\lambda_0 - \Delta\lambda_-, \lambda_0 + \Delta\lambda_+]$ the function $\Phi(\lambda)$ is given by (7). The algorithm finds $\Delta\lambda_-$ and $\Delta\lambda_+$ as well.

The main idea of the Dijkstra algorithm is the following. The algorithm assigns a label l_i and integer j_i to each node $i \in \mathcal{N}$, where l_i and j_i represent the cost and the preceding node of the current shortest path from s to i, respectively. The algorithm maintains a list LIST of nodes to be considered at next iterations. Initially, LIST contains only the source node s, $l_s = 0$ and $l_i = \infty$ for $i \neq s$. At each iteration, the algorithm selects node $i \in$ LIST with the smallest label l_i. For each node j such that there exists arc $(i, j) \in \mathcal{A}$, the algorithm tests if $l_i + l_{ij} < l_j$ and if so, updates the label l_j and the preceding node j_j and places node j into LIST. Then the algorithm eliminates the node i from the list. The algorithm stops, when all nodes are considered, that is when LIST becomes empty. The label l_t represents the cost of the shortest path from the source node s to the sink node t. An optimal path is then recovered by a backtracking procedure.

The Extended Shortest Path (ESP) Algorithm. The ESP is similar to the Dijkstra algorithm. For each node i it finds two shortest paths P_{min} and

P_{max} from s to i with minimal and maximal total weights correspondingly. It assigns additional labels w_i^-, w_i^+ and integers j_i^- and j_i^+ to each node $i \in N$, where pairs (w_i^-, j_i^-) and (w_i^+, j_i^+) represent the weight and preceding node of P_{min} and P_{max}, respectively. Like the Dijkstra algorithm, at each iteration, the ESP considers node $i \in$ LIST with the smallest l_i and every arc $(i,j) \in A$ incident from i. If $l_j > l_i + l_{ij}$, labels l_i are updated similarly to the Dijkstra algorithm. The case $l_j = l_i + l_{ij}$ means that there are two paths with the same cost. In this case, the pair (w_j^-, j_j^-), corresponding to the path from s to j with the cost l_j and the least weight, is updated if $w_j^- > w_i^- + w_{ij}$. The pair (w_j^+, j_j^+) corresponds to the path from s to j with the cost l_j and the greatest weight and is updated if $w_j^+ < w_i^+ + w_{ij}$. The algorithm stops when all nodes are considered. The label l_t represents the cost of the shortest path from the source node s to the sink node t. Labels w_t^- and w_t^+ represent the weights of paths P_{min} and P_{max} with minimal and maximal total weights, correspondingly. These paths are then recovered by a backtracking procedure. The paths P_- and P_+ in (7) are then $P_- = P_{max}$ and $P_+ = P_{min}$.

The ESP algorithm finds $\Delta\lambda_-$ and $\Delta\lambda_+$ by solving the following optimization problems

$$\Delta\lambda_- = \min_{(i,j)\in A} \frac{(l_i + l_{ij}) - l_j}{(w_i^+ + w_{ij}) - w_j^+} \tag{8}$$
$$\text{s.t. } (w_i^+ + w_{ij}) - w_j^+ > 0$$

and

$$\Delta\lambda_+ = \min_{(i,j)\in A} \frac{(l_i + l_{ij}) - l_j}{w_j^- - (w_i^- + w_{ij})} \tag{9}$$
$$\text{s.t. } w_j^- - (w_i^- + w_{ij}) > 0$$

The ESP algorithm is presented in a pseudo-code form in Appendix A.

Proposition 2 below proves that ESP algorithm is correct, i.e., values P_-, P_+, $\Delta\lambda_-$ and $\Delta\lambda_+$ calculated by the ESP algorithm satisfy condition (7). The proof of the proposition relies on the following statements.

Statement 1 *When node i is selected from LIST, its labels l_i, w_i^-, j_i^-, w_i^+ and j_i^+ represent, respectively:*

- *the length of the shortest path from s to i,*
- *the weight of a shortest path with minimal weight P_-,*
- *the preceding node of the path P_-,*
- *the weight of a shortest path with maximal weight P_+,*
- *the preceding node of the path P_+.*

Proof: Consider a particular iteration of the algorithm, at which node i is selected from LIST. Let S denote the set of all nodes that have already been considered by the algorithm and eliminated from LIST. Let $\tilde{S}$ be the set of all other nodes, so that $\tilde{S} = \mathcal{N}\backslash S$.

We will prove the statement by mathematical induction using the cardinality of S. The base of the induction corresponds to the very first iteration of the algorithm. At this iteration, $i = s$ and the only shortest path from s to i is the path consisting from just one node s, and thus, the labels $l_s = 0$, $w_s^- = 0$ and $w_s^+ = 0$ are correct.

Suppose, that at some iteration the statement holds for every node $k \in S$. Namely, every $k \in S$ had correct labels when it was eliminated from LIST. Since the labels of a node are updated only when a "better" path is found, this implies that the labels of k were never updated since then and the node did not enter LIST again.

Consider an arbitrary node $j \in$ LIST. LIST contains all the nodes from $\tilde{S}$ incident to some node in S. According to the algorithm, $l_j = \min_{\substack{k \in S \\ (k,j) \in \mathcal{A}}} (l_k + l_{kj})$,

i.e., l_j is the length of a shortest path from s to j among those whose pre-last node belongs to S. Likewise, w_j^- and w_j^+, equal to the smallest and the largest total weight among such shortest paths and j_j^- and j_j^+, contain the corresponding preceding nodes.

Now consider the node i selected from LIST at current iteration. To prove the induction step, we will show that any shortest path from s to i has its pre-last node from S. Indeed, consider a path $\mathcal{P}$ from s to i whose pre-last node does not belong to S. Then it contains an arc (k, j), such that $k \in S$, $j \in$ LIST. We can split the path $\mathcal{P}$ into two paths: a path $\mathcal{P}_1$ from s to j and a path $\mathcal{P}_2$ from j to i. The length of $\mathcal{P}_1$ is such that $length(\mathcal{P}_1) \geq l_j \geq l_i$. According to our assumption, the arc (k, j) cannot be the last arc in $\mathcal{P}$. This means, that $length(\mathcal{P}_2) > 0$. Thus, $length(\mathcal{P}) = length(\mathcal{P}_1) + length(\mathcal{P}_2) > l_i$, and $\mathcal{P}$ cannot be a shortest path from s to i.

Therefore, the paths from s to i, shortest among those whose pre-last node belongs to S, are shortest among all the paths from s to i. Consequently, the statement holds for the node i. This concludes the induction and finishes the proof of the statement. ∎

Statement 2 *The final values of the labels calculated by the ESP algorithm for $\lambda = \lambda_0$ are such that for every node j and for every $\lambda \in [\lambda_0 - \Delta\lambda_-, \lambda_0]$, the length of the shortest path from s to j is $l_j - (\lambda_0 - \lambda)w_j^+$; for every $\lambda \in [\lambda_0, \lambda_0 + \Delta\lambda_+]$ the length of the shortest path from s to j is $l_j + (\lambda - \lambda_0)w_j^-$.*

Proof: Consider an arbitrary $\lambda \in [\lambda_0 - \Delta\lambda_-, \lambda_0]$. For every node j, the path from s to j defined by the labels j_j^+ has length $l_j - (\lambda_0 - \lambda)w_j^+$. So, it is left to prove that there is no path shorter than that.

We will prove the statement by induction using the number of arcs in a path. For the path consisting of just one node s the statement is obvious. Suppose that the statement is true for every path containing $r - 1$ arcs. Consider an arbitrary path $\mathcal{P}$ containing r arcs. Let its last arc be (i, j). Then its length

$length(\mathcal{P}) \geq l_i - (\lambda_0 - \lambda)w_i^+ + l_{ij} - (\lambda_0 - \lambda)w_{ij} = l_i + l_{ij} - (\lambda_0 - \lambda)(w_i^+ + w_{ij})$.
According to Statement 1, $l_j \leq l_i + l_{ij}$. If $w_j^+ \geq w_i^+ + w_{ij}$, then $length(\mathcal{P}) \geq$
$l_i + l_{ij} - (\lambda_0 - \lambda)(w_i^+ + w_{ij}) \geq l_j - (\lambda_0 - \lambda)w_j^+$. If $w_j^+ < w_i^+ + w_{ij}$, then
$length(\mathcal{P}) - (l_j - (\lambda_0 - \lambda)w_j^+) \geq l_i + l_{ij} - l_j - (\lambda_0 - \lambda)(w_i^+ + w_{ij} - w_j^+) \geq$
$l_i + l_{ij} - l_j - \Delta\lambda_-(w_i^+ + w_{ij} - w_j^+) \geq 0$. This proves the induction.

The case of $\lambda \in [\lambda_0, \lambda_0 + \Delta\lambda_+]$ is proved similarly.   ∎

Proposition 2 (correctness of the ESP algorithm). *Suppose that paths*
$\mathcal{P}_-, \mathcal{P}_+, \Delta\lambda_-$ *and* $\Delta\lambda_+$ *are calculated by the ESP algorithm for some* $\lambda = \lambda_0$.
Then $\Delta\lambda_-$ *and* $\Delta\lambda_+$ *are positive. Moreover, the path* $\mathcal{P}_-$ *is a shortest path
from* s *to* t *for every* $\lambda \in [\lambda_0 - \Delta\lambda_-, \lambda_0]$, *and the path* $\mathcal{P}_+$ *is a shortest path
from* s *to* t *for every* $\lambda \in [\lambda_0, \lambda_0 + \Delta\lambda_+]$.

Proof: Suppose that paths $\mathcal{P}_-$ and $\mathcal{P}_+$ are calculated by the ESP algorithm
for some $\lambda = \lambda_0$.

From Statement 1, it follows, that inequality $w_j^+ < w_i^+ + w_{ij}$ implies that
a shortest path from s to j cannot pass through the arc (i, j), i.e., $l_j < l_i + l_{ij}$.
Consequently, $\Delta\lambda_-$ defined by (8) is strictly positive. Likewise, $\Delta\lambda_+$ is also
strictly positive.

From Statement 2 it follows that the path $\mathcal{P}_-$ is a shortest path from s
to t for every $\lambda \in [\lambda_0 - \Delta\lambda_-, \lambda_0]$, and the path $\mathcal{P}_+$ is a shortest for every
$\lambda \in [\lambda_0, \lambda_0 + \Delta\lambda_+]$.   ∎

Proposition 3. *The complexity of the ESP algorithm is given by*

$$O(m \log n) \sim K\, m \log_2 n, \tag{10}$$

where K is a constant.

Proof: At each iteration, the algorithm finds a node with the smallest label l_i
and updates labels of all nodes adjacent to it. If the list of nodes is maintained
efficiently[1], the selection of the node with smallest label l_i requires no more
than $\log_2 n$ operations. Since each node is considered only once, the number
of iterations is n. The total number of label updating operations equals to
the number of arcs, m. If the label of a node changes, the node should be
repositioned in the list, which requires $O(\log_2 n)$ operations. Consequently,
the complexity of the algorithm is $O(n \log_2 n + m \log_2 n) \sim O(m \log_2 n) \sim$
$K m \log_2 n$ for some constant K.   ∎

4 General Iterative Procedure

A typical iterative algorithm for solving an optimization problem usually consists of several parts, including initialization, choice of a point to calculate

---

[1]for example, as an almost well-balanced binary tree

objective function at, objective function calculation, recalculation of variables between iterations, and stopping criteria. An algorithm for calculating the objective function in problem (5) was discussed in the previous section. In this section, we focus on the iteration framework: initialization, recalculation of variables between iterations and stopping criteria. How to choose λ to calculate the objective function at each iteration at, is left for the following sections.

For the iterative procedure, developed in this section, we prove its correctness and finiteness regardless of the choice of λ, at which the objective function is calculated. However, the complexity of the algorithm depends on this choice significantly, and is discussed in the following section.

4.1 Algorithm Formulation

To find the maximum for the piecewise linear concave function $\Phi(\lambda)$ in (5) we suggest the following iterative procedure. At each iteration, we consider two points λ_- and λ_+ such that a solution λ^* belongs to the interval $[\lambda_-, \lambda_+]$. We will use values of function $\Phi(\lambda)$ and its supergradients at these points to choose a point $\lambda_* \in [\lambda_-, \lambda_+]$ at which the dual function will be calculated. The problem of choosing λ_* will be a subject for the following sections. Now we consider iterative procedure in general with an arbitrary choice of $\lambda_* \in [\lambda_-, \lambda_+]$.

Given λ_*, we will then use the ESP algorithm for $\lambda = \lambda_*$ to calculate $\Phi(\lambda_*)$, $g_-(\lambda_*)$, $g_+(\lambda_*)$, $\Delta\lambda_-(\lambda_*)$ and $\Delta\lambda_+(\lambda_*)$. According to Proposition 2, the value of the function $\Phi(\lambda)$ on the interval $[\lambda_* - \Delta\lambda_-, \lambda_* + \Delta\lambda_+]$ will be given by (7). At the next iteration, we consider the interval $[\lambda_-, \lambda_* - \Delta\lambda_-]$ or $[\lambda_* + \Delta\lambda_+, \lambda_+]$ depending on the sign of supergradient at point λ_*. Notice that at each iteration, we shift additionally by $\Delta\lambda_-$ or $\Delta\lambda_+$. This fact will be used to find an exact solution in finitely many iterations for any choice of λ_*.

For the values of λ_- and λ_+ at the first iteration, we use the following expressions

$$\lambda_-^1 = 0, \tag{11}$$

$$\lambda_+^1 = \frac{f(\mathcal{P}_0) - f(\mathcal{P}_\infty)}{g(\mathcal{P}_\infty)}, \tag{12}$$

where $\mathcal{P}_0$ is a path with the smallest cost, and $\mathcal{P}_\infty$ is a path with the smallest weight.

The algorithm will stop when it finds an exact solution. This implies several stopping criteria:

- if $\lambda_- = \lambda_+$ then $\lambda^* = \lambda_- = \lambda_+$ is the solution
- if $g_-(\lambda_*) \geq 0$ and $g_+(\lambda_*) \leq 0$ then $\lambda^* = \lambda_*$ is the solution
- if $\Phi(\lambda_-) = \Phi(\lambda_+) + (\lambda_- - \lambda_+)g_-(\lambda_+)$ then $\lambda^* = \lambda_-$ is the solution
  or if $\Phi(\lambda_+) = \Phi(\lambda_-) + (\lambda_+ - \lambda_-)g_+(\lambda_-)$ then $\lambda^* = \lambda_+$ is the solution

The last criterion follows from concavity of $\Phi(\lambda)$.

We call this algorithm the general iterative procedure. It is presented in a pseudo-code form in Appendix B.

Proposition 4. *If the general iterative procedure stops, then it yields exactly one of the following results:*

1. *it shows, that the CSPP is infeasible,*
2. *it finds a solution to the CSPP,*
3. *it finds a solution to the dual problem.*

Proposition 4 is proved in Appendix B.

4.2 Algorithm Finiteness

In this section, we will prove that the general iterative procedure finds an exact solution in finitely many iterations for any choice of λ_*.

To consider the algorithm's finiteness and complexity we introduce the following values

$$
\begin{aligned}
W_M &= \max_{(i,j) \in \mathcal{A}} (w_{ij}), \\
C_M &= \max_{(i,j) \in \mathcal{A}} (c_{ij}).
\end{aligned}
\tag{13}
$$

Then for any path $\mathcal{P}$ without loops we have that its total cost $f(\mathcal{P})$ and total weight $\bar{g}(\mathcal{P})$ satisfy the following inequalities

$$
\begin{aligned}
f(\mathcal{P}) &\leq C_M \cdot n, \\
\bar{g}(\mathcal{P}) &\leq W_M \cdot n.
\end{aligned}
\tag{14}
$$

The following proposition establishes an important property of $\Delta\lambda_-$ and $\Delta\lambda_+$, which will be used to prove finiteness and complexity of the general iterative procedure.

Proposition 5. *For a given graph $\mathcal{G}$ with arc costs $c_{ij} \geq 0$ and arc weights $w_{ij} \geq 0$ and any $\lambda > 0$, both $\lambda - \Delta\lambda_-(\lambda)$ and $\lambda + \Delta\lambda_+(\lambda)$ assume only values from a finite set $U = \{b_1, b_2, \ldots, b_u\}$, $b_1 < b_2 < \cdots < b_u$. Moreover, if parameters c_{ij}, w_{ij} and w are nonnegative integers, then*

$$
b_{i+1} - b_i \geq \frac{1}{(nW_M)^2} \quad \forall i,
$$

where W_M is defined by (13).

Proof: From the definition (8) of $\Delta\lambda_-$, there exists an arc $(i,j) \in \mathcal{A}$, such that

$$
\Delta\lambda_- = \frac{(l_i + l_{ij}) - l_j}{(w_i^+ + w_{ij}) - w_j^+}.
$$

Here w_j^+ is the total weight of some path $\mathcal{P}_j^+$ from node s to node j, $l_j = c_j^+ + \lambda w_j^+$, where c_j^+ is the total cost of the path $\mathcal{P}_j^+$. l_{ij} is defined to be $l_{ij} = c_{ij} + \lambda w_{ij}$. That is why

$$\lambda - \Delta\lambda_- = \frac{-(c_i^+ + c_{ij}) + c_j^+}{(w_i^+ + w_{ij}) - w_j^+}. \tag{15}$$

As there are only finite number m of arcs, c_{ij} and w_{ij} can assume only a finite number of values. c_i^+ and c_j^+ are sums of no more than m of c_{ij}, and hence can only assume a finite number of values. Likewise, w_i^+ and w_j^+ are sums of w_{ij} and can only assume a finite number of values. That is why $\lambda - \Delta\lambda_-$ can only assume a finite number of values.

The similar statement about $\lambda + \Delta\lambda_+$ is proved similarly.

Suppose that all c_{ij} and w_{ij} are nonnegative integers. Then $\lambda - \Delta\lambda_-$ given by (15) is a rational number. Notice that mentioned paths $\mathcal{P}_i^+$, $\mathcal{P}_j^+$ are paths without loops, so (14) holds and the denominator in (15) is no greater than nW_M. Similarly, $\lambda + \Delta\lambda_+$ can also be represented as a rational number with denominator no greater than nW_M. A difference of any two such numbers, if positive, cannot be less than $\frac{1}{(nW_M)^2}$. ∎

The statement about algorithm's finiteness follows from the above proposition.

Proposition 6. *The general iterative procedure finds a solution in a finite number of steps.*

Proof: According to Proposition 5 for any $\lambda > 0$, both $\lambda - \Delta\lambda_-(\lambda)$ and $\lambda + \Delta\lambda_+(\lambda)$ assume only values from finite set U. Let $d_1 < d_2 < \cdots < d_v$ be the ordered elements of the set $U \cup \{\lambda_-^1, \lambda_+^1\}$. Then at each iteration k we have that $\lambda_- = d_i$, $\lambda_+ = d_j$ for some i and j. The function $l(k) = j - i$ is then nonnegative and assumes only integer values. Moreover, according to the algorithm, λ_* is chosen such that $\lambda_- \leq \lambda_* \leq \lambda_+$. From Proposition 2, we obtain $\Delta\lambda_- > 0$ and $\Delta\lambda_+ > 0$. Consequently, at each iteration either λ_+ decreases or λ_- increases. Therefore, the function $l(k)$ is strictly decreasing. Consequently, the algorithm is finite. ∎

Complexity of the general iterative procedure depends on a particular choice of λ_* at each iteration. In the following section, we will estimate the complexity under some assumptions on this choice.

5 Contractive Algorithms

5.1 Definitions

In the previous section, we considered the general iterative procedure for solving the dual problem. This procedure allows us to choose the next point

$\lambda_* \in [\lambda_-, \lambda_+]$ arbitrarily. A method for choosing $\lambda_* \in [\lambda_-, \lambda_+]$ determines a particular implementation of the general iterative procedure. In this section, we will consider several properties for λ_*, which if satisfied, guarantee a particular complexity for the general iterative procedure.

At each iteration of the algorithm, the following data are available: λ_-, λ_+, $\Phi(\lambda_-)$, $\Phi(\lambda_+)$, and supergradients $g_-(\lambda_-)$ and $g_+(\lambda_+)$ of Φ at λ_- and λ_+. These data are used to estimate the region of a solution to the dual problem. One approach to evaluate the complexity of an algorithm is to estimate the speed at which the size of this region decreases. This is the approach we take here.

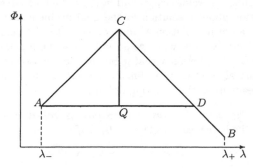

Figure 3 A region of possible solutions to the dual problem

Suppose at some iteration k of the general iterative procedure, we have some λ_-^k, $\lambda_+^k > \lambda_-^k$, $\Phi(\lambda_-^k)$, $\Phi(\lambda_+^k)$, $g_-^k > 0$ and $g_+^k < 0$. Figure 3 shows lines AC and BC, determined by points $A(\lambda_-^k, \Phi(\lambda_-^k))$, $B(\lambda_+^k, \Phi(\lambda_+^k))$ and slopes g_-^k and g_+^k, respectively. There, AD is the horizontal line $\Phi = \Phi(\lambda_-^k)$, and CQ is is the height from C onto AD. Figure 3 depicts the case when $\Phi(\lambda_-^k) \geq \Phi(\lambda_+^k)$.

Denote a solution to the problem as λ^*. Due to concavity of the function $\Phi(.)$, the point $(\lambda^*, \Phi(\lambda^*))$ lies in the triangle ACD. Expressing this condition in coordinate form, we obtain that λ^* belongs to the closed interval AD or $\Delta_x(k)$ defined by

$$\Delta_x(k) = \begin{cases} \lambda_-^k \leq \lambda^* \leq \lambda_+^k - \frac{1}{g_+}(\Phi(\lambda_+^k) - \Phi(\lambda_-^k)), & \Phi(\lambda_-^k) \geq \Phi(\lambda_+^k) \\ \lambda_-^k + \frac{1}{g_-}(\Phi(\lambda_+^k) - \Phi(\lambda_-^k)) \leq \lambda^* \leq \lambda_+^k, & \Phi(\lambda_-^k) \leq \Phi(\lambda_+^k) \end{cases} \quad (16)$$

and $\Phi(\lambda^*)$ belongs to the closed interval CQ or $\Delta_y(k)$ defined by

$$\begin{cases} \Phi(\lambda^*) \geq \max\{\Phi(\lambda_-^k), \Phi(\lambda_+^k)\} \\ \Phi(\lambda^*) \leq \left(\frac{1}{g_-^k} - \frac{1}{g_+^k}\right)^{-1} \left(\lambda_+^k - \lambda_-^k + \frac{\Phi(\lambda_-^k)}{g_-^k} - \frac{\Phi(\lambda_+^k)}{g_+^k}\right) \end{cases} \quad (17)$$

Denote the length of the interval $\Delta_x(k)$, i.e., AD, by $\sigma_x(k)$,

$$\sigma_x(k) = \begin{cases} \lambda_+^k - \lambda_-^k - \frac{1}{g_+}(\Phi(\lambda_+^k) - \Phi(\lambda_-^k)), & \Phi(\lambda_-^k) \geq \Phi(\lambda_+^k) \\ \lambda_+^k - \lambda_-^k - \frac{1}{g_-}(\Phi(\lambda_+^k) - \Phi(\lambda_-^k)), & \Phi(\lambda_-^k) \leq \Phi(\lambda_+^k) \end{cases} \qquad (18)$$

and the length of the interval $\Delta_y(k)$, i.e., CQ, by $\sigma_y(k)$,

$$\sigma_y(k) = \left(\frac{1}{g_-^k} - \frac{1}{g_+^k}\right)^{-1} \left(\lambda_+^k - \lambda_-^k + \frac{\Phi(\lambda_-^k)}{g_-^k} - \frac{\Phi(\lambda_+^k)}{g_+^k}\right)$$
$$- \max\{\Phi(\lambda_-^k), \Phi(\lambda_+^k)\} \qquad (19)$$

We can consider the values $\sigma_x(k)$ and $\sigma_y(k)$ to be measures of uncertainty in our information about a solution to the dual problem. After k iterations, we do not know an exact solution λ^*, but we know that it belongs to a closed interval of length $\sigma_x(k)$. Likewise, we do not know the maximal value of the dual function $\Phi(\lambda)$, but we know that it belongs to a closed interval of length $\sigma_y(k)$. Thus, the algorithms providing fast decrease in these values are of interest. This leads us to the following definitions.

Definition 1. *The general iterative procedure is called σ_x-contractive (or domain-contractive) with coefficient $\rho \in (0,1)$ if*

$$\lambda_*^k \in \Delta_x(k) \quad \forall k \in N,$$

and

$$\sigma_x(k+1) \leq \rho\sigma_x(k), \quad \forall k \in N. \qquad (20)$$

Definition 2. *The general iterative procedure is called σ_y-contractive (or range-contractive) with coefficient $\rho \in (0,1)$ if*

$$\sigma_y(k+1) \leq \rho\sigma_y(k), \quad \forall k \in N. \qquad (21)$$

Notice that in the above definitions, ρ depends neither on the structure of the graph G nor on arc costs and weights. In the following subsection, we will estimate the complexity of σ_x-contractive and σ_y-contractive implementations of the general iterative procedure.

5.2 Complexity of Contractive Algorithms

In this section, we consider the case in which all costs and weights are nonnegative integers.

Theorem 1. *Suppose that all costs and weights are nonnegative integers. Let the general iterative procedure be σ_x-contractive with coefficient ρ. Then the upper bound estimate for the complexity of the algorithm is given by*

$$Km\log_2 n \cdot \log_{1/\rho}(C_M W_M^2 n^3),$$

where K is the constant defined in (10), and W_M and C_M are defined in (13).

Proof: At the first iteration of the algorithm, from (12) and (14), we have

$$\sigma_x(1) = \lambda_+^1 = \frac{f(P_0) - f(P_\infty)}{g(P_\infty)} \leq f(P_0) \leq C_M \cdot n.$$

Since the algorithm is σ_x-contractive, according to (20), at the k-th iteration, we have $\sigma_x(k) \leq \rho^{k-1}\sigma_x(1)$, and for

$$k = [\log_{1/\rho}((C_M \cdot n)(W_M \cdot n)^2)] + 1,$$

where [.] denotes the integer part of a number, we have

$$\sigma_x(k) \leq \frac{1}{(W_M \cdot n)^2}.$$

Based on Proposition 5, there will be at most one element from the set U, which belongs to the interval $\Delta_x(k)$. Consequently, algorithm will stop no later than in $k + 1$ iterations. It is easy to estimate that the complexity of each iteration of the general iterative procedure is bounded by $Km \log_2 n$. Thus, the statement of the theorem holds.   ∎

Theorem 2. *Suppose that all costs and weights are nonnegative integers. Let the general iterative procedure be σ_y-contractive with coefficient ρ. Then the upper bound estimate for the complexity of the algorithm is given by*

$$Km \log_2 n \cdot \log_{1/\rho}(C_M W_M^2 n^3),$$

where K is the constant defined in (10), and W_M and C_M are defined in (13).

To prove the theorem, we need the following proposition.

Proposition 7. *Suppose that p_{ij}, w_{ij} and w are nonnegative integers. Then at each iteration k of the general iterative procedure,*

$$either \quad \sigma_y(k) = 0 \quad or \quad \sigma_y(k) \geq \frac{1}{(W_M \cdot n)^2} \quad holds\ true. \qquad (22)$$

Proof: For an arbitrary iteration k, there exists a path $P_- \in \mathcal{G}$ such that $\Phi(\lambda_-^k) = f(P_-) + \lambda_-^k g(P_-)$ and $g(P_-) = g_-^k$. Likewise, there exists a path $P_+ \in \mathcal{G}$ such that $\Phi(\lambda_+^k) = f(P_+) + \lambda_+^k g(P_+)$ and $g(P_+) = g_+^k$.

Substituting these into (19) and making some simplifications, we obtain

$$\sigma_y(k) = \frac{f(P_-)g_+^k - f(P_+)g_-^k}{g_+^k - g_-^k} - \max\{f(P_-) + \lambda_-^k g(P_-), f(P_+) + \lambda_+^k g(P_+)\}.$$
$$(23)$$

Since p_{ij}, w_{ij} and w are nonnegative integers, both $f(.)$ and $g(.)$ assume only integer values. That is why the first term in (23) is always rational. From (14) and (3) we can conclude that its denominator $|g_+^k - g_-^k| \leq W_M \cdot n$. From Proposition 5 and the iteration procedure (see (42) and (43) in Appendix

B) we can conclude that λ_-^k and λ_+^k are also rational with denominators no greater than $W_M \cdot n$. Thus, $\sigma_y(k)$ is a difference of two rational numbers with denominators no greater than $W_M \cdot n$, and so, it is rational with a denominator no greater than $(W_M \cdot n)^2$. And since $\sigma_y(k) \geq 0$, we conclude that either $\sigma_y(k) = 0$ or $\sigma_y(k) \geq \frac{1}{(W_M \cdot n)^2}$ holds true.   ∎

Now we can prove the theorem. Representing $\sigma_y(1)$ in the form of (23), we obtain

$$\sigma_y(1) = \frac{f(P_0)g(P_1) - f(P_1)g(P_0)}{g(P_1) - g(P_0)} - \max\{f(P_0), f(P_1) + \lambda_+^1 g(P_1)\}.$$

If the algorithm does not stop at the first step, then $g(P_0) > 0$ and $g(P_1) < 0$. Taking into account that $f(P_0) = \min_P f(P) \leq f(P_1)$, we obtain

$$\sigma_y(1) \leq \frac{f(P_0)g(P_1) - f(P_1)g(P_0)}{g(P_1) - g(P_0)} \leq f(P_1) \leq C_M \cdot n.$$

Since the algorithm is σ_y-contractive, from (21) we have $\sigma_y(k + 1) \leq \rho^k \sigma_y(1)$, and for

$$k = \lceil \log_{1/\rho}((C_M \cdot n)(W_M \cdot n)^2) \rceil + 2,$$

we obtain $\sigma_y(k) < \frac{1}{(W_M \cdot n)^2}$. But according to Proposition 7, this implies that $\sigma_y(k) = 0$.

Consequently, the algorithm will stop no later than in $(k + 1)$ iterations. To complete the proof, notice that the complexity of each iteration of the algorithm is bounded by $Km \log_2 n$.

Notice that if all costs and weights are nonnegative binary fractions with no more than M digits in the mantissa, then we can multiply all costs and weights by 2^M and obtain the case in which costs and weights are nonnegative integers.

6 Optimal Contractive Algorithms

In the previous section, we saw that the faster σ_x or σ_y decrease at each iteration, the faster the algorithm finds a solution. From this point of view, an optimal strategy for choosing λ_* would be such that minimizes σ_x or σ_y at the next iteration. In this section, we develop these two strategies. We derive closed-form expressions for corresponding λ_*'s and show that such choosing λ_* lead to σ_x-contractive and σ_y-contractive algorithms, whose complexity was evaluated in the previous section.

6.1 Definitions

Consider the general iterative procedure for some arbitrary graph $\mathcal{G}$. Suppose at some k-th iteration, we are provided the following data: λ_-^k, $\lambda_+^k > \lambda_-^k$,

$\Phi(\lambda_-^k)$, $\Phi(\lambda_+^k)$, $g_-^k > 0$ and $g_+^k < 0$. Our aim is to calculate λ_*^k as a function of these data

$$\lambda_*^k = \phi\left(\lambda_-^k, \lambda_+^k, \Phi(\lambda_-^k), \Phi(\lambda_+^k), g_-^k, g_+^k\right) \tag{24}$$

by minimizing $\sigma_x(k+1)$ or $\sigma_y(k+1)$. In the following we consider the same iteration and therefore may omit the upper index k.

Notice that the function $\phi(.)$ in (24) should be the same for all possible functions $\Phi(.)$. However, $\sigma_x(k+1)$ and $\sigma_y(k+1)$ depend not only on the arguments of $\phi(.)$ and our choice of λ_*, but also on the value of the objective function $\Phi(\lambda_*)$ and its supergradients $g_-(\lambda_*)$ and $g_+(\lambda_*)$ at this point. We cannot make our choice of λ_* based only on these values. Thus, for each possible choice of λ_*, we consider the worst-case $\sigma_x(k+1)$ and $\sigma_y(k+1)$, denoted by $h_x(\lambda_*)$ and $h_y(\lambda_*)$ respectively, i.e.,

$$h_x(\lambda_*) = \max_{\Phi(\lambda_*), g_-(\lambda_*), g_+(\lambda_*)} \sigma_x(k+1), \tag{25}$$

$$h_y(\lambda_*) = \max_{\Phi(\lambda_*), g_-(\lambda_*), g_+(\lambda_*)} \sigma_y(k+1). \tag{26}$$

We then define domain-optimal and range-optimal implementations of the general iterative procedure as the following.

Definition 3. *The general iterative procedure is called* domain-optimal *if at each iteration*

$$\lambda_*^k = \arg\min_{\lambda \in \Delta_x(k)} h_x(\lambda), \tag{27}$$

where $\Delta_x(k)$ is defined by (16), and is $h_x(\lambda)$ defined by (25).

Definition 4. *The general iterative procedure is called* range-optimal *if at each iteration*

$$\lambda_*^k = \arg\min_{\lambda \in [\lambda_-^k, \lambda_+^k]} h_y(\lambda), \tag{28}$$

where $h_y(\lambda)$ is defined by (26).

In the rest of this section, we will derive explicit expressions for λ_* and develop corresponding algorithms.

Further we assume that

$$\Phi(\lambda_-) \geq \Phi(\lambda_+). \tag{29}$$

In the case of $\Phi(\lambda_-) \leq \Phi(\lambda_+)$, all the results are obtained by similar reasoning.

6.2 Domain-Optimal Algorithm

In this section, we find an explicit expression for λ_*, optimal in the sense of Definition 3. That is, we solve the min-max problem (27), (25). First of all, notice that according to (16), $\Delta_x(k) = [\lambda_-, \mu]$, where $\mu = \lambda_- + \sigma_x(k)$. Let the algorithm chooses some $\lambda_* \in \Delta_x(k)$. Then

$$\sigma_x(k+1) \leq \max\{\lambda_* - \lambda_-, \mu - \lambda_*\},$$

and it is easy to show that this bound is tight. This means that $h_x(\lambda_*) = \max\{\lambda_* - \lambda_-, \mu - \lambda_*\}$. That is why, according to Definition 3, a domain-optimal choice of λ_*^k is the midpoint of the interval $\Delta_x(k)$. Such λ_*^k guarantees to have

$$\sigma_x(k+1) \leq \frac{1}{2}\sigma_x(k). \tag{30}$$

Similarly, if $\Phi(\lambda_-^k) < \Phi(\lambda_+^k)$, then the optimal choice of λ_*^k is also the midpoint of the interval $\Delta_x(k)$.

This leads us to the following implementation of the general iterative procedure.

Algorithm 1 (Domain-optimal algorithm) *Run the general iterative procedure, using the following expression for λ_*^k*

$$\lambda_*^k = \begin{cases} \frac{1}{2}\left(\lambda_-^k + \lambda_+^k - \frac{1}{g_+}(\Phi(\lambda_+^k) - \Phi(\lambda_-^k))\right) & \Phi(\lambda_-^k) \geq \Phi(\lambda_+^k) \\ \frac{1}{2}\left(\lambda_-^k + \lambda_+^k + \frac{1}{g_-}(\Phi(\lambda_+^k) - \Phi(\lambda_-^k))\right) & \Phi(\lambda_-^k) < \Phi(\lambda_+^k) \end{cases} \tag{31}$$

Since Algorithm 1 is an implementation of the general iterative procedure, its correctness and finiteness follow from Propositions 4 and 6. An upper bound estimate for the algorithm's complexity in the case of integer costs and weights follows from (30) and Theorem 1 with $\rho = 1/2$, and equals

$$Km\log_2 n \cdot \log_2(C_M W_M^2 n^3).$$

6.3 Range-Optimal Algorithm

In this section, we develop a procedure for calculating λ_* optimal in the sense of Definition 4. Namely, we consider a particular iteration k with some λ_-, λ_+, $\Phi(\lambda_-)$, $\Phi(\lambda_+)$, g_- and g_+ given, and solve min-max problem (26), (28). Finding an optimal λ_* is illustrated in Figure 4.

Figure 4 shows lines AC and BC, passing through points $A(\lambda_-, \Phi(\lambda_-))$ and $B(\lambda_+, \Phi(\lambda_+))$ and having slopes g_- and g_+, respectively. Horizontal lines AD and B_1B are determined by equations $\Phi = \Phi(\lambda_-)$ and $\Phi = \Phi(\lambda_+)$, respectively.

As the first step in solving problem (26), (28), we derive an expression for the function $h_y(\lambda_*)$ in (26). Suppose that some λ_* is chosen, and $\Phi(\lambda_*)$, $g_-(\lambda_*)$, $g_+(\lambda_*)$, $\Delta\lambda_-(\lambda_*)$ and $\Delta\lambda_+(\lambda_*)$ are calculated. Concavity of $\Phi(.)$ implies that point $E(\lambda_*, \Phi(\lambda_*))$ lies in triangle ABC (see Figure 4). Let AE intersect BC at point G, and let BE intersect AC at point F.

To derive an explicit expression for the function $h_y(\lambda_*)$, we solve (26) with respect to $g_-(\lambda_*)$ and $g_+(\lambda_*)$. Consider two cases: $\Phi(\lambda_*) \geq \Phi(\lambda_-)$ and $\Phi(\lambda_*) < \Phi(\lambda_-)$.

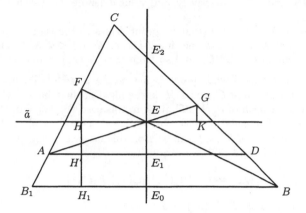

Figure 4 Finding λ_* optimal in the sense of Definition 4

If $\Phi(\lambda_*) \geq \Phi(\lambda_-) = \max\{\Phi(\lambda_-), \Phi(\lambda_+)\}$, then the point E lies in the triangle ACD. Consider a horizontal line $\tilde{a}$ passing through E, and lines FH and GK orthogonal to $\tilde{a}$. Figure 4 shows that if $g_-(\lambda_*) < 0$ then $\sigma_y(k+1) \leq |FH|$, and if $g_+(\lambda_*) > 0$ then $\sigma_y(k+1) \leq |GK|$. (Notice that if $g_-(\lambda_*) \geq 0$ and $g_+(\lambda_*) \leq 0$, then λ_* is a solution to the dual problem, and the algorithm stops at this iteration.)

In the second case, if $\Phi(\lambda_*) < \max\{\Phi(\lambda_-), \Phi(\lambda_+)\}$ then the point E lies in the triangle ADB. Let line FH' be orthogonal to AD, then in this case, we have $\sigma_y(k+1) \leq |FH'|$.

Therefore, in any case we obtain the following upper bound estimate for $\sigma_y(k+1)$

$$\sigma_y(k+1) \leq \bar{h} = \begin{cases} \max\{|FH|, |GK|\}, & \Phi(\lambda_*) \geq \max\{\Phi(\lambda_-), \Phi(\lambda_+)\} \\ |FH'|, & \Phi(\lambda_*) < \max\{\Phi(\lambda_-), \Phi(\lambda_+)\} \end{cases} \quad (32)$$

Moreover, this estimate is tight.

Values $\lambda_-, \lambda_+, \Phi(\lambda_-), \Phi(\lambda_+), g_-$ and g_+ are fixed, and $|FH|, |GK|, |FH'|$ and $\bar{h}$ are functions of λ_* and $\Phi(\lambda_*)$. From (26),

$$h_y(\lambda_*) = \max_{\Phi(\lambda_*)} \bar{h}. \quad (33)$$

To derive an explicit expression for $h_y(\lambda_*)$, we solve (33).

To simplify calculations, denote $x = \lambda_+ - \lambda_*$, $y = \lambda_* - \lambda_-$, $d = \Phi(\lambda_-) - \Phi(\lambda_+)$, $x_0 = -\frac{d}{g_+}$, $c = \lambda_+ - \lambda_-$, $a = c + \frac{d}{g_-}$, $b = c - x_0$. According to assumption (29), $d \geq 0$, and so $x_0 \geq 0$. In Figure 4, values a and b denote lengths of BB_1 and AD, respectively, and value d denote the distance between AD and BB_1.

Assume that $\lambda_* \geq \lambda_- + b$ (that is, point E lies to the right of point D). In this case point E lies below the line AD, i.e., $\Phi(\lambda_*) < \Phi(\lambda_-)$. Therefore, from (32) we obtain $\bar{h} = |FH'|$, and so from (33), $h_y(\lambda_*) = \max_{\Phi(\lambda_*)} |FH'|$, and the maximum is achieved when point E lies on the line BC. Then the point F coincides with the point C, and FH' is the hight of the triangle ACD from point C with the length of $\sigma_y(k)$. That is, in this case, $h_y(\lambda_*) = \sigma_y(k)$.

Further we will consider only the interval $\lambda_* \in [\lambda_-, \lambda_- + b]$, and we will see that the minimum of $h_y(\lambda_*)$ achieved on this interval is less than $\sigma_y(k)$.

According to (32) and (33), we can write

$$h_y(\lambda_*) = \max\{h_1(\lambda_*), h_2(\lambda_*), h_3(\lambda_*)\},$$

where $h_1(\lambda_*) = \max_{\Phi(\lambda_*)} |FH|$, $h_2(\lambda_*) = \max_{\Phi(\lambda_*)} |GK|$ and $h_3(\lambda_*) = \max_{\Phi(\lambda_*)} |FH'|$.

Notice that for any λ we have $h_3(\lambda) \leq |FH|\big|_{\Phi(\lambda)=\Phi(\lambda_-)} \leq h_1(\lambda)$, and thus,

$$h_y(\lambda) = \max\{h_1(\lambda), h_2(\lambda), h_3(\lambda)\} = \max\{h_1(\lambda), h_2(\lambda)\}. \tag{34}$$

We will express $h_1(\lambda)$ as a function of $x = \lambda_+ - \lambda_*$. Let $\alpha = \angle FBB_1 = \angle EBB_1$, i.e., $\alpha = \arctan \frac{\Phi(\lambda_*) - \Phi(\lambda_+)}{x}$, and let $\gamma_1 = \angle BB_1 C = \arctan(g_-)$. Let the vertical line, determined by λ, intersect BB_1 at point E_0, AD at point E_1 and BC at point E_2. Let FH_1 be a hight from F to BB_1. Then

$$|FH| = |FH_1| - |HH_1| = \frac{a}{\cot \gamma_1 + \cot \alpha} - x \tan \alpha.$$

To find $h_1(\lambda)$, we should maximize $|FH|$ over those α, which correspond to E inside the triangle ACD, i.e., E lying between E_1 and E_2. This corresponds to $\alpha \in [\phi_1, \phi_2]$, where $\phi_1 = \angle E_1 BB_1$ and $\phi_2 = \angle E_2 BB_1 = \arctan(-g_+)$.

Taking the derivative we obtain that the unconstrained maximum is achieved at

$$\alpha = \alpha_0 = \arctan\left(\left(\sqrt{\frac{a}{x}} - 1\right)\tan \gamma_1\right).$$

Three cases are possible:

1. $\alpha_0 \in [\phi_1, \phi_2]$. In this case $h_1(x) = |FH|\big|_{\alpha=\alpha_0} = (\sqrt{a} - \sqrt{x})^2 g_-$. In terms of x, the condition $\alpha_0 \leq \phi_2$ is equivalent to $x \geq a\frac{g_-^2}{(g_- - g_+)^2}$. From the condition $\alpha_0 \geq \phi_1$, we obtain the quadratic inequality $(\sqrt{x})^2 - \sqrt{a}\sqrt{x} + \frac{d}{g_-} \leq 0$, which yields $a - 4\frac{d}{g_-} \geq 0$ and $x_1 \leq x \leq x_2$, where

$$x_1 = \left(\frac{\sqrt{a} - \sqrt{a - 4\frac{d}{g_-}}}{2}\right)^2 \qquad x_2 = \left(\frac{\sqrt{a} + \sqrt{a - 4\frac{d}{g_-}}}{2}\right)^2. \tag{35}$$

2. $\alpha_0 > \phi_2$. In this case, $h_1(x) = |FH| \big|_{\alpha = \phi_2} = a\frac{g - g_+}{g_+ - g_-} + xg_+$. In terms of x, the condition $\alpha_0 > \phi_2$ is equivalent to $x < a\frac{g_-^2}{(g_- - g_+)^2}$.

3. $\alpha_0 < \phi_1$. In this case, $h_1(x) = |FH| \big|_{\alpha = \phi_1} = \frac{c - x}{1/g_- + x/d}$. This is true for all x if $a - 4\frac{d}{g_-} < 0$. If $a - 4\frac{d}{g_-} \geq 0$, then it holds for $x \notin (x_1, x_2)$, where x_1 and x_2 are defined by (35).

Combining all these cases, for $\lambda \in [\lambda_-, \lambda_- + b]$ we obtain

$$
h_1(\lambda) = h_1(\lambda_+ - x)
$$

$$
= \begin{cases}
a\frac{g - g_+}{g_+ - g_-} + xg_+, & -\frac{d}{g_+} \leq x \leq a\frac{g_-^2}{(g_- - g_+)^2} \\
(\sqrt{a} - \sqrt{x})^2 g_-, & x \geq a\frac{g_-^2}{(g_- - g_+)^2} \text{ and } x \in [x_1, x_2] \\
\frac{c - x}{1/g_- + x/d}, & x \notin (x_1, x_2), \text{ or } a - 4\frac{d}{g_-} < 0
\end{cases}
\tag{36}
$$

The function $h_2(\lambda)$ is calculated similarly, except that if we draw a figure for h_2 similar to Figure 4, we will see that the case of E^* lying below E_1 (which is now equivalent to $\alpha_0 < 0$) is impossible. Hence, the final formula, expressed in terms of $y = \lambda - \lambda_-$, takes the form

$$
h_2(\lambda) = h_2(\lambda_- + y) = \begin{cases}
b\frac{g - g_+}{g_+ - g_-} - yg_-, & 0 \leq y \leq y_1 = b\frac{g_+^2}{(g_- - g_+)^2} \\
-g_+(\sqrt{b} - \sqrt{y})^2, & y_1 \leq y \leq b
\end{cases}
\tag{37}
$$

A closed-form expression for $h_y(\lambda)$ is obtained based on formulas (36), (37) and (34). Then λ_*, optimal in the sense of Definition 4, is given by

$$
\lambda_* = \arg \min_{\lambda \in [\lambda_-, \lambda_- + b]} \max\{h_1(\lambda), h_2(\lambda)\}.
$$

From (36) and (37) we see that $h_1(\lambda)$ and $h_2(\lambda)$ are continuous on the closed interval $[\lambda_-, \lambda_- + b]$. Since $h_1(\lambda)$ decreases with respect to x, it increases as a function of λ. Likewise, since $h_2(\lambda)$ decreases with with respect to y, it decreases as a function of λ. Figure 4 show that if E lies on the vertical line passing through A, then it coincides with A, and $|FH| = 0$. Thus, $h_1(\lambda_-) = 0$. From (37), $h_2(\lambda_- + b) = 0$. Taking into account monotonicity and continuity of $h_1(\lambda)$ and $h_2(\lambda)$, we formulate the following proposition

Proposition 8. *There exists a unique λ such that $h_1(\lambda) = h_2(\lambda)$. This λ solves the problem*

$$
\min_{\lambda} \max\{h_1(\lambda), h_2(\lambda)\}.
$$

Hence, to find the optimal λ_*, we solve the equation

$$
h_1(\lambda) = h_2(\lambda),
\tag{38}
$$

where functions $h_1(\lambda)$ and $h_2(\lambda)$ are determined by (36) and (37), respectively. Since $h_1(\lambda)$ and $h_2(\lambda)$ are monotonic, equation (38) can be solved by the dichotomy method.

A solution λ_* to (38) will depend only on $\lambda_+ - \lambda_-$, $\Phi(\lambda_-) - \Phi(\lambda_+)$, g_- and g_+. Consequently,

$$y = \lambda_* - \lambda_- = F(\lambda_+ - \lambda_-, \Phi(\lambda_-) - \Phi(\lambda_+), g_-, g_+), \qquad (39)$$

where the function F is derived in Appendix C analytically, and an algorithm for calculating this function is presented in Appendix D. For λ_* we will then have

$$\lambda_* = \lambda_- + F(\lambda_+ - \lambda_-, \Phi(\lambda_-) - \Phi(\lambda_+), g_-, g_+).$$

Notice, however, that all these results were obtained for the case $\Phi(\lambda_-) \geq \Phi(\lambda_+)$ The case $\Phi(\lambda_-) < \Phi(\lambda_+)$ reduces to this case by substitution $\lambda = \text{const} - \lambda$. After this substitution, $\lambda_+ - \lambda_-$ is unchanged; $\Phi(\lambda_-)$ becomes $\Phi(\lambda_+)$, and $\Phi(\lambda_+)$ becomes $\Phi(\lambda_-)$; g_- becomes $-g_+$, and g_+ becomes $-g_-$. Further, $\lambda_+ - \lambda_*$ becomes $\lambda_* - \lambda_- = y$, and can be calculated by Algorithm 5. Therefore, in this case, an optimal λ_* is computed by

$$\lambda_* = \lambda_+ - F(\lambda_+ - \lambda_-, \Phi(\lambda_+) - \Phi(\lambda_-), -g_+, -g_-).$$

Thus, we obtain the following algorithm, optimal in $\Phi(\lambda)$ in the sense of Definition 4.

Algorithm 2 (Range-optimal algorithm) *Run the general iterative procedure, using the following expression for λ_**

$$\lambda_* = \begin{cases} \lambda_- + F(\lambda_+ - \lambda_-, \Phi(\lambda_-) - \Phi(\lambda_+), g_-, g_+), & \Phi(\lambda_-) \geq \Phi(\lambda_+) \\ \lambda_+ - F(\lambda_+ - \lambda_-, \Phi(\lambda_+) - \Phi(\lambda_-), -g_+, -g_-), & \Phi(\lambda_-) < \Phi(\lambda_+) \end{cases}$$
$$(40)$$

where F is given by (39).

Since Algorithm 2 is one of possible implementations of the general iterative procedure, its correctness and finiteness follow from Propositions 4 and 6. To evaluate algorithm complexity, we need the following proposition

Proposition 9. *If the general iterative procedure is range-optimal according to Definition 4, then it is σ_y-contractive with coefficient $q = 1/2$. That is, for each iteration of the algorithm*

$$\sigma_y(k+1) \leq \frac{1}{2}\sigma_y(k).$$

Proof: Consider a sub-optimal $\lambda_* = \lambda_1$, where λ_1 is the midpoint of the interval $\Delta_x(k)$, i.e., AD (see Figure 4). In this case, we have $|FH| \leq \frac{1}{2}|CQ| = \frac{1}{2}\sigma_y(k)$, and similarly $|GK| \leq \frac{1}{2}\sigma_y(k)$. Therefore, according to (34), $h_y(\lambda_1) \leq \frac{1}{2}\sigma_y(k)$.

For the optimal λ_*, we have $h_y(\lambda_*) = \min_\lambda h_y(\lambda) \leq h_y(\lambda_1) \leq \frac{1}{2}\sigma_y(k)$, which proves the proposition. ∎

The upper bound estimate for algorithm's complexity in the case of integer costs and weights follows from this proposition and Theorem 2 with $\rho = 1/2$. It is given by

$$Km \log_2 n \cdot \log_2(C_M W_M^2 n^3).$$

7 Conclusions

We have developed a general iterative procedure for solving the dual to the constrained shortest path problem (CSPP). The procedure couples the extended shortest path algorithm with one of the contractive algorithms: domain-optimal or range-optimal. The domain-optimal and range-optimal algorithms are based on the idea of "greedy" algorithms and find an optimal solution to the dual by minimizing the region of feasible solutions in the domain-range plane of the Lagrange function for the CSPP. At a particular iteration, the domain-optimal and range-optimal algorithms minimize the domain and range of the Lagrange function, correspondingly, and find an optimal λ for the next iteration analytically. We have proved that the complexity of any of the suggested contractive algorithms is linear in the number of arcs in the graph, logarithmic in the number of nodes, and logarithmic in the maximum value for arcs' costs and the maximum value for arcs' weights.

Appendix

A Extended Shortest Path Algorithm

Algorithm 3   *1. Initialization:*

$$l_{ij} := c_{ij} + \lambda w_{ij}, \quad (i,j) \in \mathcal{A}$$
$$l_i := \infty \; w_i^- := \infty \; w_i^+ := \infty \quad i \in \mathcal{N}, i \neq s$$
$$l_i := 0 \; w_i^- := 0 \; w_i^+ := 0 \quad i = s$$
$$LIST := \{s\}.$$

2. While $LIST \neq \emptyset$ do:
   a) Select $i \in Q$ with the smallest l_i.
   b) $LIST := LIST \setminus \{i\}$
   c) For each j, $(i,j) \in \mathcal{A}$
   - *If $l_j > l_i + l_{ij}$, then*

   $$l_j := l_i + l_{ij} \; w_j^- := w_i^- + w_{ij} \; w_j^+ := w_i^+ + w_{ij} \; j_j^- := i \; j_j^+ := i$$

   If $j \notin LIST$ then $LIST := LIST \cup \{j\}$
   - *If $l_j = l_i + l_{ij}$*
     - *If $w_j^- > w_i^- + w_{ij}$, then $w_j^- := w_i^- + w_{ij}; \; j_j^- := i$; if $j \notin LIST$ then $LIST := LIST \cup \{j\}$.*
     - *If $w_j^+ < w_i^+ + w_{ij}$, then $w_j^+ := w_i^+ + w_{ij}; \; j_j^+ := i$; if $j \notin LIST$ then $LIST := LIST \cup \{j\}$.*

3.

$$\Delta\lambda_- := \min_{(i,j)\in\mathcal{A}} \frac{(l_i + l_{ij}) - l_j}{(w_i^+ + w_{ij}) - w_j^+}$$

$$s.t.\ (w_i^+ + w_{ij}) - w_j^+ > 0$$

$$\Delta\lambda_+ := \min_{(i,j)\in\mathcal{A}} \frac{(l_i + l_{ij}) - l_j}{w_j^- - (w_i^- + w_{ij})}$$

$$s.t.\ w_j^- - (w_i^- + w_{ij}) > 0$$

4. *Recover path $\mathcal{P}_-$ by backtracking procedure starting from node t using labels j_i^+. Recover path $\mathcal{P}_+$ by backtracking procedure starting from node t using labels j_i^-. Return $\Phi(\lambda) = l_t - \lambda w$, $\mathcal{P}_-$, $\mathcal{P}_+$, $\Delta\lambda_-$ and $\Delta\lambda_+$.*
END

B General Iterative Procedure

Algorithm 4   *1. Run the ESP for $\lambda = 0$. $\mathcal{P}_0 := \mathcal{P}_+$.*
For $(i,j) \in \mathcal{A}$ do $c'_{ij} := w_{ij}$, $w'_{ij} := c_{ij}$.
Run the ESP for graph with arc costs c'_{ij} and arc weights w'_{ij} and for $\lambda = 0$. $\mathcal{P}_\infty := \mathcal{P}_+$.
Calculate $g(\mathcal{P}_0)$, $g(\mathcal{P}_\infty)$ according to (3).
If $g(\mathcal{P}_\infty) > 0$ then return "CSPP is infeasible". STOP.
If $g(\mathcal{P}_\infty) = 0$ then return $\mathcal{P}_\infty$ as a solution to the CSPP. STOP.
If $g(\mathcal{P}_0) \leq 0$ then return $\mathcal{P}_0$ as a solution to the CSPP. STOP.
If $f(\mathcal{P}_0) = f(\mathcal{P}_\infty)$ then return $\mathcal{P}_\infty$ as a solution to the CSPP. STOP.

$$\lambda_-^1 := 0$$

$$\lambda_+^1 := \frac{f(\mathcal{P}_0) - f(\mathcal{P}_\infty)}{g(\mathcal{P}_\infty)}$$

Run the ESP for $\lambda = \lambda_+^1$. $\mathcal{P}_1 := \mathcal{P}_-$.
If $g(\mathcal{P}_1) = 0$ then return $\mathcal{P}_1$ as a solution to the CSPP. STOP.

$$g_-^1 := g(\mathcal{P}_0);\ \Phi(\lambda_-^1) := f(\mathcal{P}_0) + \lambda_-^1 \cdot g(\mathcal{P}_0);$$
$$g_+^1 := g(\mathcal{P}_1);\ \Phi(\lambda_+^1) := f(\mathcal{P}_1) + \lambda_+^1 \cdot g(\mathcal{P}_1) \tag{41}$$

$$k := 1$$

2. If $\lambda_-^k = \lambda_+^k$ then return $\lambda^* = \lambda_-^k$. STOP.
If $\Phi(\lambda_-^k) = \Phi(\lambda_+^k) + (\lambda_-^k - \lambda_+^k)g_+^k$ then return $\lambda^* = \lambda_-^k$. STOP.
If $\Phi(\lambda_+^k) = \Phi(\lambda_-^k) + (\lambda_+^k - \lambda_-^k)g_-^k$ then return $\lambda^* = \lambda_+^k$. STOP.

3. *Choose λ_*^k such that $\lambda_-^k \leq \lambda_*^k \leq \lambda_+^k$.*
4. *Run the ESP algorithm to calculate $\Phi(\lambda_*^k)$, $g_-(\lambda_*^k)$, $g_+(\lambda_*^k)$, $\Delta\lambda_-(\lambda_*^k)$, $\Delta\lambda_+(\lambda_*^k)$.*
5. *If $g_+(\lambda_*^k) \leq 0 \leq g_-(\lambda_*^k)$ then return $\lambda^* = \lambda_*^k$. STOP.*
   If $g_-(\lambda_*^k) < 0$ then
   begin

$$\lambda_-^{k+1} := \lambda_-^k; \qquad g_-^{k+1} := g_-^k; \qquad \Phi(\lambda_-^{k+1}) := \Phi(\lambda_-^k)$$
$$\lambda_+^{k+1} := \lambda_*^k - \Delta\lambda_-(\lambda_*^k); \qquad g_+^{k+1} := g_-(\lambda_*^k) \qquad (42)$$
$$\Phi(\lambda_+^{k+1}) := \Phi(\lambda_*^k) - \Delta\lambda_-(\lambda_*^k)g_-(\lambda_*^k)$$

$$k := k + 1$$

   Go to step 2.
   end
   If $g_+(\lambda_*^k) > 0$ then
   begin

$$\lambda_+^{k+1} := \lambda_+^k; \qquad g_+^{k+1} := g_+^k; \qquad \Phi(\lambda_+^{k+1}) := \Phi(\lambda_+^k)$$
$$\lambda_-^{k+1} := \lambda_*^k + \Delta\lambda_+(\lambda_*^k); \qquad g_-^{k+1} := g_+(\lambda_*^k) \qquad (43)$$
$$\Phi(\lambda_-^{k+1}) := \Phi(\lambda_*^k) + \Delta\lambda_+(\lambda_*^k)g_+(\lambda_*^k)$$

   Go to step 2.
   end

Now we prove Proposition 4 about correctness of this algorithm.

Proposition 4 *If the general iterative procedure stops, then it yields exactly one of the following results:*

1. *it shows, that the CSPP is infeasible,*
2. *it finds a solution to the CSPP,*
3. *it finds a solution to the dual problem.*

Proof: If the algorithm stops at the first step, then the statement of the Theorem obviously holds. Let us prove, that it also holds if the algorithm stops at the second step.

First of all, let us prove that for every $k \in N$, if Algorithm 4 does not stop before step 2 at the k-th iteration, then the following holds

$$g_-^k > 0, \qquad g_+^k < 0. \qquad (44)$$

Indeed, from concavity of function $\Phi(.)$ and from the definition of $\mathcal{P}_1$ (see step 1 of Algorithm 4) we can conclude that $g(\mathcal{P}_1) \leq 0$. If $g(\mathcal{P}_1) = 0$, the algorithm finds an optimal solution and stops. Also, according to the algorithm, if $g(\mathcal{P}_0) \leq 0$ then we will stop at step 1. Therefore, if algorithm does not stop at step 1, then from (41) we have $g_-^1 = g(\mathcal{P}_0) > 0$ and $g_+^1 = g(\mathcal{P}_1) < 0$, which proves the induction base. If for some k condition (44) holds, then according to (42) and (43), condition (44) holds also for $k+1$. By induction it holds for every k.

Notice also that $g(\mathcal{P}_\lambda) \in \partial(\varPhi(\lambda))$. Then from (41) we have that g^1_- and g^1_+ are supergradients of $\varPhi(\lambda)$ at $\lambda = \lambda^1_-$ and $\lambda = \lambda^1_+$, respectively. Taking into account (42) and (43), we can conclude by induction that for every k values g^k_- and g^k_+ remain supergradients for $\varPhi(\lambda)$ at $\lambda = \lambda^k_-$ and $\lambda = \lambda^k_+$, respectively.

Taking into account concavity of $\varPhi(\lambda)$ and (44), we obtain $\lambda^k_- \leq \lambda^k_+$, and there always exists a solution to the dual problem in the interval $[\lambda^k_-, \lambda^k_+]$. In particular, if $\lambda^k_- = \lambda^k_+$, then $\lambda^* = \lambda^k_- = \lambda^k_+$ is a solution.

Notice also, that from Proposition 2 we can conclude that $\varPhi(\lambda^{k+1}_-)$ and $\varPhi(\lambda^{k+1}_+)$, calculated according to (42) and (43), respectively, are indeed values for $\varPhi(.)$ for the corresponding λ's. Thus, from concavity of $\varPhi(\lambda)$ we have that if $\varPhi(\lambda^k_+) = \varPhi(\lambda^k_-) + (\lambda^k_+ - \lambda^k_-)g^k_-$ then λ^k_+ is a solution to the dual problem. Analogously, if $\varPhi(\lambda^k_-) = \varPhi(\lambda^k_+) + (\lambda^k_- - \lambda^k_+)g^k_+$, then λ^k_- is indeed a solution to the dual problem.

Therefore, if the algorithm stops at the second step, then the statement of the theorem also holds.

If at step 5, $g_+(\lambda^k_*) \leq 0 \leq g_-(\lambda^k_*)$, then λ^k_* maximizes $\varPhi(\lambda)$ by the properties of concave functions and the fact that $g_+(\lambda^k_*)$ and $g_-(\lambda^k_*)$ are supergradients of $\varPhi(\lambda)$ at $\lambda = \lambda^k_*$. ∎

C Analytic Derivation of F

To find an analytic solution to (38) we should consider six cases.

1. λ_* is such that $x = \lambda_+ - \lambda_0$ and $y = \lambda_* - \lambda_-$ satisfy conditions $-\frac{d}{g_+} \leq x \leq a\frac{g^2_-}{(g_- - g_+)^2}$ and $b\frac{g^2_+}{(g_- - g_+)^2} \leq y \leq b$. Notice that $x + y = \lambda_+ - \lambda_- = c$. In this case from (36), (37), (38) we have the following system

$$\begin{cases} a\frac{g_- - g_+}{g_+ - g_-} + xg_+ = -g_+(\sqrt{b} - \sqrt{y})^2 \\ x + y = c \end{cases}$$

Recall that $a = c + \frac{d}{g_-}$ and $b = c + \frac{d}{g_+}$. Solving the system above with respect to x and y, we obtain

$$y = \left(1 - \frac{g_-}{2(g_- - g_+)}\right)^2 \left(c + \frac{d}{g_+}\right) \tag{45}$$

2. If λ_* is such that $x \geq a\frac{g^2_-}{(g_- - g_+)^2}$, $x \in [x_1, x_2]$ and $0 \leq y \leq b\frac{g^2_+}{(g_- - g_+)^2}$, where x_1 and x_2 are defined by (35), then from (36) and (37) we have

$$h_1(\lambda_*) = (\sqrt{a} - \sqrt{x})^2 g_- = g_- \left(\sqrt{c + \frac{d}{g_-}} - \sqrt{x} \right)^2 ,$$

$$h_2(\lambda_*) = b \frac{g_- - g_+}{g_+ - g_-} - y g_- = \left(c + \frac{d}{g_+} \right) \frac{g_- - g_+}{g_+ - g_-} - (c - x) g_- .$$

Solving equation (38) for this case, we obtain

$$y = c - \left(1 + \frac{g_+}{2(g_- - g_+)} \right)^2 \left(c + \frac{d}{g_-} \right) . \tag{46}$$

3. If λ_* is such that $x \geq a \frac{g_-^2}{(g_- - g_+)^2}$, $x \in [x_1, x_2]$ and $b \frac{g_+^2}{(g_- - g_+)^2} \leq y \leq b$, then from (36), (37) and (38) we have a system

$$\begin{cases} g_- \left(\sqrt{c + \frac{d}{g_-}} - \sqrt{x} \right)^2 = -g_+ \left(\sqrt{c + \frac{d}{g_+}} - \sqrt{y} \right)^2 \\ x + y = c \end{cases}$$

Denoting $\xi = \sqrt{x}$, $\eta = \sqrt{y}$ and solving this system with respect to ξ and η, we obtain

$$\eta_{1,2} = \frac{1}{g_- - g_+} \left\{ \sqrt{-g_+} \left(\sqrt{-g_+} \sqrt{c + \frac{d}{g_+}} - \sqrt{g_-} \sqrt{c + \frac{d}{g_-}} \right) \right.$$
$$\left. \pm \sqrt{2g_-} \sqrt[4]{-g_+ g_-} \left(c + \frac{d}{g_-} \right) \left(c + \frac{d}{g_+} \right) . \right\} \tag{47}$$

Notice that we should choose a solution such that $\eta \geq 0$, $\xi \geq 0$, and corresponding x and y satisfy the condition above. Proposition 8 guarantees that there exists only one such a solution.

4. If λ_* is such that for corresponding x and y, $0 \leq y \leq b \frac{g_+^2}{(g_- - g_+)^2}$ and either $a - 4 \frac{d}{g_-} < 0$ or $x \notin (x_1, x_2)$, then from (36), (37) and (38) we obtain

$$\frac{y}{\frac{1}{g_-} + \frac{c - y}{d}} = b \frac{g_+ g_-}{g_+ - g_-} - y g_- .$$

Solutions to this quadratic equation are given by

$$y_{1,2} = \frac{cg_- + d}{2g_-} \left(\frac{2g_+ - g_-}{g_+ - g_-} \mp \sqrt{\left(\frac{g_-}{g_+ - g_-} \right)^2 + \frac{4d}{cg_- + d}} \right) . \tag{48}$$

It is not difficult to prove that y_2, corresponding to the plus sign, is greater than a. But a solution should satisfy $y \leq b \frac{g_+^2}{(g_- - g_+)^2} < b \leq a$. Therefore, the solution to (38) is $y = y_1$.

5. If λ_* is such that for corresponding x and y, $b\frac{g_+^2}{(g_- - g_+)^2} \leq y \leq b$ and either $a - 4\frac{d}{g_-} < 0$ or $x \notin (x_1, x_2)$, then from (36), (37) and (38) we have

$$\frac{y}{\frac{1}{g_-} + \frac{c-y}{d}} = -g_+ \left(\sqrt{c + \frac{d}{g_+}} - \sqrt{y} \right)^2.$$

If we denote $z = \sqrt{y}$, then the equation above reduces to a quartic equation

$$z^4 - 2\sqrt{c + \frac{d}{g_+}} z^3 - \frac{d}{g_-} z^2 + 2\left(c + \frac{d}{g_-} \right)\sqrt{c + \frac{d}{g_+}} z$$
$$- \left(c + \frac{d}{g_-} \right)\left(c + \frac{d}{g_+} \right) = 0, \tag{49}$$

whose roots can be found by Ferrari's formulas. Among the four roots of the equation, we should choose those that are real, nonnegative and for which $y = z^2$ satisfies the conditions for this case. From Proposition 8, it follows that the equation $h_1(\lambda_*) = h_2(\lambda_*)$ has only one solution. Consequently, we have only one z satisfying all the requirements. The solution will be then $y = z^2$.

6. In the last case, we consider λ_* such that $-\frac{d}{g_+} \leq x \leq a\frac{g_-^2}{(g_- - g_+)^2}$ and $0 \leq y \leq b\frac{g_+^2}{(g_- - g_+)^2}$. But as long as $g_- > 0$ and $g_+ < 0$, this implies that

$$x \leq a\left(\frac{g_-}{g_- - g_+} \right)^2 < a\frac{g_-}{g_- - g_+} = \left(c + \frac{d}{g_-} \right)\frac{g_-}{g_- - g_+}$$

$$y \leq b\left(\frac{-g_+}{g_- - g_+} \right)^2 < b\frac{-g_+}{g_- - g_+} = \left(c + \frac{d}{g_+} \right)\frac{-g_+}{g_- - g_+}$$

and

$$x + y < c\left(\frac{g_-}{g_- - g_+} + \frac{-g_+}{g_- - g_+} \right) + d\left(\frac{1}{g_- - g_+} + \frac{-1}{g_- - g_+} \right) = c$$

which contradicts the condition $x + y = c$. Thus, there is no λ that corresponds to x and y satisfying simultaneously $-\frac{d}{g_+} \leq x \leq a\frac{g_-^2}{(g_- - g_+)^2}$ and $0 \leq y \leq b\frac{g_+^2}{(g_- - g_+)^2}$. Consequently, this case is impossible.

Formulas (45)-(48) and the root of equation (49) determine y and the optimal $\lambda_* = \lambda_- + y$.

D Algorithm for Calculation of F

Function $F(c, d, g_1, g_2)$ will return optimal y given $c = \lambda_+ - \lambda_-$, $d = \Phi(\lambda_-) - \Phi(\lambda_+)$, $g_1 = g_-$, $g_2 = g_+$ for the case $\Phi(\lambda_-) \geq \Phi(\lambda_+)$.

Algorithm 5 (Calculation of F)

1.

$$a := c + \frac{d}{g_1}, \quad b := c + \frac{d}{g_2}$$

If $c - 3\frac{d}{g_2} \geq 0$ then

$$y_1 := c - \left(\frac{\sqrt{c + \frac{d}{g_1}} + \sqrt{c - 3\frac{d}{g_1}}}{2} \right)^2$$

$$y_2 := c - \left(\frac{\sqrt{c + \frac{d}{g_1}} - \sqrt{c - 3\frac{d}{g_1}}}{2} \right)^2$$

else

$$y_1 := -1 \quad y_2 := -1$$

end if.

$$y_3 := c - a\frac{g_1^2}{(g_1 - g_2)^2}$$

$$y_4 := b\frac{g_2^2}{(g_2 - g_1)^2}$$

2.

$$k := 0$$

For $i = 1, 2, 3, 4$ do
if $y_i \in [0, b]$ then $k := k + 1$, $d_k := y_i$.
Sort $d_1, \ldots, d_k$ in the increasing order.

$$d_0 := 0 \quad d_{k+1} := b$$

3.

$$j := \max\{j \mid h_1(c - d_j) < h_2(d_j)\}$$

where $h_1(x)$ is given by (36), $h_2(y)$ is given by (37).

4.

$$t := \frac{d_j + d_{j+1}}{2}$$

If $t \geq y_3$ then $i_1 := 1$.
If $t \notin [y_1, y_2]$ then $i_1 := 3$.
If $t < y_3$ and $t \in (y_1, y_2)$ then $i_1 := 2$.
If $t \leq y_4$ then $i_2 := 1$ else $i_2 := 2$.

5. Depending on i_1 and i_2 do

a) If $i_1 = 1$, $i_2 = 2$ then return

$$F = \left(1 - \frac{g_1}{2(g_1 - g_2)}\right)^2 \left(c + \frac{d}{g_2}\right)$$

STOP

b) If $i_1 = 2$, $i_2 = 1$ then return

$$F = c - \left(1 + \frac{g_2}{2(g_1 - g_2)}\right)^2 \left(c + \frac{d}{g_1}\right)$$

STOP

c) If $i_1 = 2$, $i_2 = 2$ then

$$\eta_1 := \frac{-\sqrt{-g_2}(\sqrt{g_1}\sqrt{a} - \sqrt{-g_2}\sqrt{b}) - \sqrt{2g_1}\sqrt[4]{-g_1 g_2 ab}}{g_1 - g_2}$$

$$\eta_2 := \frac{-\sqrt{-g_2}(\sqrt{g_1}\sqrt{a} - \sqrt{-g_2}\sqrt{b}) + \sqrt{2g_1}\sqrt[4]{-g_1 g_2 ab}}{g_1 - g_2}$$

For $i = 1, 2$ do
if $\eta_i \geq 0$ and $\eta_i^2 \in [d_j, d_{j+1}]$ then return $F = \eta_i^2$.
STOP.

d) If $i_1 = 3$, $i_2 = 1$ then return

$$F = \frac{cg_1 + d}{2g_1}\left(\frac{2g_2 - g_1}{g_2 - g_1} - \sqrt{\left(\frac{g_1}{g_2 - g_1}\right)^2 + \frac{4d}{cg_1 + d}}\right)$$

STOP.

e) If $i_1 = 3$, $i_2 = 2$ then

$$a_3 := -2\sqrt{c + \frac{d}{g_2}}, \quad a_2 := -\frac{d}{g_1}, \quad a_1 := 2d\sqrt{c + \frac{d}{g_2}}\left(\frac{1}{g_1} + \frac{c}{d}\right),$$

$$a_0 := \left(-\frac{d^2}{g_2} - cd\right)\left(\frac{1}{g_1} + \frac{c}{d}\right)$$

$$b_2 := -a_2, \quad b_1 := a_1 a_3 - 4a_0, \quad b_0 := 4a_2 a_0 - a_1^2 - a_3^2 a_0,$$

$$Q := \frac{3b_1 - b_2^2}{9}$$

$$R := \frac{9b_2 b_1 - 27b_0 - 2b_2^3}{54}$$

$$D := Q^3 + R^3$$

$$S := \sqrt[3]{R + \sqrt{D}}$$

$$T := \sqrt[3]{R - \sqrt{D}}$$

$$u := -\frac{1}{3}b_2 + S + T$$

$$H := \sqrt{\frac{1}{4}a_3^2 - a_2 + u}$$

$$G := \begin{cases} \sqrt{\frac{3}{4}a_3^2 - H^2 - 2a_2 + \frac{1}{4}(4a_3a_2 - 8a_1 - a_3^3)H^{-1}} & \text{for } H \neq 0 \\ \sqrt{\frac{3}{4}a_3^2 - 2a_2 + 2\sqrt{u^2 - 4a_0}} & \text{for } H = 0 \end{cases}$$

$$E := \begin{cases} \sqrt{\frac{3}{4}a_3^2 - H^2 - 2a_2 - \frac{1}{4}(4a_3a_2 - 8a_1 - a_3^3)H^{-1}} & \text{for } H \neq 0 \\ \sqrt{\frac{3}{4}a_3^2 - 2a_2 - 2\sqrt{u^2 - 4a_0}} & \text{for } H = 0 \end{cases}$$

$$z_1 := -\frac{1}{4}a_3 + \frac{1}{2}H + \frac{1}{2}G$$

$$z_2 := -\frac{1}{4}a_3 + \frac{1}{2}H - \frac{1}{2}G$$

$$z_3 := -\frac{1}{4}a_3 - \frac{1}{2}H + \frac{1}{2}E$$

$$z_4 := -\frac{1}{4}a_3 - \frac{1}{2}H - \frac{1}{2}E$$

For $i = 1, 2, 3, 4$ do
if $z_i \in \mathbb{R}$ and $z_i \geq 0$ and $z_i^2 \in [d_j, d_{j+1}]$ then return $F = z_i^2$.
STOP.

Let us prove the correctness of this algorithm. We should prove that algorithm calculates optimal $y = \lambda_* - \lambda_-$ given $c = \lambda_+ - \lambda_-$, $d = \Phi(\lambda_-) - \Phi(\lambda_+)$, $g_1 = g_-$, $g_2 = g_+$ for the case $\Phi(\lambda_-) \geq \Phi(\lambda_+)$.

We proved in Proposition 8 that optimal λ_* is the root of the equation $h_1(\lambda_*) = h_2(\lambda_*)$. To calculate $h_1(.)$ based on (36), we consider three cases, and to calculate $h_1(.)$ based on (37), we consider two cases.

In Appendix C, we considered all these six cases and obtained formulas (45)-(48) and equation (49) that provide an optimal y in each of these cases.

At step 1 of the algorithm, we compute a, b, y_1, y_2, y_3 and y_4, where a and b are the same as in (36) and (37); and $y_1, \ldots, y_4$ are the values of y, which split the set of all possible values of y into the intervals corresponding to different cases in computing h_1 and h_2.

At step 2, we order $y_1, \ldots, y_4$ to be d_i-s, so that for each interval $[d_i, d_{i+1}]$, only one case for computing h_1 and h_2 is possible.

Monotonicity of $h_1(.)$ and $h_2(.)$ implies that optimal y lies in the interval $[d_j, d_{j+1}]$ determined at step 3.

Values i_1 and i_2, introduced at step 4, are used to calculate $h_1(.)$ and $h_2(.)$ on the selected interval in expressions (36) and (37), and calculate optimal y in step 5. For example, $i_1 = 1$, $i_2 = 2$ correspond to the case 1 in Appendix C. Therefore, formula (45) was used for this case to calculate optimal y. Likewise, for different values of i_1 and i_2, an optimal y is calculated according to the cases considered in Appendix C. For the case $i_1 = 3$, $i_2 = 2$, values a_0, a_1, a_2 and a_3 are the coefficients in quartic equation (49), which is subsequently solved by Ferrari formulas.

Consequently, Algorithm 5 provides an optimal y in the case $\Phi(\lambda_-) \geq \Phi(\lambda_+)$.

References

1. Y.P. Aneja, V. Aggarwal and K.P.K. Nair, Shortest Chain subject to Side Constraints. *Networks*, **13**, pp. 295–302, 1983.
2. J.E. Beasley and N. Christofides, An Algorithm for the Resource Constrained Shortest Path Problem, *Networks*, **19**, pp. 379–394, 1989.
3. J.F. Benders, Partitioning procedures for solving mixed-variables programming problems, *Numerische Mathematik*, **4**, pp. 238–252, 1962.
4. W.M. Carlyle and R.K. Wood, Lagrangian Relaxation and Enumeration for Solving Constrained Shortest-Path Problems, 38th Annual ORSNZ Conference, University of Waikato, Hamilton, New Zealand, 21-22 November 2003.
5. M. Desrochers and F. Soumis, A Generalized Permanent Labeling Algorithm for the Shortest Path Problem with Time Windows, *INFOR*, **26**, pp. 191–212, 1988.
6. G. Dooms, Y. Deville and P. Dupont, Constrained path finding in biochemical networks, *Proceedings of JOBIM 2004*, pp. JO–40, 2004.
7. I. Dumitrescu, Constrained Path and Cycle Problems, Ph.D. Thesis, Department of Mathematics and Statistics, University of Melbourne, 2002.
8. I. Dumitrescu and N. Boland, Algorithms for the Weight Constrained Shortest Path Problem. *ITOR*, **8**, pp. 15–29, 2001.
9. I. Dumitrescu and N. Boland, Improved Preprocessing, Labeling and Scaling Algorithms for the Weight-Constrained Shortest Path Problem, *Networks*, **42**, pp. 135–153, 2003.
10. M.R. Garey and D.S. Johnson, *Computers and intractability: A guide to the theory of NP-completeness*, Freeman, San Francisco, 1979.
11. R.E. Gomory, An Algorithm for Integer Solutions to Linear Programs, in R.L. Graves and P. Wolfe (eds), *Recent Advances in Mathematical Programming*, McGraw-Hill, New York, 1963.
12. G.Y. Handler and I. Zang, A Dual Algorithm for the Constrained Shortest Path Problem, *Networks*, **10**, pp. 293–309, 1980.
13. R. Hassin, Approximation Schemes for the Restricted Shortest Path Problem, *Mathematics of Operations Research*, **17**, pp. 36–42, 1992.
14. H. Joksch, The shortest route problem with constraints, *Journal of Mathematical Analysis and Application*, **14**, pp. 191–197, 1966.
15. J.E. Kelley, The Cutting-Plane Method for Solving Convex Programs, *Journal of SIAM*, **8**(4), pp. 703–712, 1960.
16. T. Korkmaz and M. Krunz, Multi-Constrained Optimal Path Selection, *Proceedings of the IEEE INFOCOM 2001 Conference*, **2**, pp. 834–843, 2001.
17. J.L. Latourell, B.C. Wallet and B. Copeland, Genetic Algorithm to Solve Constrained Routing Problems with Applications for Cruise Missile Routing, in *Applications and Science of Computational Intelligence, Proceedings of SPIE*, Vol. 3390, Society of Photo-optical Instrumentation Engineers, Bellingham, Washington, USA, pp. 490–500, 1998.
18. C. Lemarchal, Lagrangian Relaxation, in M. Junger and D. Naddef (eds), *Computational Combinatorial Optimization: Optimal or Provably Near-Optimal Solutions*, vol. 2241/2001, Springer-Verlag, Heidelberg, 2001.

19. D.H. Lorenz and D. Raz, A simple efficient approximation scheme for the restricted shortest path problem, *Operations Research Letters*, **28**(5), pp. 213−219, 2001.

20. K. Mehlhorn and M. Ziegelmann, Resource Constraint Shortest Paths, 7th Annual European Symposium on Algorithms (ESA 2000), LNCS 1879, pp. 326−337, 2000.

21. M. Minoux and C.C. Ribeiro, Solving Hard Constrained Shortest Path Problems by Lagrangean Relaxation and Branch-and-Bound Algorithms, *Methods of Operations Research*, **53**, pp. 305−316, 1986.

22. C. Ribeiro and M. Minoux, A Heuristic Approach to Hard Constrained Shortest Path Problems. *Discrete Applied Mathematics*, **10**, pp. 125−137, 1986.

23. J.-P. Vial, J.-L. Goffin and O. du Merle, On the Comparative Behavior of Kelley's Cutting Plane Method and the Analytic Center Cutting Plane Method, Technical Report 1996.4, Department of Management Studies, University of Geneva, Switzerland, 1996.

24. J.Y. Yen, Finding the k-shortest, Loopless Paths in a Network, *Management Science*, **17**, pp. 711−715, 1971.

25. M. Zabarankin, S. Uryasev and R. Murphey, Aircraft Routing under the Risk of Detection, *Naval Research Logistics*, Vol. 53, Issue 8, pp. 728−747, 2006.

26. M. Zabarankin, S. Uryasev and P. Pardalos, Optimal Risk Path Algorithms, in R. Murphey and P. Pardalos (eds), Cooperative Control and Optimization, Kluwer Academic Publishers, pp. 271−303, 2002.

27. M. Ziegelmann, Constrained Shortest Paths and Related Problems, Ph.D. Thesis, Universität des Saarlandes, Saarbrücken, 2001.

8 Towards an Irreducible Theory of Complex Systems

Victor Korotkikh

Summary. In the chapter we present results to develop an irreducible theory of complex systems in terms of self-organization processes of prime integer relations. Based on the integers and controlled by arithmetic only the self-organization processes can describe complex systems by information not requiring further explanations. Important properties of the description are revealed. It points to a special type of correlations that do not depend on the distances between parts, local times and physical signals and thus proposes a perspective on quantum entanglement. Through a concept of structural complexity the description also ·computationally suggests the possibility of a general optimality condition of complex systems. The computational experiments indicate that the performance of a complex system may behave as a concave function of the structural complexity. A connection between the optimality condition and the majorization principle in quantum algorithms is identified. A global symmetry of complex systems belonging to the system as a whole, but not necessarily applying to its embedded parts is also presented. As arithmetic fully determines the breaking of the global symmetry, there is no further need to explain why the resulting gauge forces exist the way they do and are not even slightly different.

1 Introduction

By networking different components a new generation of technologies is being developed in control of human operations. As a result it becomes increasingly important to have confidence in the theory of complex systems. This calls for clear explanations why the foundations of the theory are valid in the first place. The ideal situation would be to have an irreducible theory of complex systems not requiring a deeper explanatory base in principle. But the question arises: where such a theory could come from, when even the concept of spacetime is questioned [1] as a fundamental entity.

As a possible answer it is suggested that the concept of integers may take responsibility in the search for an irreducible theory of complex systems [2]. It is shown that complex systems can be described in terms of self-organization processes of prime integer relations [2], [3]. Based on the integers and controlled by arithmetic only the self-organization processes can describe complex systems by information not requiring further explanations. In the chapter we present further results to progress in this direction.

2 Invariant Quantities of a Complex System and Underlying Correlations

To understand a complex system we consider the dynamics of the elementary parts and focus on the correlations preserving certain quantities of the complex system [2], [3].

Let I be an integer alphabet and

$$I_N = \{x = x_1...x_N, x_i \in I, i = 1, ..., N\}$$

be the set of sequences of length $N \geq 2$. We consider N elementary parts $P_i, i = 1, ..., N$ with the state of an element P_i in its reference frame specified by a generalized coordinate $x_i \in I, i = 1, ..., N$ (for example, the position of the element P_i in space) and the state of the elements by a sequence $x = x_1...x_N \in I_N$.

For a geometric representation of the sequences we use piecewise constant functions. Let $\varepsilon > 0$ and $\delta > 0$ be length scales of a two-dimensional lattice. Let

$$\rho_{m\varepsilon\delta} : x \to f$$

be a mapping that realizes the geometric representation of a sequence $x = x_1...x_N \in I_N$ by associating it with a function $f \in W_{\varepsilon\delta}[t_m, t_{m+N}]$, denoted $f = \rho_{m\varepsilon\delta}(x)$, such that

$$f(t_m) = x_1\delta, \ f(t) = x_i\delta, \ t \in (t_{m+i-1}, t_{m+i}], \ i = 1, ..., N,$$

$$t_i = i\varepsilon, \ i = m, ..., m + N$$

and whose integrals $f^{[k]}$ satisfy the condition

$$f^{[k]}(t_m) = 0, \ k = 1, 2, ... ,$$

where m is an integer. The sequence $x = x_1...x_N$ is called a code of the function f and denoted $c(f)$.

By using the geometric representation for a state $x = x_1...x_N \in I_N$ of the elements $P_i, i = 1, ..., N$ we define quantities of the complex system as the integrals

$$f^{[k]}(t_{m+N}) = \int_{t_m}^{t_{m+N}} f^{[k-1]}(t)dt, \ f^{[0]} = f, \ k = 1, 2, ...$$

of a function $f^{[0]} = f = \rho_{m\varepsilon\delta}(x) \in W_{\varepsilon\delta}[t_m, t_{m+N}]$ [2].

Remarkably, the integer code series [4] expresses the definite integral

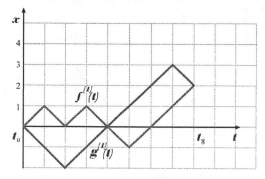

Fig. 1. For states $x = +1-1+1-1-1+1+1+1$ and $x' = -1-1+1+1+1+1+1-1$ the first integrals are equal $f^{[1]}(t_8) = g^{[1]}(t_8)$, where $f = \rho_{m\varepsilon\delta}(x), g = \rho_{m\varepsilon\delta}(x')$ and $m = 0, \varepsilon = 1, \delta = 1$.

$$f^{[k]}(t_{m+N}) = \sum_{i=0}^{k-1} \alpha_{kmi}((m+N)^i x_1 + ... + (m+1)^i x_N)\varepsilon^k \delta \qquad (1)$$

of a function $f \in W_{\varepsilon\delta}([t_m, t_{m+N}])$ by using the code $c(f) = x_1...x_N$ of the function f, powers

$$(m+N)^i, ..., (m+1)^i, \ i = 0, ..., k-1$$

of integers $(m+N), ..., (m+1)$ and combinatorial coefficients

$$\alpha_{kmi} = ((-1)^{k-i-1}(m+1)^{k-i} + (-1)^{k-i}m^{k-i})/(k-i)!i!,$$

where $k \geq 1$ and $i = 0, ..., k-1$.

We consider the correlations conserving the quantities

$$f^{[k]}(t_{m+N}) = g^{[k]}(t_{m+N}), \quad k = 1, ..., C(x, x'), \qquad (2)$$

$$f^{[C(x,x')+1]}(t_{m+N}) \neq g^{[C(x,x')+1]}(t_{m+N}), \qquad (3)$$

as elements $P_i, i = 1, ..., N$ change from one state $x = x_1...x_N \in I_N$ to another $x' = x'_1...x'_N \in I_N$, where $f = \rho_{m\varepsilon\delta}(x), g = \rho_{m\varepsilon\delta}(x')$ [2]. The conditions (2) and (3) suggest that, as a complex system changes, its parts are correlated to preserve $C(x, x')$ of the quantities (Figures 1 and 2).

It is worthwhile to mention that the conditions (2) and (3) specify a symmetry transformation that could possibly determine the equation of motion of the complex system. Therefore, once the number $C(x, x')$ of invariants is viewed as a variable to describe an order, we may get a scheme classifying equations of motion.

By using (1) it is proved [2] that $C(x, x') \geq 1$ of the quantities of a complex system remain invariant, if and only if $C(x, x')$ equations take place

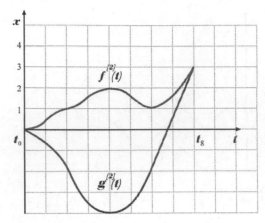

Fig. 2. For the states $x = +1-1+1+1-1-1+1+1+1$ and $x' = -1-1+1+1+1+1+1+1-1$ the second integrals are also equal $f^{[2]}(t_8) = g^{[2]}(t_8)$, but the third integrals are not $f^{[3]}(t_8) \neq g^{[3]}(t_8)$. Thus, two quantities remain invariant and $C(x,x') = 2$.

$$(m+N)^{C(x,x')-1}\Delta x_1 + (m+N-1)^{C(x,x')-1}\Delta x_2 + ... + (m+1)^{C(x,x')-1}\Delta x_N = 0$$

$$\cdot \qquad \cdot \qquad \cdot \qquad \cdot \qquad \cdot$$

$$(m+N)^1 \Delta x_1 + (m+N-1)^1 \Delta x_2 + ... + (m+1)^1 \Delta x_{\dot{N}} = 0$$
$$(m+N)^0 \Delta x_1 + (m+N)^0 \Delta x_2 + ... + (m+1)^0 \Delta x_N = 0 \qquad (4)$$

characterizing in view of an inequality

$$(m+N)^{C(x,x')}\Delta x_1 + (m+N-1)^{C(x,x')}\Delta x_1 + ... + (m+1)^{C(x,x')}\Delta x_N \neq 0, \quad (5)$$

the correlations between the parts of the complex system, where $\Delta x_i = x'_i - x_i$ are the changes of the elements $P_i, i = 1, ..., N$ in their reference frames and m is an integer.

The coefficients of the system of linear equations (4) become the entries of the Vandermonde matrix

$$\begin{pmatrix} (m+N)^{N-1} & (m+N-1)^{N-1} & ... & (m+1)^{N-1} \\ & ... & & \\ (m+N)^1 & (m+N-1)^1 & ... & (m+1)^1 \\ (m+N)^0 & (m+N-1)^0 & ... & (m+1)^0 \end{pmatrix}$$

if the number of the equations is N. This fact is important in order to prove that the number $C(x,x')$ of the conserved quantities of a complex system satisfies the condition $C(x,x') < N$ [2].

The equations (4) present a special type of correlations that do not have reference to the distances between parts, local times and physical signals. The space and non-signaling aspects of the correlations are known from explanations of quantum correlations through entanglement [5]. The time aspect of the correlations may suggest something new into the agenda.

The system of equations (4) may bring interesting associations. For example:

1. It is suggested that the system of equations (4) may be connected with a smooth projective curve over a finite field and the number $C(x, x')$ of the invariants with the genus of the curve [2].
2. The system of equations (4) can be written for

$$s = 0, -1, ..., -C(x, x') + 1$$

and $m = 0$ as

$$\sum_{n=1}^{N} \frac{\Delta x_{N-n+1}}{n^s} = 0$$

to have a naive resemblance with the Dirichlet zeta function

$$L(s, \chi) = \sum_{n=1}^{\infty} \frac{\chi_n}{n^s},$$

where χ_n are some coefficients and s is defined for proper complex numbers. This points to a possible link of the equations (4) with the zeroes of the Dirichlet zeta function

$$L(s, \chi) = \sum_{n=1}^{\infty} \frac{\chi_n}{n^s} = 0$$

and the zeroes of the Riemann zeta function

$$\zeta(s) = \sum_{n=1}^{\infty} \frac{1}{n^s} = 0$$

in particular.

3 Self-Organization Processes of Prime Integer Relations and Correlation Structures of Complex Systems

The analysis of the equations (4) and inequality (5) reveals hierarchical structures of prime integer relations in the description of complex systems [2], [3] (Figure 3). As the equations can be viewed as identities, this result may be useful to investigate whether the Ward identities and their generalizations [6] could be presented in a more explicit form.

The hierarchical structures underlying equations (4) may be in a certain superposition with each other. Namely, a prime integer relation may simultaneously belong to a number of the hierarchical structures. The measurements specifying the behavior of the part controlled by the prime integer relation can propagate through the superposition to associated prime integer relations and thus effect other parts of the complex system.

Through the hierarchical structures a new type of processes, i.e., the self-organization processes of prime integer relations, is revealed [2]. Starting with integers as the elementary building blocks and following a single principle, a self-organization process makes up the prime integer relations of a level of a hierarchical structure from the prime integer relations of the lower level (Figure 3).

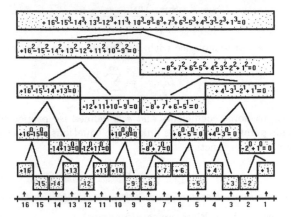

Fig. 3. One of the hierarchical structures associated with the integer relations (6) and inequality (7). The structure can be interpreted as a result of a self-organization process of prime integer relations. The process starts with integers $1, 4, 6, 7, 10, 11, 13, 16$ in the positive state and integers $2, 3, 5, 8, 9, 12, 14, 15$ in the negative state at the zero level to make up the prime integer relations at the first level. Guided by a single principle, the self-organization process forms the prime integer relations from the first to the fourth level. But the process can not reach the fifth level, because, according to arithmetic, the left side of (7) does not equal zero.

We illustrate the results by considering two states

$$x = -1 + 1 + 1 - 1 + 1 - 1 - 1 + 1 + 1 - 1 - 1 + 1 - 1 + 1 + 1 - 1,$$

$$x' = +1 - 1 - 1 + 1 - 1 + 1 + 1 - 1 - 1 + 1 + 1 - 1 + 1 - 1 - 1 + 1$$

of the elements $P_i, i = 1, ..., 16$ of a complex system. The states are specified by the Prouhet-Thue-Morse (PTM) sequences of length $N = 16$ starting with

-1 and $+1$ respectively. In this case $C(x, x') = 4$ and the equations (4) can be written as four integer relations

$$+16^0 - 15^0 - 14^0 + 13^0 - 12^0 + 11^0 + 10^0 - 9^0$$

$$-8^0 + 7^0 + 6^0 - 5^0 + 4^0 - 3^0 - 2^0 + 1^0 = 0,$$

$$+16^1 - 15^1 - 14^1 + 13^1 - 12^1 + 11^1 + 10^1 - 9^1$$

$$-8^1 + 7^1 + 6^1 - 5^1 + 4^1 - 3^1 - 2^1 + 1^1 = 0,$$

$$+16^2 - 15^2 - 14^2 + 13^2 - 12^2 + 11^2 + 10^2 - 9^2$$

$$-8^2 + 7^2 + 6^2 - 5^2 + 4^2 - 3^2 - 2^2 + 1^2 = 0,$$

$$+16^3 - 15^3 - 14^3 + 13^3 - 12^3 + 11^3 + 10^3 - 9^3$$

$$-8^3 + 7^3 + 6^3 - 5^3 + 4^3 - 3^3 - 2^3 + 1^3 = 0 \qquad (6)$$

as the inequality (5) takes the form

$$+16^4 - 15^4 - 14^4 + 13^4 - 12^4 + 11^4 + 10^4 - 9^4$$

$$-8^4 + 7^4 + 6^4 - 5^4 + 4^4 - 3^4 - 2^4 + 1^4 \neq 0, \qquad (7)$$

where $m = 0$. A common factor 2 originated from the spacetime variables $\Delta x_i, i = 1, ..., 16$ is not shown. This describes the dynamics by the signs only, but at the same time allows us to identify prime integer relations. It is worth to note that calculations in (6) and (7) and their results are completely determined by arithmetic.

There is a number of hierarchical structures of prime integer relations associated with the system of integer relations (6) and inequality (7). One of the hierarchical structures is shown in Figure 3. In the structure the relationships between the elements of neighboring levels can be interpreted as a consequence of a self-organization process of prime integer relations. The process starts with integers $1, ..., 16$ in certain states, i.e., positive or negative, and proceeds level by level following the same organizing principle:

on each level the powers of the integers in the prime integer relations are increased by 1, so that through emerging arithmetic interdependencies the prime integer relations could self-organize as the components to form prime integer relations of the higher level.

It is important to note that the formation of prime integer relations is more than the simple sum.

To indicate where the variety of the hierarchical structures comes from we give another example. The first level of this structure includes the prime integer relations

$$+16^0 - 14^0 = 0, \quad -15^0 + 13^0 = 0, \quad -12^0 + 10^0 = 0,$$

$$+11^0 - 9^0 = 0, \quad -8^0 + 6^0 = 0, \quad +7^0 - 5^0 = 0,$$

$$+4^0 - 2^0 = 0, \quad -3^0 + 1^0 = 0.$$

The second level - the prime integer relations

$$(+16^1 - 14^1) + (-15^1 + 13^1) = 0,$$

$$(-12^1 + 10^1) + (+11^1 - 9^1) = 0,$$

$$(-8^1 + 6^1) + (+7^1 - 5^1) = 0, \quad (+4^1 - 2^1) + (-3^1 + 1^1) = 0.$$

The third level - the prime integer relations

$$((+16^2 - 14^2) + (-15^2 + 13^2))$$

$$+((-12^2 + 10^2) + (+11^2 - 9^2)) = 0,$$

$$((-8^2 + 6^2) + (+7^2 - 5^2)) + ((+4^2 - 2^2) + (-3^2 + 1^2)) = 0.$$

The fourth level - the prime integer relation

$$(((+16^3 - 14^3) + (-15^3 + 13^3) + ((-12^3 + 10^3) + (+11^3 - 9^3)))$$

$$+(((-8^3 + 6^3) + (+7^3 - 5^3)) + ((+4^3 - 2^3) + (-3^3 + 1^3))) = 0.$$

Let us explain the notion of prime integer relation. A prime integer relation of the first level is made up of integers from the zero level. An integer comes to the first level in positive or negative state (Figure 3). Following the organizing principle prime integer relations of a level make up a prime integer relation of the higher level. A prime integer relation is made up as an inseparable object: if even one of the prime integer relations is not included, then the rest of the prime integer relations can not form an integer relation. In other words, each and every prime integer relation involved in the formation of a prime integer relation is critical.

For example, an integer relation

$$+7^2 - 6^2 - 5^2 + 3^2 + 2^2 - 1^2 = 0$$

is a prime integer relation. However, an integer relation

$$+7^0 - 6^0 - 5^0 + 3^0 = 0$$

is not prime, because it consists of two prime integer relations

$$+7^0 - 6^0 = 0, \quad -5^0 + 3^0 = 0.$$

For simplicity prime integer relations such as

$$+14^0 - 7^0 = 0, \quad +2 \cdot 14^0 - 2 \cdot 7^0 = 0$$

are not distinguished. Multiple 2 means that we have two integers 14 in the positive state and two integers 7 in the negative state.

The prime integer relations set limits of causality in the sense that complex systems controlled by separate prime integer relations have no effect on each other. But, if a part of a complex system changes, then, in accordance with the prime integer relation, the other parts change no matter how far they may be from each other. At the same time the prime integer relation has no causal power to influence parts of complex systems controlled by separate prime integer relations.

An analogy between the prime integer relation and the light cone may be suggested. While in spacetime events can be only observed in light cones, in our approach events can only happen through the hierarchical structure of the prime integer relations.

The dynamics Δx_i of an elementary part $P_i, i = 1, ..., N$ has the following interpretation. The absolute value of Δx_i is the number of integers $(m + N - i + 1)$ starting each of the self-organization processes, while $sign(\Delta x_i)$ determines the positive or negative state of the integers, provided that $\Delta x_i \neq 0, i = 1, ..., N$.

The correlation structures underlying the conservation of the quantities (2) in view of (3) are defined by the hierarchical structures of prime integer relations associated with the system of equations (4) and inequality (5).

We illustrate the result by using the system of integer relations (6), where we return to the symbolic form of (4) to change our perspective from the self-organization processes of prime integer relations to the dynamics of the complex system in spacetime. Starting with the prime integer relation $+16^0 - 15^0 = 0$ (Figure 3), which stands as

$$+16^0(+2) + 15^0(-2) = +16^0\Delta x_1 + 15^0\Delta x_2 = 0, \tag{8}$$

we can see that the changes Δx_1 and Δx_2 of the elements P_1 and P_2 are correlated as

$$+16^0\Delta x_1 = -15^0\Delta x_2 \tag{9}$$

and the elements P_1 and P_2 thus make a part $P_1 \leftrightarrow P_2$.

The prime integer relation (8) does not contain information about a physical signal that may realize the correlation between the elements P_1 and P_2. But, if the dynamics Δx_2 of the element P_2 is specified, then, according to (9), the dynamics Δx_1 of the element P_1 is determined and vice versa. We may also say that the correlation is nonlocal, because the prime integer relation (8) does not have any reference to the distance between the elements P_1 and P_2.

Similarly, the prime integer relation $-14^0 + 13^0 = 0$ leads to

$$+14^0\Delta x_3 + 13^0\Delta x_4 = 0,$$

which specifies the correlation

$$14^0\Delta x_3 = -13^0\Delta x_4$$

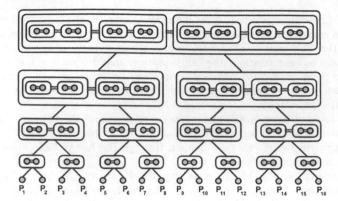

Fig. 4. A correlation structure of the complex system is defined by the hierarchical structure of prime integer relations (Figure 3). A horizontal link denotes that the parts are correlated through a prime integer relation. As arithmetic behind the prime integer relations makes them sensitive to a minor change, so does the correlation structure. If the complex system deviates from the dynamic behavior even slightly, then some of the correlation links disappear and the complex system decays.

between the elements P_3 and P_4 and describes a part $P_3 \leftrightarrow P_4$.

In its turn the prime integer relation

$$+16^1 - 15^1 - 14^1 + 13^1 = 0,$$

made up of the prime integer relations $+16^0 - 15^0 = 0$ and $-14^0 + 13^0 = 0$, corresponds to

$$(16^1 \Delta x_1 + 15^1 \Delta x_2) + (14^1 \Delta x_3 + 13^1 \Delta x_4) = 0,$$

which shows that the parts $P_1 \leftrightarrow P_2$ and $P_3 \leftrightarrow P_4$ are correlated as

$$(16^1 \Delta x_1 + 15^1 \Delta x_2) = -(14^1 \Delta x_3 + 13^1 \Delta x_4)$$

and form a larger composite part

$$(P_1 \leftrightarrow P_2) \leftrightarrow (P_3 \leftrightarrow P_4).$$

Continuing in the same manner we can associate the self-organization process of prime integer relations with the formation of a correlation structure of the complex system (Figures 3 and 4). The formation process starts with the elements $P_1, ..., P_{16}$ and combine them into parts to make up then larger parts and so on until the whole correlation structure is built.

A complex system can be described by self-organization processes of prime integer relations in a distinctive way. Information about a complex system can be given by prime integer relations, which are true statements. The prime integer relations are organized as hierarchical structures and there is no need for deeper principles to explain why the hierarchical structures exist the way they do and not otherwise.

For example, prime integer relations

$$+7^0 - 6^0 = 0, \quad -5^0 + 3^0 = 0, \quad +2^0 - 1^0 = 0$$

of level 1 form a prime integer relation

$$+7^1 - 6^1 - 5^1 + 3^1 + 2^1 - 1^1 = 0$$

of level 2, which can self-organize itself into a prime integer relation

$$+7^2 - 6^2 - 5^2 + 3^2 + 2^2 - 1^2 = 0 \tag{10}$$

of level 3. However, the prime integer relation (10), can not on its own progress to level 4, because

$$+7^3 - 6^3 - 5^3 + 3^3 + 2^3 - 1^3 \neq 0.$$

4 Geometrization of the Self-Organization Processes and its Applications

In the above considerations it was convenient to think about prime integer relations as objects of formation rather than as objects of calculation. However, integers and integer relations as abstract entities do not produce pictures we associate with formations of physical objects. They are also not familiar as geometric objects suitable for observation and measurement.

Remarkably, by using the integer code series [4], the prime integer relations can be geometrized as two-dimensional geometric patterns and the self-organization processes can be isomorphically expressed by transformations of the patterns [2]. As it becomes possible to measure a prime integer relation by an isomorphic geometric pattern, quantities of the prime integer relation and a complex system it describes can be defined by quantities of the geometric pattern such as the area and the length of its boundary curve (Figure 5).

To illustrate the results we consider a self-organization process that can progress through the hierarchical levels. The process is connected with the PTM sequence and can be specified in terms of critical point [7], [8] features [2]. This properly resonates with the fact that the PTM sequence symbolically describes the period-doubling [9] in physical systems [10].

The left side of Figure 5 shows a hierarchical structure of prime integer relations, when

$$x = 0 \ 0 \ 0 \ 0 \ 0 \ 0 \ 0 \ 0,$$

$$x' = +1 - 1 - 1 + 1 - 1 + 1 + 1 - 1$$

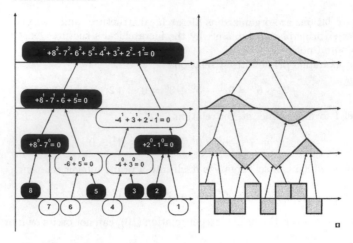

Fig. 5. The left side shows one of the hierarchical structures of prime integer relations, when a complex system has $N = 8$ elements $P_i, i = 1, ..., 8$, $x = 00000000$, $x' = +1 - 1 - 1 + 1 - 1 + 1 + 1 - 1$, $m = 0$ and $C(x, x') = 3$. The hierarchical structure is built by a self-organization process of prime integer relations and controls a correlation structure of the complex system. The right side presents an isomorphic hierarchical structure of geometric patterns. On scale level 0 eight rectangles specify the dynamics of the elements $P_i, i = 1, ..., 8$. Under the integration of the function, the geometric patterns of one level form the geometric patterns of the higher level, so we can observe the geometric patterns length scale by length scale. A prime integer relation can be measured by the area of a corresponding geometric pattern and the length of its boundary curve. The boundary curves of the geometric patterns describe the dynamics of the parts. All geometric patterns are symmetric and their symmetries are interconnected. The symmetry of a geometric pattern is global and belongs to a corresponding part as a whole.

$N = 8$ and $m = 0$. The sequence x' is the initial segment of length 8 of the PTM sequence starting with $+1$.

The right side presents an isomorphic hierarchical structure of geometric patterns as a result of the geometrization of the prime integer relations. The process can be also investigated by measuring prime integer relations in terms of corresponding geometric patterns. For example, a prime integer relation can be measured by the area of a corresponding geometric pattern and the length of its boundary curve. These quantities of prime integer relations have interesting properties.

For example, the first property shows that complex systems may be described by prime integer relations in a strong scale covariant form. Although a PTM geometric pattern at scale level $\mathcal{N} > 1$ is bounded by an intricate curve, it has nevertheless a concise and universal description working for all levels. In particular, the area S of a PTM geometric pattern at scale level $\mathcal{N} \geq 1$

can be expressed as the area of a triangle by

$$S = \frac{LH}{2}, \tag{11}$$

where L and H are the length and the height of the PTM geometric pattern (Figure 6). Consequently, the law of PTM pattern area is the same and in the simplest possible form for all scale levels. In other words, the description of S is *strongly scale covariant*, i.e.,

under the scale transformations the equation of S is preserved in the simplest possible form (11).

A PTM geometric pattern is a result of the formation, but for the description of its area S there is no need to know what happens at the lower scale levels. All information can be obtained by measuring the length L and the height H at the level of consideration. The history of the formation of a PTM geometric pattern is encoded by the boundary curve.

Importantly, the quantity of the complex system specified by the height H of the geometric pattern and denoted by M is determined by the area quantities E_l and E_r of two parts, i.e., left and right, the complex system is made of (Figure 5). Namely, the height H of the geometric pattern equals one half of the sum of the areas of the geometric patterns characterizing the parts forming the complex system from the lower scale level.

Thus, we have

$$M = \frac{E_l + E_r}{2}$$

suggesting that, as the parts on scale level $\mathcal{N} - 1$ form the complex system on scale level $\mathcal{N}$, their E quantities are converted into the M quantity of the complex system. The M quantity is proportional and as such can be seen equivalent to the E quantity of the complex system

$$E = \frac{ML}{2}.$$

Because of (11) the PTM geometric patterns have a scale invariant property dividing the hierarchical scale levels into groups of *three* successive levels. Namely, the lengths and the heights of PTM geometric patterns at scale levels $\mathcal{N} = 1, 2, 3$ and $\mathcal{N} = 4, 5, 6$ are given in terms of ε and δ as

$$(2\varepsilon, \varepsilon\delta), \ (4\varepsilon, \varepsilon^2\delta), \ (8\varepsilon, 2\varepsilon^3\delta)$$

and

$$(16\varepsilon, 8\varepsilon^4\delta), \ (32\varepsilon, 64\varepsilon^5\delta), \ (64\varepsilon, 1024\varepsilon^6\delta).$$

By using the renomalization group transformation

$$\varepsilon' = 2^3\varepsilon, \ \delta' = \varepsilon^3\delta,$$

the lengths and the heights of PTM geometric patterns at scale levels $\mathcal{N} = 4, 5, 6$ can be expressed in terms of ε' and δ' in the same way

$$(2\varepsilon', \varepsilon'\delta'), \quad (4\varepsilon', \varepsilon'^2\delta'), \quad (8\varepsilon', 2\varepsilon'^3\delta')$$

as the lengths and the heights of PTM geometric patterns are given in terms of ε and δ at scale levels $\mathcal{N} = 1, 2, 3$. The situation repeats for levels $\mathcal{N} = 7, 8, 9$ and so on.

The second property is connected with a long-standing problem to explain why constants of nature, such as the fine-structure constant α measured as

$$\alpha = 0.00729735... \ ,$$

have the values they do and are not even slightly different. Although the logic of digits in the fine-structure constant α has not been established yet, it is known that α is fragile. If the fine-structure constant α could be varied a bit, then complex physical systems would not be able to exist [11].

A prime integer relation is also an intricate entity, because it is sensitive to a minor change. However, when we read a prime integer relation, unlike a constant of nature, we can understand and accept it as a true statement. Moreover, a prime integer relation can describe a correlation structure of the complex system. In this capacity it encodes the parts of the correlation structure, the relationships between them as well as their interconnected dynamics. As a result, a minor change breaking a prime integer relation also leads to a collapse of a corresponding complex system, because some of the relationships between the parts of the correlation structure disappear.

Thus, there is an interesting analogy with the fine-structure constant, although a prime integer relation is fragile to changes of the elements, not to variations of a single number, like α. The analogy would be made stronger, if we could associate a prime integer relation with a number.

The geometrization of a prime integer relation can provide such a number. In particular, the length of the curve, i.e., a number, encodes the geometric pattern and thus the prime integer relation. With a minor change to the number the prime integer relation as well as a corresponding complex system cease to exist. Indeed, if the number changes even slightly, the boundary curve in its turn changes the geometric pattern, which leads the prime integer relation and the complex system to decay.

To show how such numbers can be obtained we consider the above example with $\varepsilon = 1$ and $\delta = 1$ (Figure 5). By using corresponding geometric patterns at levels 1, 2 and 3, we can define the numbers for prime integer relations.

For prime integer relation $+8^0 - 7^0 = 0$ we have

$$\vartheta_1 = 2 \int_0^1 \sqrt{1 + (\frac{df^{[1]}}{dt})^2} dt =$$

$$2 \int_0^1 \sqrt{1+1} dt = 2\sqrt{2} = 2 \times 1.414223562... ,$$

where $f = \rho_{011}(x')$, $x' = +1 - 1 - 1 + 1 - 1 + 1 + 1 - 1$.
For prime integer relation

$$+8^1 - 7^1 - 6^1 + 5^1 = 0$$

we obtain

$$\vartheta_2 = 4 \int_0^1 \sqrt{1 + (\frac{df^{[2]}}{dt})^2} dt =$$

$$4 \int_0^1 \sqrt{1 + t^2} dt = 4 \times 1.14779... .$$

For prime integer relation

$$+8^2 - 7^2 - 6^2 + 5^2 - 4^2 + 3^2 + 2^2 - 1^2 = 0$$

we get

$$\vartheta_3 = 4 \int_0^2 \sqrt{1 + (\frac{df^{[3]}}{dt})^2} dt =$$

$$4(\int_0^1 \sqrt{1 + \frac{t^4}{4}} dt + \int_1^2 \sqrt{1 + (\frac{-t^2}{2} + 2t - 1)^2} dt) =$$

$$4 \times (1.0242... + 1.30702...) = 4 \times 2.33122... .$$

The numerical results in the second and third cases are computed by using *Mathematica*.

We can write the prime integer relations and their corresponding numbers as

$$+8^0 - 7^0 = 0 \implies 2 \times 1.414223562...$$

$$+8^1 - 7^1 - 6^1 + 5^1 = 0 \implies 4 \times 1.14779...$$

$$+8^2 - 7^2 - 6^2 + 5^2 - 4^2 + 3^2 + 2^2 - 1^2 = 0 \implies 4 \times 2.33122... .$$

On the side of prime integer relations we have confidence in the arithmetic statements, as we can check them easily. Moreover, we know how the prime integer relations are built, can observe symmetry in their corresponding geometric patterns (Figure 5) and associate them with correlation structures of the complex system.

But on the other side it is not clear what logic digits of the numbers may follow. The situation would become intriguing, once the numbers found to be constants in physical experiments.

When a constant of nature is measured, the information from physical devices comes in the numerical form. Presenting numerical information through prime integer relations may give us a tool to understand experimental results relying on irreducible arguments. For instance, if some physical experiments identify a constant and as a number it corresponds to a prime integer relation

$$+7^2 - 6^2 - 5^2 + 3^2 + 2^2 - 1^2 = 0,$$

then it may be considered that the experiments through the constant actually reveal a true statement we can understand and agree with.

Therefore, we propose to explore the idea:

constants of nature may be numerical expressions of prime integer relations or their metrics.

5 Global Symmetry of Complex Systems and Gauge Forces

Our description presents a global symmetry of complex systems through the geometric patterns of prime integer relations and their transformations. It belongs to the system as a whole, but does not necessarily apply to its embedded parts (Figure 5). The differences between the behaviors of the parts may be interpreted through the existence of gauge forces acting in their reference frames.

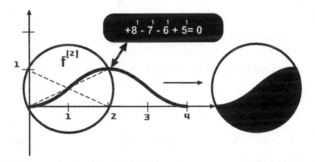

Fig. 6. The geometric pattern of the part $(P_1 \leftrightarrow P_2) \leftrightarrow (P_3 \leftrightarrow P_4)$. From above the pattern is limited by the boundary curve, i.e., the graph of the second integral $f^{[2]}(t)$, $t_0 \le t \le t_4$ of the function f defined on scale level 0 (Figure 5), where $t_i = i\varepsilon$, $i = 1, ..., 4$, $\varepsilon = 1$, $\delta = 1$, and it is restricted by the t axis from below. The geometric pattern is isomorphic to the prime integer relation $+8^1 - 7^1 - 6^1 + 5^1 = 0$ and determines the dynamics. If the part deviates from this dynamics even slightly, then some of the correlation links provided by the prime integer relation disappear and the part decays. The boundary curve has a special property ensuring that the area of the geometric pattern is given as the area of a triangle: $S = \frac{HL}{2}$, where H and L are the height and the length of the geometric pattern. In the figure $H = 1$ and $L = 4$, thus $S = 2$. The property is illustrated in yin-yang motifs.

We consider whether the description of the dynamics of parts of a scale level is invariant as through the formation they become embedded in a part of the higher level.

At scale level 2 the second integral

$$f^{[2]}(t), \ t_0 \leq t \leq t_4, \ t_i = i\varepsilon, \ i = 1, ..., 4, \ \varepsilon = 1, \delta = 1$$

characterizes the dynamics of a part

$$(P_1 \leftrightarrow P_2) \leftrightarrow (P_3 \leftrightarrow P_4).$$

This composite part is made up of elements P_1, P_2, P_3, P_4 and parts

$$P_1 \leftrightarrow P_2, \quad P_3 \leftrightarrow P_4$$

embedded by the formations within its correlation structure (Figures 5 and 6).

The description of the dynamics of elements P_1, P_2, P_3, P_4 and parts $P_1 \leftrightarrow P_2, P_3 \leftrightarrow P_4$ within the part $(P_1 \leftrightarrow P_2) \leftrightarrow (P_3 \leftrightarrow P_4)$ is invariant relative to their reference frames. In particular, the dynamics of elements P_1 and P_2 in a reference frame of the element P_1 is specified by

$$f^{[2]}(t) = f^{[2]}_{P_1}(t_{P_1}) = \frac{t^2_{P_1}}{2!},$$

$$t_0 = t_{0,P_1} \leq t_{P_1} \leq t_{1,P_1} = t_1, \tag{12}$$

$$f^{[2]}(t) = f^{[2]}_{P_1}(t_{P_1}) = -\frac{t^2_{P_1}}{2!} + 2t_{P_1} - 1,$$

$$t_1 = t_{1,P_1} \leq t_{P_1} \leq t_{2,P_1} = t_2.$$

The transition from the coordinate system of the element P_1 to a coordinate system of the element P_2 given by the transformation

$$t_{P_2} = -t_{P_1} - 2, \quad f^{[2]}_{P_2} = -f^{[2]}_{P_1} - 1$$

shows that the characterization

$$f^{[2]}_{P_2}(t_{P_2}) = \frac{t^2_{P_2}}{2!}, \quad t_{0,P_2} \leq t_{P_2} \leq t_{1,P_2} \tag{13}$$

of the dynamics of the element P_2 is invariant, if we compare (12) and (13).

The description is also invariant, when we consider the dynamics of the elements P_3 and P_4 in their reference frames. The dynamics of the element P_3 in the coordinate system of the element P_1 is specified by

$$f^{[2]}_{P_1}(t_{P_1}) = -\frac{t^2_{P_1}}{2!} + 2t_{P_1} - 1,$$

$$t_{2,P_1} \leq t_{P_1} \leq t_{3,P_1}.$$

The description of the dynamics of the element P_3 takes the same form

$$f_{P_3}^{[2]}(t_{P_3}) = \frac{t_{P_3}^2}{2!}, \quad t_{0,P_3} \leq t_{P_3} \leq t_{1,P_3} \tag{14}$$

as (12) and (13), when, using the transformation

$$t_{P_3} = t_{P_1} - 2, \quad f_{P_3}^{[2]} = -f_{P_1}^{[2]} + 1,$$

a transition to a coordinate system of the element P_3 is made.

Similarly, the dynamics of the element P_4 in the coordinate system of the element P_1 is specified by

$$f_{P_1}^{[2]}(t_{P_1}) = \frac{t_{P_1}^2}{2!} - 4t_{P_1} + 8,$$

$$t_{3,P_1} \leq t_{P_1} \leq t_{4,P_1}.$$

The transformation

$$t_{P_4} = -t_{P_1} + 4, \quad f_{P_4}^{[2]} = f_{P_1}^{[2]}$$

leads to a coordinate system of the element P_4 to demonstrate that the description of the dynamics of the element P_4 has the same form

$$f_{P_4}^{[2]}(t_{P_4}) = \frac{t_{P_4}^2}{2!}, \quad t_{0,P_4} \leq t_{P_4} \leq t_{1,P_4}$$

as (12), (13) and (14).

Furthermore, descriptions of the dynamics of the parts $P_1 \leftrightarrow P_2$ and $P_3 \leftrightarrow P_4$ are the same relative to their coordinate systems. Namely, for the dynamics of the part $P_1 \leftrightarrow P_2$ in its reference frame we have

$$f_{P_1 \leftrightarrow P_2}^{[2]}(t_{P_1 \leftrightarrow P_2}) = \begin{cases} \frac{t_{P_1 \leftrightarrow P_2}^2}{2!}, \\ t_{0,P_1 \leftrightarrow P_2} \leq t_{P_1 \leftrightarrow P_2} \leq t_{1,P_1 \leftrightarrow P_2} \\[2mm] -\frac{t_{P_1 \leftrightarrow P_2}^2}{2!} + 2t_{P_1 \leftrightarrow P_2} - 1, \\ t_{1,P_1 \leftrightarrow P_2} \leq t_{P_1 \leftrightarrow P_2} \leq t_{2,P_1 \leftrightarrow P_2} \end{cases} \tag{15}$$

For the dynamics of the part $P_3 \leftrightarrow P_4$ in the reference frame of the part $(P_1 \leftrightarrow P_2)$ we have

$$f_{P_1 \leftrightarrow P_2}^{[2]}(t_{P_1 \leftrightarrow P_2}) = \begin{cases} -\frac{t_{P_1 \leftrightarrow P_2}^2}{2!} + 2t_{P_1 \leftrightarrow P_2} - 1, \\ t_{2,P_1 \leftrightarrow P_2} \leq t_{P_1 \leftrightarrow P_2} \leq t_{3,P_1 \leftrightarrow P_2} \\[2mm] \frac{t_{P_1 \leftrightarrow P_2}^2}{2!} - 4t_{P_1 \leftrightarrow P_2} + 8, \\ t_{3,P_1 \leftrightarrow P_2} \leq t_{P_1 \leftrightarrow P_2} \leq t_{4,P_1 \leftrightarrow P_2} \end{cases} \tag{16}$$

The description (16) takes the same form

$$f^{[2]}_{P_3 \leftrightarrow P_4}(t_{P_3 \leftrightarrow P_4}) = \begin{cases} \dfrac{t^2_{P_3 \leftrightarrow P_4}}{2!}, \\ t_{0,P_3 \leftrightarrow P_4} \leq t_{P_3 \leftrightarrow P_4} \leq t_{1,P_3 \leftrightarrow P_4} \\[2mm] -\dfrac{t^2_{P_3 \leftrightarrow P_4}}{2!} + 2t_{P_3 \leftrightarrow P_4} - 1, \\ t_{1,P_3 \leftrightarrow P_4} \leq t_{P_3 \leftrightarrow P_4} \leq t_{2,P_3 \leftrightarrow P_4} \end{cases}$$

as (15), if under the transformation

$$t_{P_3 \leftrightarrow P_4} = t_{P_1 \leftrightarrow P_2} + 2, \quad f^{[2]}_{P_3 \leftrightarrow P_4} = -f^{[2]}_{P_1 \leftrightarrow P_2} + 1,$$

we make a transition from the reference frame of the part $P_1 \leftrightarrow P_2$ to a reference frame of the part $P_3 \leftrightarrow P_4$.

Thus, as the perspective is changed from the reference frame of the part $P_1 \leftrightarrow P_2$ to the reference frame of the part $P_3 \leftrightarrow P_4$ the description of the dynamics remains invariant.

However, at scale level 3 the description of the dynamics is not invariant. In particular, the dynamics of elements P_1 and P_2 within the part

$$((P_1 \leftrightarrow P_2) \leftrightarrow (P_3 \leftrightarrow P_4)) \leftrightarrow ((P_5 \leftrightarrow P_6) \leftrightarrow (P_7 \leftrightarrow P_8))$$

relative to a coordinate system of the element P_1 can be specified accordingly by (Figure 5)

$$f^{[3]}_{P_1}(t_{P_1}) = \frac{t^3_{P_1}}{3!}, \quad t_{0,P_1} \leq t \leq t_{1,P_1}, \tag{17}$$

$$f^{[3]}_{P_1}(t_{P_1}) = -\frac{t^3_{P_1}}{3!} + t^2_{P_1} - t_{P_1} + \frac{1}{3}, \quad t_{1,P_1} \leq t_{P_1} \leq t_{2,P_1}.$$

The transitions from the coordinate systems of the element P_1 to the coordinate systems of the element P_2 do not preserve the form (17). For example, if under the transformation

$$t_{P_2} = t_{P_1} + 2, \quad f^{[3]}_{P_2} = -f^{[3]}_{P_1} + 1$$

the perspective is changed from the coordinate system of the element P_1 to a coordinate system of the element P_2, then it turns out that the description of the dynamics (17) is not invariant

$$f^{[2]}(t) = f^{[3]}_{P_2}(t_{P_2}) = \frac{t^3_{P_2}}{3!} - t_{P_2}, \quad t_{1,P_2} \leq t_{P_1} \leq t_{2,P_2}$$

due to the additional linear term $-t_{P_2}$.

Therefore, on scale level 3 arithmetic determines the different dynamics of the elements P_1 and P_2. The information about the difference may be obtained from observers positioned at the coordinate system of the element P_1 and the coordinate system of the element P_2 respectively. As one observer reports about the dynamics of the element P_1 and the other about the dynamics of the element P_2, we may interpret the difference in dynamics through the existence of a gauge force F acting on the element P_2 in its coordinate system to the effect of the linear term $\chi(F) = -t_{P_2}$

$$f_{P_2}^{[3]}(t_{P_2}) = \frac{t_{P_2}^3}{3!} - \chi(F),$$

$$t_{0,P_2} \le t_{P_2} \le t_{1,P_2}.$$

The introduction of the gauge force F restores the local symmetry

$$f_{P_2}^{[3]}(t_{P_2}) = \frac{t_{P_2}^3}{3!}, \quad t_{0,P_2} \le t_{P_2} \le t_{1,P_2} \qquad (18)$$

as we can see comparing (17) and (18).

The results can be schematically expressed as follows

Arithmetic →

Prime integer relations in control

of correlation structures of complex systems ↔

Global symmetry: geometric patterns in control

of the dynamics of complex systems →

→ *Not locally invariant descriptions*

of embedded parts of complex systems ↔

↔ *Gauge forces to restore local symmetries*

We consider whether the hierarchical scale levels in our description could be subdivided into groups of successive scale levels so that the description of complex systems would be group scale invariant. If that were the case then the classification of the gauge forces could be made possible by focusing on any group of scale levels. Every other group would have the same classification although expressed in its own terms.

Remarkably, it turns out that the scale levels can be subdivided into groups of *three* successive levels, where the description of complex systems remains the same. In particular, using the renormalization group transformation applied to ε, δ

$$\varepsilon' = 2^3 \varepsilon, \quad \delta' = \varepsilon^3 \delta,$$

and elements

$$P_1' = (((P_1 \leftrightarrow P_2) \leftrightarrow (P_3 \leftrightarrow P_4)) \leftrightarrow ((P_5 \leftrightarrow P_6) \leftrightarrow (P_7 \leftrightarrow P_8))),$$

$$P_2' = (((P_9 \leftrightarrow P_{10}) \leftrightarrow (P_{11} \leftrightarrow P_{12})) \leftrightarrow ((P_{13} \leftrightarrow P_{14}) \leftrightarrow (P_{15} \leftrightarrow P_{16}))),$$

$$\cdot \quad \cdot \quad \cdot \quad \cdot \quad \cdot \quad \cdot$$

$$P'_k =$$

$$(((P_{8(k-1)+1} \leftrightarrow P_{8(k-1)+2}) \leftrightarrow (P_{8(k-1)+3} \leftrightarrow P_{8(k-1)+4})) \leftrightarrow$$

$$((P_{8(k-1)+5} \leftrightarrow P_{8(k-1)+6}) \leftrightarrow (P_{8(k-1)+7} \leftrightarrow P_{8(k-1)+8}))),$$

we can obtain the same description of complex systems on scale levels 4, 5, 6 as for complex systems on scale levels 1, 2, 3. The difference is that the description on scale levels 4, 5, 6 is given in terms of ε' and δ' with parts $P'_1, P'_2, \ldots$ as the elements, while on scale levels 1, 2, 3 it is given in terms of ε, δ and elements $P_1, P_2, \ldots$. The situation repeats for scale levels 7, 8, 9 and so on [3].

6 Structural Complexity in Optimality Condition of Complex Systems and Optimal Quantum Algorithms

Despite their different origins complex systems have much in common and are investigated to satisfy universal laws. Our description points out that the universal laws may originate not from forces in spacetime, but through arithmetic.

There are many notions of complexity introduced in the search to communicate the universal laws into theory and practice. The concept of structural complexity is defined to measure the complexity of a system in terms of self-organization processes of prime integer relations [2]. In particular, as self-organization processes of prime integer relations progress from a level to the higher level, the system becomes more complex, because its parts at the level are combined to make up more complex parts at the higher level. Therefore, the higher the level self-organization processes progress to, the greater is the structural complexity of a corresponding complex system.

Existing concepts of complexity do not explain in general how the performance of a complex system may depend on its complexity. To address the situation we conducted computational experiments to investigate whether the concept of structural complexity could make a difference [12], [13]. A special optimization algorithm, as a complex system, was developed to minimize the average distance in the travelling salesman problem. Remarkably, for each problem tested the performance of the algorithm behaved as a concave function. As a result, the algorithm and a problem were characterized by a single performance optimum. The analysis of the performance optimums for all problems revealed a relationship between the structural complexity of the algorithm and the structural complexity of the problem approximating it well enough by a linear function [13].

The results of the computational experiments suggest the possibility of a general optimality condition of complex systems:

A complex system demonstrates the optimal performance for a problem, when the structural complexity of the system is in a certain relationship with the structural complexity of the problem.

The optimality condition presents the structural complexity of a system as a key to its optimization. Indeed, according to the optimality condition the optimal result can be obtained as long as the structural complexity of the system is properly related with the structural complexity of the problem. From this perspective the optimization of a system should be primarily concerned with the control of the structural complexity of the system to match the structural complexity of the environment.

The computational results also indicate that the performance of a complex system may behave as a concave function of the structural complexity. Once the structural complexity controlled as a single entity, the optimization of a complex system could be potentially reduced to a one-dimensional concave optimization irrespective of the number of variables involved its description.

In the search to identify a mathematical structure underlying optimal quantum algorithms the majorization principle emerges as a necessary condition for efficiency in quantum computational processes [14]. We find a connection between the optimality condition and the majorization principle in quantum algorithms.

According to the majorization principle in an optimal quantum algorithm the probability distribution associated to the quantum state has to be step-by-step majorized until it is maximally ordered. This means that an optimal quantum algorithm works in such a way that the probability distribution p_{k+1} at step $k + 1$ majorizes $p_k \prec p_{k+1}$ the probability distribution p_k at step k. There are special conditions in place for the probability distribution p_{k+1} to majorize the probability distribution p_k with intuitive meaning that the distribution p_k is more disordered than p_{k+1} [14].

In our description the algorithm revealing the optimality condition also uses a similar principle, but based on the structural complexity. The algorithm tries to work in such a way that the structural complexity $\mathbf{C}_{k+1}$ of the algorithm at step $k + 1$ majorizes $\mathbf{C}_k \prec \mathbf{C}_{k+1}$ its structural complexity $\mathbf{C}_k$ at step k. The concavity of the algorithm's performance suggests efficient means to find optimal solutions [13].

7 Conclusions

In the chapter we have presented results to develop an irreducible theory of complex systems in terms of self-organization processes of prime integer relations. Based on the integers and controlled by arithmetic only the self-organization processes can describe complex systems by information not requiring further explanations. The following properties of the description have been revealed.

First, the description points to a special type of correlations that do not depend on the distances between parts, local times and physical signals. Apart from the time aspect such correlations are known in quantum physics and attributed to quantum entanglement [5]. Thus, the description proposes a perspective on quantum entanglement suggesting to include the time aspect into the agenda.

Second, through a concept of structural complexity the description computationally reveals the possibility of a general optimality condition of complex systems. The experiments indicate that the performance of a complex system may behave as a concave function of the structural complexity. Therefore, once the structural complexity were controlled as a single entity, the optimization of a complex system could be potentially reduced to a one-dimensional concave optimization irrespective of the number of variables involved in its description.

A connection between the majorization principle in quantum algorithms and the optimality condition has been identified. While the quantum majorization principle suggests that the computational process should stop when the probability distribution is maximally ordered, it does not however specify what this order actually means in the context of a particular problem. At the same time our approach is clear on this matter: to obtain the performance maximum the computational process should stop when its structural complexity is in a certain relationship with the structural complexity of the problem.

Third, the description introduces a global symmetry of complex systems that belongs to the system as a whole, but does not necessarily apply to its embedded parts. The breaking of the global symmetry may be interpreted through the existence of gauge forces. There is no further need to explain why the resulting gauge forces exist the way they do and are not even slightly different as their existence is fully determined by arithmetic.

References

1. L. Smolin, *The Case for Background Independence*, arXiv:hep-th/0507235.
2. V. Korotkikh, *A Mathematical Structure for Emergent Computation*, Kluwer Academic Publishers, Dordrecht, 1999.
3. V. Korotkikh and G. Korotkikh, *Description of complex systems in terms of self-organization processes of prime integer relations*, in M. M. Novak (ed.) Complexus Mundi: Emergent Patterns in Nature, World Scientific, New Jersey/London, pp. 63-72, 2006.
4. V. Korotkikh, *Integer code series with some applications in dynamical systems and complexity*, Computing Centre of the Russian Academy of Sciences, Moscow, 1993.
5. N. Gisin, *Can Relativity be Considered Complete? From Newtonian Nonlocality to Quantum Nonlocality and Beyond*, arXiv:quant-ph/0512168, and references therein.
6. G. Sardanashvily, *Green function identities in Euclidean quantum field theory*, arXiv:hep-th/0604003.
7. L. Kadanoff, W. Gotze, D. Hamblen, R. Hecht, E.A.S. Lewis, V.V. Palciaukas, M. Rayl and J. Swift, Rev. Mod. Phys., **39**(2), 395, 1967.
8. K. G. Wilson and J. Kogut, Phys. Rep. C, **12**, 75, 1974.
9. M. Feigenbaum, Los Alamos Sci. **1**, 4, 1980.

8. K. G. Wilson and J. Kogut, Phys. Rep. C, **12**, 75, 1974.
9. M. Feigenbaum, Los Alamos Sci. **1**, 4, 1980.
10. J. P. Allouche and M. Cosnard, in *Dynamical Systems and Cellular Automata*, Academic Press, 1985.
11. J.D. Barrow, *The Constants of Nature: From Alpha to Omega*, Jonathan Cape, London, 2002 and Pantheon, New York, 2002; J.D. Barrow and J.K. Webb, Sci. Amer. **292**, 6, 33, 2005.
12. V. Korotkikh, *The search for the universal concept of complexity and a general optimality condition of cooperative agents*, in S. Butenko, R. Murphey and P. M. Pardalos (eds), Cooperative Control: Models, Applications and Algorithms, Kluwer, Dordrecht, pp. 129-163, 2003.
13. V. Korotkikh, G. Korotkikh and D. Bond, *On Optimality condition of complex systems: computational evidence*, arXiv:cs.CC/0504092.
14. R. Orus, J. Latorre and M. A. Martin-Delgado, *Systematic Analysis of Majorization in Quantum Algorithms*, arXiv:quant-ph/0212094.

9 Complexity of a System as a Key to its Optimization

Victor Korotkikh and Galina Korotkikh

Summary. Computational results presented in the chapter show that the performance of a complex system may behave as a concave function of its structural complexity and reveal the possibility of an optimality condition. The optimality condition could potentially reduce the optimization of a complex system to a one-dimensional concave optimization. This would become possible irrespective of the number of variables involved in the description of the complex system, because the structural complexity of the system, considered as one variable, could control its performance in such a remarkable way. The optimality condition presents the structural complexity of a system as a key to its optimization. An important connection between the optimality condition and the majorization principle in quantum algorithms is also discussed.

1 Introduction

Management of complex systems and their optimization become increasingly important. However, it is still unknown whether a general condition determining the optimal performance of a complex system may exist [1]. To address the situation we show that the concept of structural complexity [2] computationally reveals the possibility of such an optimality condition [3] - [5], [8].

In particular, a special optimization algorithm, as a complex system, has been developed to minimize the average distance in the traveling salesman problem. Remarkably, for each problem tested the performance of the algorithm is concave. As a result, the algorithm and a problem are characterized by a single optimum. The analysis of the performance optimums reveals a relationship between the structural complexity of the algorithm and the structural complexity of the problem approximating it well enough by a linear function.

The results of the computational experiments suggest the possibility of a general optimality condition of complex systems:

A complex system demonstrates the optimal performance for a problem, when the structural complexity of the system is in a certain relationship with the structural complexity of the problem.

The optimality condition presents the structural complexity of a system as a key to its optimization. Indeed, according to the optimality condition the optimal result can be obtained as long as the structural complexity of the system is properly related with the structural complexity of the problem. From this perspective the optimization of a system should be primarily concerned with the control of the structural complexity of the system to match the structural complexity of the problem.

The computational results also indicate that the performance of a complex system may behave as a concave function of the structural complexity. Therefore, once the structural complexity controlled as a single entity, the optimization of a complex system could be reduced to a one-dimensional concave optimization irrespective of the number of variables involved in the description of the complex system.

We find an important connection between the optimality condition of complex systems and the majorization principle in quantum algorithms. In the search to identify a mathematical structure underlying optimal quantum algorithms the majorization principle emerges as a necessary condition for efficiency in quantum computational processes [9]. In our description the algorithm revealing the optimality condition also uses a majorization principle, but based on the structural complexity.

Perception-based information [10] may also play a new role. It was specified through a parameter aimed to translate changes in perceptions about the problem into monotonic control of the structural complexity of the algorithm. This suggests that perception-based information as human knowledge about the problem should be coherently presented with the control of the structural complexity of the system.

2 Structural Complexity in Irreducible Description of Complex Systems

In this section we give a brief introduction into the concept of structural complexity [2].

The concept of structural complexity is based on the description of complex systems in terms of self-organization processes of prime integer relations. These processes determine correlation structures of complex systems and may characterize their dynamics in a strong scale covariant form. Controlled by arithmetic only, the self-organization processes of prime integer relations can describe complex systems by information not requiring further explanations [2], [7].

To understand a complex system we consider the dynamics of the elementary parts and focus on the correlations preserving certain quantities of the complex system.

Let I be an integer alphabet and

$$I_N = \{x = x_1...x_N, x_i \in I, i = 1, ..., N\}$$

be the set of sequences of length $N \geq 2$. We consider N elementary parts $P_i, i = 1, ..., N$ with the state of an element P_i in its reference frame specified by a generalized coordinate $x_i \in I, i = 1, ..., N$ (for example, the position of the element P_i in space) and the state of the elements by a sequence $x = x_1...x_N \in I_N$.

It is proved that $C(x, x') \geq 1$ of the quantities of a complex system remain invariant, if and only if $C(x, x')$ equations take place

$$(m+N)^{C(x,x')-1}\Delta x_1 + (m+N-1)^{C(x,x')-1}\Delta x_2 + ... + (m+1)^{C(x,x')-1}\Delta x_N = 0$$

$$\cdot \quad \cdot \quad \cdot \quad \cdot \quad \cdot \quad \cdot \quad \cdot \quad \cdot \quad \cdot$$

$$(m+N)^1\Delta x_1 + (m+N-1)^1\Delta x_2 + ... + (m+1)^1\Delta x_N = 0$$
$$(m+N)^0\Delta x_1 + (m+N-1)^0\Delta x_2 + ... + (m+1)^0\Delta x_N = 0 \qquad (1)$$

characterizing in view of an inequality

$$(m+N)^{C(x,x')}\Delta x_1 + (m+N-1)^{C(x,x')}\Delta x_2 + ... + (m+1)^{C(x,x')}\Delta x_N \neq 0,$$

the correlations between the parts of the complex system, where $\Delta x_i = x'_i - x_i$, $x' = x'_1...x'_N, x = x_1...x_N$, $x'_i, x_i \in I, i = 1, ..., N$ are the changes of the elements $P_i, i = 1, ..., N$ in their own reference frames and m is an integer. The coefficients of the system of linear equations become the entries of the Vandermonde matrix, if the number of the equations is N. This fact is important in order to prove that $C(x, x') < N$ [2].

The equations (1) introduce a special type of correlations. These correlations do not depend on the distances between the parts, physical signals and local times. Therefore, under the equations and underlying assumptions it is possible to imagine that the parts of a complex system may be correlated as one event irrespective of the distances between them, their own local times and without signaling.

The space and non-signaling aspects of the correlations resonate well with explanations of quantum correlations through entanglement [11].

Analysis of the equations (1) reveals hierarchical structures of prime integer relations in the description of complex systems (Figure 1). Through the hierarchical structures we discovered a new type of processes - the self-organization processes of prime integer relations [2]. Starting with integers as the elementary building blocks and following a single principle, a self-organization process makes up the prime integer relations of a level of a hierarchical structure from the prime integer relations of the lower level (Figure 1).

The concept of structural complexity is defined to measure the complexity of a system in terms of self-organization processes of prime integer relations [2]. In particular, as self-organization processes of prime integer relations progress from a level to the higher level, the system becomes more complex, because its parts at the level are combined to make up more complex parts at the higher

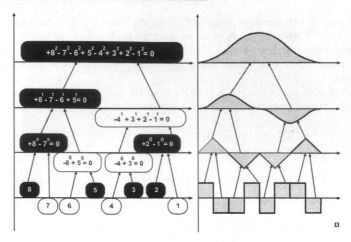

Fig. 1. The left side shows one of the hierarchical structures of prime integer relations, when a complex system has $N = 8$ elements $P_i, i = 1, ..., 8$, $x = 00000000$, $x' = +1 - 1 - 1 + 1 + 1 - 1 + 1 + 1 - 1$, $m = 0$ and $C(x, x') = 3$. The hierarchical structure is built by a self-organization process of prime integer relations and controls a correlation structure of the complex system. The right side presents an isomorphic hierarchical structure of geometric patterns. On scale level 0 eight rectangles specify the dynamics of the elements $P_i, i = 1, ..., 8$. The boundary curves of the geometric patterns describe the dynamics of the parts. All geometric patterns are symmetric and their symmetries are interconnected. The symmetry of a geometric pattern is global and belongs to a corresponding part as a whole.

level. Therefore, the higher the level self-organization processes progress to, the greater is the structural complexity of a corresponding complex system.

By using the integer code series [12] the prime integer relations can be geometrized as two-dimensional geometric patterns and the self-organization processes can be isomorphically expressed by transformations of the patterns [2]. As it becomes possible to measure a prime integer relation by an isomorphic geometric pattern, quantities of the prime integer relation and a complex system it describes can be defined by quantities of the geometric pattern, such as the area and the length of its boundary curve.

Due to the isomorphism, the structure and the dynamics of a complex system in our description are combined. As self-organization processes determine the correlation structure of a complex system, transformations of corresponding geometric patterns may characterize its dynamics in a strong scale covariant form [6], [7].

Many notions of complexity have been introduced, but they do not explain how the performance of a complex system may depend on its complexity. The idea of the experiments presented in the next sections was to investigate whether the concept of structural complexity could make a difference [3], [8].

3 Optimization Algorithm as a Complex System

An optimization algorithm $\mathcal{A}$, as a complex system of $N \geq 2$ computational agents, that minimizes the average distance in the travelling salesman problem is considered. All agents start in the same city and at each step an agent visits the next city by using one of the two strategies:

- a random strategy, i.e., visit the next city at random, or
- the greedy strategy, i.e, visit the next closest city.

It is defined that all agents start with the random strategy. The state of the agents visiting n cities of the problem at step $j = 1, ..., n-1$ is described by a binary sequence $x_j = x_{1j}...x_{Nj}$, where $x_{ij} = +1$, if agent $i = 1, ..., N$ to visit the next city uses the random strategy and $x_{ij} = -1$, if the agent uses the greedy strategy. The dynamics of the agents is determined by their choice of strategies and can be encoded by an $N \times (n-1)$ binary strategy matrix

$$S = \{x_{ij}, \ i = 1, ..., N, \ j = 1, ..., n-1\}.$$

A parameter v to control the dynamics of the agents in a specific manner is introduced. The parameter v also aims to translate changes in perceptions about the problem into monotonic control of the structural complexity of the algorithm $\mathcal{A}$.

Let D_{ij} be the distance travelled by agent $i = 1, ..., N$ after $j = 1, ..., n-1$ steps and

$$D_j^- = \min_{i=1,...,N} D_{ij}, \quad D_j^+ = \max_{i=1,...,N} D_{ij}.$$

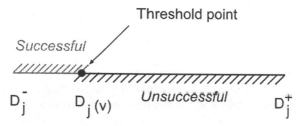

Fig. 2. At each step the range of the current distances is divided into two parts to specify agents with successful and unsuccessful strategies.

All distances travelled by the agents after $j = 1, ..., n-1$ steps belong to the interval $[D_j^-, D_j^+]$. The control parameter v specifies a threshold point (Figure 2)

$$D_j(v) = D_j^+ - v(D_j^+ - D_j^-), \quad 0 \leq v \leq 1.1$$

dividing the interval into two parts, i.e., successful $[D_j^-, D_j(v)]$ and unsuccessful $(D_j(v), D_j^+]$. If the distance D_{ij} travelled by agent $i = 1, ..., N$ after

$j = 1, ..., n - 1$ steps belongs to the successful interval, then the agent's last strategy is perceived as successful. If the distance D_{ij} belongs to the unsuccessful interval, then the agent's last strategy is perceived as unsuccessful.

To control the structural complexity of the algorithm $\mathcal{A}$ monotonically we try to mimic the period doubling route to chaos. For this purpose we use the parameter v to determine the dynamics of the agents following an optimal fuzzy if-then rule [2] to choose the next strategy. The rule relies on the Prouhet-Thue-Morse (PTM) sequence

$$+1 - 1 - 1 + 1 - 1 + 1 + 1 - 1 - 1 + 1 + 1 - 1 + 1 - 1 - 1 + 1 \ ... \ ,$$

which gives a symbolic description of chaos resulting from the period-doubling [13] in complex systems [14]. It can be also associated with a special self-organization process of prime integer relations that can progress through the hierarchical levels demonstrating critical point [15], [17] features [2].

The optimal fuzzy if-then rule has the following description:

1. *If your last strategy is successful, continue with the same strategy.*
2. *If your last strategy is unsuccessful, consult PTM generator which strategy to use next.*

Each agent has its own PTM generator and a pointer attached to it. The pointer starts with the first bit of the PTM sequence and after each consultation moves one step further, so that the next bit of the PTM sequence can be used, if the strategy is unsuccessful. The control is realized by changing the parameter v from 0 to 1.1 and thus changing the perceptions about successful and unsuccessful strategies. For the extreme values of the parameter v, the algorithm $\mathcal{A}$ produces the following dynamics of the agents.

When $v = 0$, then $D_j(v) = D_j^+$. Therefore, the successful interval

$$[D_j^-, D_j(v)]$$

coincides with the whole interval $[D_j^-, D_j^+]$ at each step $j = 1, ..., n-1$ and the last strategy of each agent is successful. This means that each agent always uses the random strategy and the strategy matrix becomes

$$S = \begin{pmatrix} +1 & +1 & ... & +1 & +1 \\ +1 & +1 & ... & +1 & +1 \\ . & . & ... & . & . \\ +1 & +1 & ... & +1 & +1 \\ +1 & +1 & ... & +1 & +1 \end{pmatrix}$$

In the opposite limit, when $v = 1.1$, then $D_j(v) < D_j^-$ and the unsuccessful interval $(D_j(v), D_j^+]$ covers the whole interval $[D_j^-, D_j^+]$. This means that the last strategy of each agent is always unsuccessful and, according to the rule, at each step $j = 1, ..., n - 1$ an agent asks the PTM generator which strategy to use next. As a result, the binary sequence of each agent becomes the initial segment of length $(n-1)$ of the PTM sequence and the strategy matrix turns into

$$S = \begin{pmatrix} +1 & -1 & ... & -1 & +1 \\ +1 & -1 & ... & -1 & +1 \\ . & . & ... & . & . \\ +1 & -1 & ... & -1 & +1 \\ +1 & -1 & ... & -1 & +1 \end{pmatrix}$$

The control parameter v can be also seen from the perspective of correlations or cooperative behavior of the agents. When $v = 0$, the agents are independent, because the behavior of an agent does not depend on the behaviors of the other agents. However, as the parameter v gets larger, the successful interval gets smaller, and the behaviors of the agents become more correlated through the PTM sequence. To stay independent, an agent has to show a result approaching the current minimum. Consequently, the rule urges an agent to follow the PTM sequence more strongly and, as a result, the presence of the PTM sequence in the strategy matrix becomes more evident. When $v = 1.1$, every row of the strategy matrix becomes the PTM sequence of length $n - 1$.

Extensive computational experiments have been conducted [3] - [5], [8] to test the problems of a class

$$P = \{eil76, eil101, st70, rat195, lin105, kroC100, kroB100,$$

$$kroA100, kroD100, d198, kroA150, pr107, u159, pr144,$$

$$pr144, pr152, pr226, pr136, pr76, ts225\}.$$

selected from the benchmark travelling salesman problems [16].

Let $D_i(p, v)$ be the distance travelled by agent $i = 1, ..., N$ for a problem $p \in P$ and value v of the control parameter. The performance of the algorithm $\mathcal{A}$ is characterized by the average distance travelled by N agents

$$D(p, v) = \sum_{i=1}^{N} D_i(p, v)/N,$$

which is minimized.

In the experiments the algorithm $\mathcal{A}$ has been applied to each problem $p \in P$ with the number of agents $N = 2000$ and values

$$v_i = i \Delta v, \quad i = 0, 1, 2, ..., 20, 22$$

of the control parameter v, where $\Delta v = 0.05$. To eliminate randomness, the computations have been repeated in a series of ten tests and the performance functions

$$\{D_k(p, v), \quad k = 1, ..., 10\}$$

have been averaged into $\bar{D}(p, v)$.

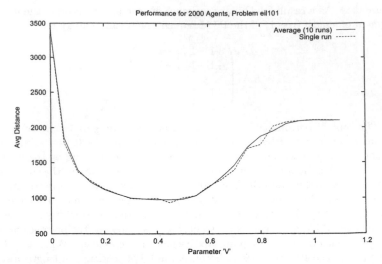

Fig. 3. The average distance obtained by the algorithm $\mathcal{A}$ for problem eil101 as a convex function of the parameter v.

Remarkably, for each problem $p \in P$ it has been found that the performance $\bar{D}(p, v)$ of the optimization algorithm $\mathcal{A}$ behaves as a concave function of the parameter v with the only optimum at $v^*(p)$. (Figures 3 and 4 show convex, not concave functions of the control parameter v, because the algorithm $\mathcal{A}$ minimizes the average distance). Consequently, it has been investigated whether the performance optimums could be characterized in terms of an optimality condition of the algorithm $\mathcal{A}$.

4 Optimality Condition of Complex Systems

It has been considered whether there could be a connection between the structural complexities of the algorithm and the problem [8].

For this purpose the structural complexity of the algorithm $\mathcal{A}$ is approximated by traces of variance-covariance matrices derived from the dynamics of the algorithm as a complex system. The structural complexity of the problem is approximated by a trace of the distance matrix. Remarkably, the computational analysis of the performance optimums reveals a relationship between the structural complexities of the algorithm and the problem by approximating it well enough with a linear function.

In particular, the performance optimums

$$\{v^*(p), \ p \in P\}$$

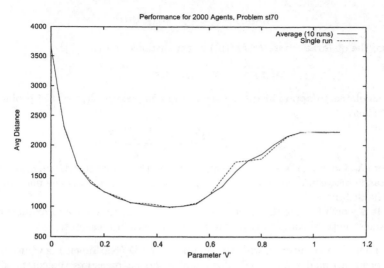

Fig. 4. The average distance obtained by the algorithm $\mathcal{A}$ for problem st70 as a convex function of the parameter v.

have been considered to specify the structural complexity of the algorithm. For a problem $p \in P$ a set of strategy matrices

$$S_k(v^*(p)) = \{s_{ij}(k),\ i = 1, ..., N,\ j = 1, ..., n-1\},\ k = 1, ..., 10$$

has been obtained as a result of ten runs for the value $v^*(p)$ of the parameter. For each strategy matrix

$$S_k(v^*(p)),\ k = 1, ..., 10$$

the variance-covariance matrix

$$V(S_k(v^*(p))) = \{V_{ij}(k),\ i, j = 1, ..., N\}, k = 1, ..., 10,$$

and its quadratic trace

$$tr(V^2(S_k(v^*(p)))) = \sum_{i=1}^{N} \lambda_{ik}^2$$

are computed, where $V_{ij}(k)$ is the linear correlation coefficient between agents $i = 1, ..., N$ and $j = 1, ..., N$ and $\lambda_{ik}, i = 1, ..., N,\ k = 1, ..., 10$ are the eigenvalues of the variance-covariance matrix $V(S_k(v^*(p)))$.

The average trace $tr(V^2(S(v^*(p)))$ is computed to approximate the structural complexity $C(\mathcal{A}(p))$ of the algorithm $\mathcal{A}$ for the problem p as

$$C(\mathcal{A}(p)) = \frac{1}{N^2} tr(V^2(S(v^*(p)))).$$

Also, the quadratic trace $tr(M^2(p))$ of the distance matrix

$$M(p) = \{d_{ij}/d_{max}, i, j = 1, ..., n\}$$

is calculated to approximate the structural complexity $C(p)$ of the problem $p \in P$ as

$$C(p) = \frac{1}{n^2} tr(M^2(p)) = \frac{1}{n^2} \sum_{i=1}^{n} \lambda_i^2,$$

where $\lambda_i, i = 1, ..., n$ are the eigenvalues of the distance matrix $M(p)$, d_{ij} is the distance between cities $i = 1, ..., n$ and $j = 1, ..., n$ and d_{max} is the maximum of the distances.

It is worth to note that there are connections between our approximations to the structural complexity and different correlation measures:

1. There is a connection with the number of KLD (Karhunen-Loeve decomposition) modes D_{KLD} [8]. The quantity D_{KLD} measures the complexity of spatiotemporal data [19], [20], [21] and a correlation length ξ_{KLD} based on D_{KLD} can characterize high-dimensional inhomogeneous spatiotemporal chaos [22].
2. The degree of correlation [23], [24] is inversely proportional to the sum of the squares of the eigenvalues of the 1-particle-electron statistical operator, while the wavefunction norm is 1, so that the trace of the operator equals the number of electrons [25].
3. The correlation index

$$\mathcal{E} = \frac{1}{N_1 N_2} tr(\Lambda \Lambda^T)$$

is used to study the correlation between finite quantum systems and applies to two systems with state vector Hilbert spaces of dimension N_1 and N_2 accordingly, where Λ is the covariance matrix of local observables [26].

An optimality condition of the algorithm $\mathcal{A}$ has been sought through a possible relationship between the structural complexity $C(\mathcal{A}(p))$ of the algorithm $\mathcal{A}$ and the structural complexity $C(p)$ of the problem. For this purpose, the points with coordinates

$$\{x = C(p), y = C(\mathcal{A}(p)), \ p \in P\}.$$

are considered.

The result of the analysis shown in Figure 5 suggests a possible linear relationship between the structural complexities. The regression line is calculated as

$$y = \alpha x + \beta = 0.67x + 0.33,$$

$$\alpha = 0.67 \pm 0.01, \ \beta = 0.33 \pm 0.01, \tag{2}$$

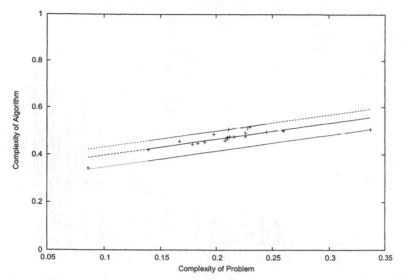

Fig. 5. The regression line of the relationship between the structural complexity of the algorithm $\mathcal{A}$ and the structural complexity of the problem with the bounds including all points.

where the standard error of estimate is 0.09 and the absolute value of the maximal individual error is 0.05. The coefficient of determination of 0.71 shows that 71 percent of the variation in the complexity of the algorithm $\mathcal{A}$ can be explained by the regression line.

The experiment has been repeated a number of times and each time the results confirmed the consistency of the regression line. Therefore, within the accuracy of the linear regression, it is possible to formulate an optimality condition of the algorithm:

If the algorithm $\mathcal{A}$ shows its best performance for a problem p, then the structural complexity $C(\mathcal{A}(p))$ of the algorithm $\mathcal{A}$ is in the linear relationship with the structural complexity $C(p)$ of the problem p

$$C(\mathcal{A}(p)) = 0.67 \times C(p) + 0.33. \tag{3}$$

To test the optimality condition further we have used a different algorithm $\mathcal{B}$. It works exactly in the same manner as the algorithm $\mathcal{A}$, except it consults a random generator instead of the PTM generator. To find a possible connection between the best performances and the relationship between the complexities, the algorithms $\mathcal{A}$ and $\mathcal{B}$ have been compared.

First, for each problem $p \in \mathcal{P}$ the best performance of the algorithm $\mathcal{A}$ has been compared with the best performance of the algorithm $\mathcal{B}$. Figure 6 shows

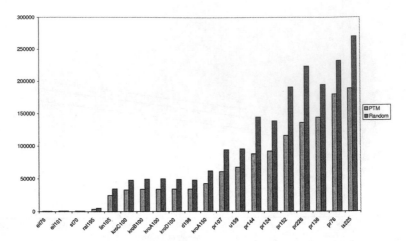

Fig. 6. The best performances of the algorithm $\mathcal{A}$ using PTM generator are compared with the best performances of the algorithm $\mathcal{B}$ using random generator. The problems are ordered as in the description of the class P.

that for majority of the problems the algorithm $\mathcal{A}$ demonstrates significantly better results than the algorithm $\mathcal{B}$.

Second, the relationship between the structural complexities of the algorithm $\mathcal{B}$ and the problem (Figure 7) has been examined in the same manner as for the algorithm $\mathcal{A}$ (Figure 5). The situation deteriorates and the relationship is less consistent for the algorithm $\mathcal{B}$. Formally, the calculated regression line

$$y = \alpha' x + \beta' = 0.48x + 0.23,$$

$$\alpha' = 0.48 \pm 0.02, \quad \beta' = 0.23 \pm 0.01 \tag{4}$$

is a less accurate estimator of the relationship between the structural complexities of the algorithm $\mathcal{B}$ and the problem. The standard error of estimate of 0.18 and the absolute value of the maximal individual error of 0.1 of the regression line (4) are doubled in comparison with the regression line (2). The coefficient of determination is only 0.26 in comparison with 0.71 for the algorithm $\mathcal{A}$.

Therefore, computationally, we have revealed a connection between the best performances and the relationship between the structural complexities. In particular, the better performance of the algorithm $\mathcal{A}$ (Figure 6) corresponds to the fact that the relationship between the structural complexities of the algorithm $\mathcal{A}$ and the problem (Figure 5) appears more consistent than it is in the case of the algorithm $\mathcal{B}$ (Figure 7).

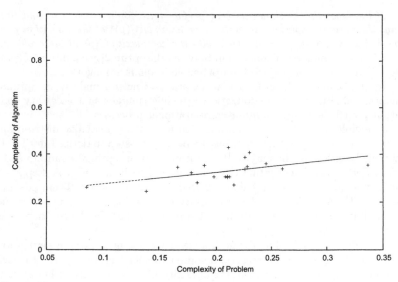

Fig. 7. The regression line of the relationship between the structural complexity of the algorithm B and the structural complexity of the problem.

Notably, the optimality condition of the algorithm A combines structure and dynamics. The structure is given by the distance network specified by the distance matrix. The dynamics of the agents is realized as they choose strategies step by step. According to the optimality condition the performance is optimal when dynamical properties of the algorithm expressed by the structural complexity are in the linear relationship (2) with structural properties of the network also expressed by the structural complexity.

By using this result potential implications of the optimality condition of the algorithm A could be considered. The observed connection suggests a possible sequence of algorithms whose best performances converge to the optimal solutions of the problems as the relationship between the structural complexities converges to a certain function g. The algorithms A and B might be among the elements of the sequence with the algorithm B followed by the algorithm A.

The convergence of the sequence would link the optimal performance and the relationship between the complexities, so that we could speculate about an algorithm G. The best performance of the algorithm G for a problem could actually provide the optimal solution. An optimality condition of the algorithm G might extend the optimality condition of the algorithm A and explain the optimal performance in terms of the relationship between the complexities:

If the algorithm $\mathcal{G}$ shows its best performance for a problem p, and thus finds the optimum solution, then the complexity $C(\mathcal{G}(p))$ of the algorithm is in the relationship $C(\mathcal{G}(p)) = g(C(p))$ with the complexity $C(p)$ of the problem.

We find an important connection between the optimality condition of complex systems and the majorization principle in quantum algorithms.

In the search to identify a mathematical structure underlying optimal quantum algorithms the majorization principle emerges as a necessary condition for efficiency in quantum computational processes [9]. According to the principle in an optimal quantum algorithm the probability distribution associated to the quantum state has to be step-by-step majorized until it is maximally ordered. The majorization means that an optimal quantum algorithm works in such a way that the probability distribution p_{k+1} at step $k+1$ majorizes $p_k \prec p_{k+1}$ the probability distribution p_k at step k. There are special conditions in place for the probability distribution p_{k+1} to majorize the probability distribution p_k with intuitive meaning that the distribution p_k is more disordered than p_{k+1} [9].

The algorithm revealing the optimality condition also uses a majorization principle, but based on the structural complexity. The algorithm tries to work in such a way that the structural complexity $\mathbf{C}_{k+1}$ of the algorithm at step $k + 1$ majorizes $\mathbf{C}_k \prec \mathbf{C}_{k+1}$ its structural complexity $\mathbf{C}_k$ at step k. In our description the optimality condition determines the optimal result in a clear way, i.e., when the structural complexity of the algorithm is in the linear relationship with the structural complexity of the problem.

5 Conclusions

Based on a concept of structural complexity we have presented results of computational experiments suggesting the possibility of an optimality condition of complex systems:

A complex system demonstrates the optimal performance for a problem, when the structural complexity of the system is in a certain relationship with the structural complexity of the problem.

The optimality condition presents the structural complexity of a system as a key to its optimization. According to the condition the optimal performance of a system can be obtained as long as the structural complexity of the system is properly related with the structural complexity of the problem. Therefore, the optimization of a system should be primarily concerned with the control its structural complexity in order to match the structural complexity of the problem. The ability to evaluate the structural complexity of the problem before the actual computations would be especially helpful. In this case, the optimization of a system could be reduced to the tuning of the structural complexity of the system to match already known structural complexity of the problem.

Moreover, the computational results have indicated that the performance of a complex system may behave as a concave function of the structural complexity. Therefore, the optimality condition could reduce the optimization of a complex system to a one-dimensional concave optimization. This would become possible irrespective of the number of variables involved in the description of a complex system, because the structural complexity of the system, considered as one variable, could control the performance of the system in such a remarkable way.

The optimality condition has been successfully used in the development of the Mackay Transport Technology [27]. The technology utilizes a fleet of various size vehicles and blends the low cost of bus service with the flexibility of taxi service by offering the anywhere to anywhere convenience on demand. By using satellite/wireless communications and digital maps it operates through a control system optimizing in real time the performance of the vehicles in uncertain and dynamic environment.

References

1. P. Ball, *Transitions Still to be Made*, Nature 402, 2, 1999, c73-c76.
2. V. Korotkikh, *A Mathematical Structure for Emergent Computation*, Kluwer Academic Publishers, Dordrecht, 1999.
3. V. Korotkikh, *The Search for the universal concept of complexity and a general optimality condition of cooperative agents*, in S. Butenko, R. Murphey and P. M. Pardalos (eds), Cooperative Control: Models, Applications and Algorithms, Kluwer, Dordrecht, 2003, pp. 129–163.
4. G. Korotkikh and V. Korotkikh, *On the role of nonlocal correlations in optimization*, in P. Pardalos and V. Korotkikh (eds), Optimization and Industry: New Frontiers, Kluwer Academic Publishers, Dordrecht/Boston/London, 2003, pp. 181–220.
5. V. Korotkikh, *An Approach to the mathematical theory of preception-based Iinformation*, in M. Nikravesh, L. Zadeh and V. Korotkikh (eds), Fuzzy Partial Differential Equations and Relational Equations, Springer, Berlin, 2004, pp. 80-115.
6. V. Korotkikh, *Integers: irreducible guides in the search for a unified theory*, Brazilian Journal of Physics, Special Issue on Decoherence, Information, Complexity and Entropy, **35**(2B), 509, 2005.
7. V. Korotkikh and G. Korotkikh, *Description of complex systems in terms of self-organization processes of prime integer relations*, in M. M. Novak (ed.), Complexus Mundi: Emergent Patterns in Nature, World Scientific, New Jersey/London, pp. 63-72, 2006.
8. V. Korotkikh, G. Korotkikh and D. Bond, *On Optimality Condition of Complex Systems: Computational Evidence*, arXiv:cs.CC/0504092.
9. R. Orus, J. Latorre and M. A. Martin-Delgado, *Systematic Analysis of Majorization in Quantum Algorithms*, arXiv:quant-ph/0212094.
10. L. Zadeh, From computing with numbers to computing with words - from manipulation of measurements to manipulation of perceptions, IEEE Transactions on Circuits Systems, 45 (1999) 105-119.

11. N. Gisin, *Can Relativity be Considered Complete? From Newtonian Nonlocality to Quantum Nonlocality and Beyond*, arXiv:quant-ph/0512168.
12. V. Korotkikh, *Integer Code Series with Some Applications in Dynamical Systems and Complexity*, Computing Centre of the Russian Academy of Sciences, Moscow, 1993.
13. M. Feigenbaum, *Universal Behaviour in Nonlinear Systems*, Los Alamos Science, vol. 1, 1980, 4.
14. J. P. Allouche and M. Cosnard, *Sequences generated by automata and dynamical systems*, in Dynamical Systems and Cellular Automata, Academic Press, 17, 1985.
15. L. Kadanoff, W. Gotze, D. Hamblen, R. Hecht, E. A. S. Lewis, V. V. Palciaukas, M. Rayl and J. Swift, Rev. Mod. Phys. 39, 2, 395, 1967.
16. G. Reinelt, TSPLIB Version 1.2, ftp://ftp.wiwi.uni-frankfurt.de/pub/TSPLIB 1.2 (Accessed 28/11/2000).
17. K. G. Wilson and J. Kogut, Phys. Rep. C12, 75, 1974.
18. V. Korotkikh and G. Korotkikh, *On a new quantization in complex systems*, in P. Pardalos, C. Sackellares, P. Carney and L. Iasemidis (eds), Quantitative Neuroscience: Models, Algorithms, Diagnostics and Therapeutic Applications, Kluwer Academic Publishers, Dordrecht, pp. 71–91, 2004.
19. L. Sirovich and A. E. Deane, Journal Fluid Mech. 222, 251, 1991.
20. S. Ciliberto and Nikolaenko, Europhys. Lett. 14, 303, 1991.
21. R. Vautard and M. Ghil, Physica 35D, Amsterdam, 395, 1989.
22. S. M. Zoldi and H. S. Greenside, Phys. Rev. Lett. 78, 1687, 1997.
23. R. Grobe, K. Rzazewski and J. H. Eberly, *Measure of electron-electron correlation in atomic physics*, J. Phys. B, 27, pp. L503-L508, 1994.
24. W. C. Liu, J. H. Eberly, S. L. Haan and R. Grobe, *Correlation effects in two-electron model atoms in intense laser fields*, Phys. Rev. Letters 83, 3, pp. 520–523, 1999.
25. A. D. Gottlieb and N. J. Mauser, *New Measure of Electron Correlation*, arXiv:quant-ph/0503098.
26. D. Ellinas and E. G. Floratos, *Prime Decomposition and Correlation Measure of Finite Quantum Systems*, arXiv:quant-ph/9806007.
27. M. McBride and V. Korotkikh, *Mackay Transport Technology*, Mackay Taxi Holdings Ltd and Central Queensland University, 2004.

10 GRASP with Path-Relinking for the Cooperative Communication Problem on Ad hoc Networks

Clayton W. Commander, Paola Festa, Carlos A. S. Oliveira,
Panos M. Pardalos, Mauricio G. C. Resende and Marco Tsitselis

Summary. Ad hoc networks are a new paradigm for communications systems in which wireless nodes can freely connect to each other without the need of a pre-specified structure. Difficult combinatorial optimization problems are associated with the design and operation of these networks. In this chapter, we consider the problem of maximizing the connection time between a set of mobile agents in an ad hoc network. Given a network and a set of wireless agents with starting nodes and target nodes, the objective is to find a set of trajectories for the agents that maximizes connectivity during their operation. This problem, referred to as the CO-OPERATIVE COMMUNICATION ON AD HOC NETWORKS is known to be NP-hard. We look for heuristic algorithms that are able to efficiently compute high quality solutions to instances of large size. We propose the use of a greedy randomized adaptive search procedure (GRASP) to compute solutions for this problem. The GRASP is enhanced by applying a path-relinking intensification procedure. Extensive experimental results are presented, demonstrating that the proposed strategy provides near optimal solutions for the 900 instances tested.

1 Introduction

Cooperative control has received a significant amount of attention by researchers [6, 7, 21, 25]. In many situations, multiple "agents" collaborate to achieve a shared goal. Cooperation between the agents is important to improve the efficiency and effectiveness by which their goal is reached. In wireless networks, groups of agents are often employed to perform a number of cooperative tasks, including the synchronization of information among a set of users and the accomplishment of missions in remote areas. In such situations, it is useful to maintain collaboration among the agents performing the cooperative tasks to maximize the probability of success.

There are many applications of this described system. These include situations where communication in a region is required, but no topologically fixed transmission system exists. Specific examples include emergency/rescue operations, disaster relief, and battlefield operations [27]. In each of these examples the goals and objectives are fixed in advance and communication is important for the attainment of these goals. The current technologies used in these types of applications allow improved communication systems that rely on ad hoc wireless protocols. However, it is computationally difficult to decide how to maintain communication for the maximum possible time when faced with the inherent restrictions of wireless systems.

Ad hoc networks are composed of a set of wireless units that can communicate directly, without the use of a pre-established server infrastructure. In this respect, ad hoc systems are fundamentally different from traditional cellular systems, where each user has an assigned base station which connects it to the wired telephony system. In an ad hoc network, each client has the capacity of accessing network nodes that are within its reach. This connectivity model allows for the existence of networks without a predefined topology, reaching a different state every time a node changes its position. Due to this inherent variability, ad hoc networks present serious challenges for the design of efficient protocols. The optimization of activities in such networks is subject to the lack of global information. Furthermore, optimal solutions may be short-lived, due to the dynamics of the agents' positions and connectivity status.

In this chapter, we study a problem involving the coordination of wireless users involved in a mission of tasks that requires each user to go from an initial location to a target location. The problem consists in maximizing the amount of connectivity among the set of users, subject to constraints on the maximum distance traveled by each user, as well as restrictions on what types of movements can be performed. This problem is referred to as the COOPERATIVE COMMUNICATION PROBLEM ON MOBILE AD HOC NETWORKS (CCPM) [13, 27].

The CCPM is a path-planning problem and has numerous military applications. For example, suppose a force is planning a reconnaissance mission over a battle space and will be using a set of so-called unmanned aerial vehicles (UAVs) for this task. It may also be required that the UAVs maintain connectivity with each other in order to share information that is being gathered. Before the mission begins, the mission planner could solve an instance of the CCPM using the knowledge of the battle space to determine a set of optimal paths for the UAVs. In the CCPM, a set of paths is said to be optimal if the communication between all pairs of agents is maximized, subject to the agents arriving at their respective destinations within the specified time horizon. After briefly reviewing some related work from the literature, we will provide a mathematical model of the CCPM and describe an effective algorithm for solving large instances.

1.1 Related work

Ad hoc networks represent an extremely active area of research [30]. Several problems related to routing, power control, and accurate position update have been studied in the last few years [26]. In terms of routing, one of the main problems in ad hoc networks is the computation of a network backbone. The objective is to find a subset of nodes with a small number of elements that can be used to send routing information. Such a structure is useful to simplify the management tasks required by a routing protocol. The backbone computation problem can be modeled as a CONNECTED DOMINATING SET (CDS) problem. Here, the objective is to find a set of minimum size forming a connected backbone, with the additional property that each network client can directly reach this set. The CDS problem, which can be modeled using unit graphs, has several approximation algorithms [4, 5, 9], all of which are based on approximation properties of the MAXIMUM INDEPENDENT SET problem on planar unit graphs [3]. The use of optimization techniques to maximize connectivity in ad hoc systems was studied by Oliveira and Pardalos in [27]. Until now, only local search heuristics have been applied to the CCPM [13].

Given the difficulty of solving the CCPM exactly, we are interested in studying the application of metaheuristic methods for the problem. In particular, we propose a greedy randomized adaptive search procedure (GRASP) for the CCPM. In addition, path-relinking [20] is applied to increase the effectiveness of GRASP. We show that the resulting algorithm is able to efficiently provide high quality solutions for large scale instances of the problem.

1.2 Problem definition

An ad hoc network is composed of a set of autonomous clients that can connect to each other using their own wireless capabilities. This includes using scarce resources such as computational processing and battery power. We model this situation using a special type of graph called a unit graph [10]. A unit graph is a planar graph $G = (V, E)$ with associated positions for each node $v \in V$. In a unit graph an edge occurs between nodes $v, w \in V$ if $dist(v, w) \leq 1$, where $dist : V \times V \to \mathbb{R}$ is the Euclidean distance. Unit graphs occur as a natural model in ad hoc networks and are used in this chapter to represent the set of configurations of nodes that share connections.

Let $G = (V, E)$ be a graph representing the set of valid positions for network clients. Suppose this graph has the property that each node is connected only to nodes that can be reached in one unit of time. Therefore, the graph can be used to represent all possible trajectories of a node. Each such trajectory is a path $\mathcal{P} = \{v_1, \ldots, v_k\}$ where $v_1 \in V$ is the starting node and $v_k \in V$ is the destination node. We also consider a set U of wireless units, a set of initial positions $S \subseteq V$, such that $|S| = |U|$, and a set of destinations $D \subseteq V$, with $|D| = |U|$. We assume that, to perform its task, each wireless unit $u_i \in U$ starts from a position $s_i \in S$, and ends at position $d_i \in D$. We are given a time limit T such that all units must reach their destinations by time T.

The trajectory of users in the system occurs as follows. For $v \in V$, let $\eta(v) \subseteq V$ be the set of all nodes in the neighborhood of v, i.e. the set of nodes $w \in V$ such that $(v, w) \in E$. Let $p_t : U \to V$ be a function returning the position of a wireless unit at time t. Then at each time step t, a wireless unit $u \in U$ can stay in its previous position p_{t-1} or move to one of its neighboring nodes $v \in \eta(p_{t-1})$. More formally, at time step t, position $p_t(u) \in p_{t-1}(u) \cup \eta(p_{t-1})$. Let $\{\mathcal{P}_i\}_{i=1}^{|U|}$ be the set of paths representing the trajectories of the wireless units in U (obviously, the first node of $\mathcal{P}_i$ is s_i and its last node is d_i). Let L_i, for $i \in \{1, \ldots, |V|\}$, be a threshold on the total cost of path $\mathcal{P}_i$. Define $\omega : V \times V \to \mathbb{R}$ to be the weight of the edge between two nodes. Then, for each path $\mathcal{P}_i = \{v_1, \ldots, v_{n_i}\}$, we require that

$$\sum_{i=2}^{n_i} \omega(v_{i-1}, v_i) \leq L_i. \tag{1}$$

With this, we can define the COOPERATIVE COMMUNICATION PROBLEM ON MOBILE AD HOC NETWORKS (CCPM) as follows. Given a network $G = (V, E)$, a set U of wireless agents, a set $S \subseteq V$ of starting nodes, a set $D \subseteq V$ of destination nodes, a maximum travel time T, and distance thresholds L_i, for $i \in \{1, \ldots, |V|\}$, a feasible solution for the CCPM consists of a set of positions $p_t(u)$, for $t \in \{1, \ldots, T\}$ and $u \in U$, such that the initial position satisfies $p_1(u) = s(u)$ for $u \in U$, the final position is $p_T(u) = d(u)$, a set positions given by $p_t(u) \in p_{t-1}(u) \cup \eta(p_{t-1})$, $\forall t \in \{2, \ldots, T-1\}$, and the inequalities in (1) are satisfied. The objective is to maximize the connectivity among the users in U over all time steps, which is measured by

$$\max \sum_{t=1}^{T} \sum_{\substack{u,v \in U \\ u \neq v}} c(p_t(u), p_t(v)),$$

where $c : V \times V \to \{0, 1\}$ is a function returning 1 if and only if $dist(p(u), p(v)) \leq 1$. The reader is referred to the paper by Oliveira and Pardalos [27] for additional integer programming formulations in which other objectives are considered and discussed. We finish this section by providing two results related to the computational complexity of the problem.

Theorem 1 *Finding an optimal solution for an instance of the* COOPERATIVE COMMUNICATION PROBLEM ON MOBILE AD HOC NETWORKS *is NP-hard.*

This result, due to Oliveira and Pardalos [27], follows by a reduction from MAXIMUM 3-SAT [18]. We now extend this result in the following theorem.

Theorem 2 *Consider an instance of the* CCPM, *with T as the time-horizon. Finding an optimal solution at each time-step $t \in [1, T]$ is NP-hard.*

Proof: We will show this result by reducing CLIQUE to CCPM at an arbitrary time-step. Recall that the CLIQUE problem is as follows. Given a graph $G = (V, E)$ and an integer $J \leq |V|$, does G contain a clique, or complete subgraph, of size J or more [18]?

Consider an instance of CCPM at any time step t. An optimal solution is one in which all the agents are pairwise connected. Thus, for n agents the

```
procedure GRASP(MaxIter, RandomSeed)
1    X* ← ∅
2    for i = 1 to MaxIter do
3        X ← ConstructionSolution(G, g, X)
4        X ← LocalSearch(X, MaxIterLS)
5        if f(X) ≥ f(X*) then
6            X* ← X
7        end
8    end
9    return X*
end procedure GRASP
```

Fig. 1. GRASP for maximization.

number of connections in an optimal solution is $n(n-1)/2$. Notice that this is equivalent to finding clique on n nodes of the graph. Therefore, given an instance of CLIQUE, by letting $J = n$, we have the result. Thus there is a bijection between optimal configurations of agents and cliques in the graph. ∎

Corollary 1 *For any instance of* CCPM, *an upper bound on the optimal solution is given by*

$$T \cdot \frac{u(u-1)}{2}, \tag{2}$$

where T is the time horizon and $u = |U|$ is the number of agents.

Proof: This proof follows directly from Theorem 2. If all u agents communicate at a given time, then they form a clique on u nodes. The clique will contain $u(u-1)/2$ vertices representing the communication links. If the agents maintain the clique formation over all time steps, then the number of communication connections will be

$$T \cdot \frac{u(u-1)}{2} \tag{3}$$

and the lemma is proved. ∎

2 GRASP for CCPM

A greedy randomized adaptive search procedure (GRASP) [14] is a multistart metaheuristic that has been used widely to provide solutions for several difficult combinatorial optimization problems [17], including SATISFIABILITY [29], JOB SHOP SCHEDULING [2], VEHICLE ROUTING [8], and QUADRATIC ASSIGNMENT [24, 28].

GRASP is a two-phase procedure which generates solutions through the controlled use of random sampling, greedy selection, and local search. For a given problem Π, let F be the set of feasible solutions for Π. Each solution $X \in F$ is composed of k discrete components $a_1, \ldots, a_k$. GRASP constructs a sequence $\{X_1, X_2, \ldots\}$ of solutions for Π, such that each $X_i \in F$. The algorithm returns the best solution found after all iterations. The GRASP procedure is described in the pseudo-code shown in Figure 1. The *construction phase* receives as parameters an instance G of the problem and a ranking function $g : A(X) \to \mathbb{R}$, where $A(X)$ is the domain of feasible components $a_1, \ldots, a_k$ for a partial solution X. The construction phase begins with an empty partial solution X. Assuming that $|A(X)| = k$, the algorithm creates a so-called *Restricted Candidate List* (RCL) containing the α percent best ranked components in $A(X)$. The parameter $\alpha \in [0, 1]$ can be fixed by the user or chosen randomly. An element $x \in$ RCL is uniformly chosen and the current partial solution is augmented to include x. This procedure is repeated until the solution is feasible, i.e., until $X \in F$.

The *intensification phase* consists of a local search which could include gradient descent methods, swap based methods, and even the use of other metaheuristics. Our local search consists of an implementation of a hill-climbing procedure. Given a solution $X \in F$, let $N(X)$ be the set of solutions that can found from X by changing one of the components $a \in X$. Then, $N(X)$ is called the neighborhood of X. The improvement algorithm consists in finding, at each step, the element X^* such that

$$X^* = \underset{X' \in N(X)}{\operatorname{argmax}} f(X'),$$

where $f : F \to \mathbb{R}$ is the objective function of the problem. At the end of each step we make $X^* \leftarrow X$ if $f(X) > f(X^*)$. The algorithm will eventually achieve a local optimum, in which case the solution X^* is such that $f(X^*) \geq f(X')$ for all $X' \in N(X^*)$. X^* is returned as the best solution from the iteration and the best solution from all iterations is returned as the overall GRASP solution.

We discuss in this section how the above algorithm can be specialized to solve the CCPM. In the following subsection, we describe an algorithm for the GRASP construction phase that provides initial solutions for instances of the CCPM problem. Then, in Subsection 2.2 we provide a local search algorithm for the improvement phase.

2.1 Construction Phase

The first task in a GRASP algorithm is to build good feasible solutions in terms of a given objective function. To do this, we need to specify the set A, the greedy function g, and, for $X \in F$, the neighborhood $N(X)$. The components of each solution X are feasible moves of a member of the ad hoc network from a node v to a node $w \in N(v) \cup \{v\}$. For an agent $u_i \in U$ located at node v in the graph, $\mathcal{P}_i(v)$ represents a shortest path from the current node v to the destination for agent u_i, namely node d_i.

The complete solution is constructed according to the following procedure outlined in the pseudo-code shown in Figure 2. In the pseudo-code, a_h refers to the current location of an agent. First, the solution which is initially empty is augmented to include the starting locations for all agents. Then, the time variable t is initialized to 1, and in line 6 an agent $u_i \in U$ is selected at random and routed from its along a shortest path $\mathcal{P}_i(s_i)$ from its source node s_i to its destination node d_i. If the total distance of $\mathcal{P}_i(s_i)$ is greater than L_i, then the instance is clearly infeasible and the algorithm ends. Otherwise, the procedure continues and the remaining agents are scheduled in the loop beginning at line 8. The procedure considers each feasible move (q, w, u) before scheduling an agent. A feasible move connects the final node q of a sub-path P_u, for $u \in U$, to another node w, such that the shortest path from w to d_u has distance at most $L_u - \sum_{e \in P_u} dist(e)$. The set of all feasible moves in a solution is defined as $A(X)$.

The loop in lines 12–14 ensures that a node currently at its destination remains there. Likewise, the loop in lines 15–17 schedules an agent h on a shortest path $\mathcal{P}_h(a_h)$ from its current position a_h to d_h if the maximum allowed travel time for agent h is equal to the $|\mathcal{P}_h(a_h)|$. In lines 19–21, the set $L \subseteq A(X)$ is formed and consists of all feasible moves for agents not yet scheduled. Then, in line 25, the greedy function g returns for each move $k \in L$ the number of additional connections created by that move. As described above, the construction procedure will rank the elements of L according to g. The α percent best-ranked elements are then added to the RCL and in lines 28–29, a move is selected at random and added to the solution. This is repeated until a complete feasible solution for the problem is obtained.

2.2 Improvement phase

In the local search phase, GRASP attempts to improve the solution built in the construction phase. As mentioned above, we use a hill-climbing procedure where the objective is to improve the solution as much as possible until a local optimal solution is found as described in the pseudo-code provided in Figure 3.

The local search receives the construction phase solution X and a parameter MaxIterLS as input. In each iteration, the neighborhood $N(X)$ of X is explored in search of a solution X' such that $f(X') > f(X)$. In order to explore $N(X)$, a perturbation function is defined as follows. In the loop in lines 5–21, agents are re-routed using a greedy method similar to that of the construction phase. In line 6, the current construction phase path for agent u_i is removed from the solution. Then each feasible move is considered and the move which adds the greatest increase to the objective function, $BestMove$, is added to the new path for agent u_i. This is repeated for all agents until a new feasible solution $X' \in N(X) \subseteq A(X)$ is created. If $f(X') > f(X)$, then

```
procedure ConstructionSolution(G, g, X)
1     X ← ∅
2     for i = 1 to k do
3         X ← X ∪ {sᵢ}
4     end
5     t ← 1
6     Select randomly uᵢ ∈ U and route uᵢ on shortest-path 𝒫ᵢ(sᵢ)
7     X ← X ∪ {Pᵢ(sᵢ)}
8     while t < T and ∃ uᵢ ∈ U \ X do
9         L ← ∅
10        RCL ← ∅
11        for uₕ ∈ U \ X do
12            if aₕ = dₕ do
13                X ← X ∪ {dₕ}
14            end
15            if dist(aₕ, dₕ) = Lₕ − ∑ₑ∈Pₕ dist(e) do
16                Route uₕ on shortest path 𝒫ₕ(aₕ)
17                X ← X ∪ {𝒫ₕ(aₕ)}
18            end
19            while ∃ (lq, lw, uₕ) do
20                L ← {(lq, lw, uₕ)}
21            end
22        end
23        α ← RealRandomNumber(0, 1)
24        δ ← (MaxContribute − (α ∗ (MaxContribute − MinContribute)))
25        for all (lᵢ, lⱼ, uₑ) such that g((lᵢ, lⱼ, uₑ)) ≥ δ do
26            RCL ← RCL ∪ {(lᵢ, lⱼ, uₑ)}
27        end
28        (lᵢ, lⱼ, uₑ) ← Get randomly from RCL
29        Add (lᵢ, lⱼ, uₑ) into the path of agent uₑ in the solution
30        t ← t + 1
31    end
32    return(X)
end procedure ConstructionSolution
```

Fig. 2. Greedy randomized constructor for CCPM.

in line 17, X' is set as the new current solution. The process returns to line 5 and repeats until no agent can be re-routed according to this greedy method and improve the current solution or until some maximum number of iterations MaxIterLS are completed.

2.3 Complexity of the heuristic

The following theorems address the computational complexity of the proposed algorithm.

```
procedure LocalSearch(X, MaxIterLS)
1    X' ← ∅
2    t ← 1
3    LastImprove ← 1
4    i ← 2
5    iter ← 0
6    while i ≠ LastImprove and iter < MaxIterLS do
7        Remove current path from sᵢ to dᵢ for agent uᵢ
8        while aᵢ ≠ dᵢ do
9            if dist(aᵢ, dᵢ) = Lᵢ − ∑ₑ∈Pᵢ dist(e) then
10               Route user uᵢ using its shortest path Pᵢ(aᵢ)
11           else
12               BestMove ← {(aᵢ, lw, uᵢ) | g((aᵢ, lw, uᵢ)) > g((aᵢ, lⱼ, uᵢ)), ∀ (aᵢ, lⱼ, uᵢ)}
13           end
14           Add BestMove into the new path for uᵢ in solution X'
15           t ← t + 1
16       end
17       if f(X') > f(X) then
18           X ← X'
19           LastImprove ← i
20       end
21       i ← i + 1 mod k
22       iter ← iter + 1
23   end
24   return(X')
end procedure LocalSearch
```

Fig. 3. Local search for CCPM.

Theorem 3 *The construction phase finds a feasible solution for the* CCPM *in* $\mathcal{O}(Tmu^2)$ *time, where T is the time horizon, $u = |U|$, and $m = |E|$.*

Proof: Notice that the **while** loop in lines 8–31 will require $T(|U| − 1)$ iterations to complete. Likewise, the loop in lines 11–22 requires $|U|$ iterations. Within the loop, the most time consuming step is the construction of the shortest path. However, this can be done using a breadth-first search in $\mathcal{O}(m)$ time [1]. Thus, we have the result. ∎

Theorem 4 *The time complexity of the local search phase is* $\mathcal{O}(kTu^2m)$, *where T is the time horizon, $u = |U|$, $m = |E|$, and $k = $ MaxIterLS.*

Proof: The proof is similar to Theorem 3. Notice that the **while** loop in lines 5–22 perform local improvements according the greedy re-routing scheme. Again, the most time consuming step is the construction of a shortest path which can be accomplished in $\mathcal{O}(m)$ time. Each improvement can require up to k iterations of the loop, and we have the proof. ∎

```
procedure PathRelinking(x_s, E)
1    x_g ← randSelect(y ∈ E : Δ(x_s, y) > δ)
2    f* ← max{f(x_s), f(x_g)}
3    x* ← arg max{f(x_s), f(x_g)}
4    x ← x_s
5    while Δ(x_s, x_g) ≠ ∅ do
6        m* ← arg max{f(x ⊕ m) : m ∈ Δ(x, x_g)}
7        Δ(x ⊕ m*, x_g) ← Δ(x, x_g) \ {m*}
8        x ← x ⊕ m*
9        if f(x) > f* then
10           f* ← f(x)
11           x* ← x
12       end
13   end
14   return x*
end procedure PathRelinking
```

Fig. 4. A path-relinking subroutine adapted from [31].

Corollary 2 *The overall time complexity of the proposed GRASP algorithm is $\mathcal{O}(lTu^2m(k+1))$, where where T is the time horizon, $u = |U|$, $m = |E(G)|$, $k = $ MaxIterLS, and $l = $ MaxIter is the overall number of GRASP iterations.*

Proof: The proof is immediate from Theorem 3 and Theorem 4. ∎

3 Path-relinking

First introduced by Glover in [19], path-relinking (PR) was used as an enhancement for tabu search heuristics. PR was first combined with GRASP by Laguna and Martí [23]. When applied to GRASP, path-relinking introduces a memory to the heuristic which usually results in improvements in solution quality. This is because in the standard GRASP framework, the multi-start nature of the heuristic does not include any long-term memory mechanism for saving traits of good solutions generated by the algorithm. Path-relinking allows GRASP to remember these traits and favor them in successive iterations. GRASP with path-relinking has been successfully applied to problems such as MAXIMUM CUT [16], QUADRATIC ASSIGNMENT [28], TDMA MESSAGE SCHEDULING [12], and originally for LINE CROSSING MINIMIZATION [23]. For a survey of GRASP with path-relinking, the reader is referred to [31].

Path-relinking works by maintaining a set of elite solutions $\mathcal{E}$, known as *guides* and examines point-to-point trajectories between a guiding solution and an incumbent solution in search of an optimum. Pseudo-code for a generic path-relinking procedure is provided in Figure 4. To perform path-relinking, we begin with a guiding solution $x_g \in \mathcal{E}$, and an initial starting solution x_s.

The guiding solution x_g is selected at random from the pool of elite solutions $\mathcal{E}$, so long as the symmetric difference $\Delta(x_s, x_g)$ between the two solutions x_s and x_g is sufficiently large. The symmetric difference is defined as the set of pairwise exchanges needed to transform x_s in to x_g. Recall that all solutions in $\mathcal{E}$ are local optima, and we are trying to discover solutions which are not located in the neighborhoods of x_s or x_g. Therefore this constraint prevents us from applying path-relinking to solutions which are too similar to each other, and would not likely yield an improved solution [15].

At each step, the procedure examines all moves $m \in \Delta(x, x_g)$, and greedily selects the move which results in the maximum increase in the objective of the current solution. This occurs in line 6 of the pseudo-code in which the move m^* is selected as the move which maximizes $f(x \oplus m)$, where $x \oplus m$ is the solution which results from incorporating m in to x. In line 7, the symmetric difference is updated, and if necessary the best solution is updated in lines 9-12. The procedure ends when $\Delta(x, x_g) = \emptyset$, i.e. when $x = x_g$ [31].

Path-relinking can be applied to a pure GRASP in a straightforward manner, which can be visualized in the pseudo-code of Figure 5. First, the set of elite solutions $\mathcal{E}$ is initialized to the empty set in line 2 and is built by including the solutions from the first MaxElite iterations. After a standard GRASP iteration of greedy randomized construction and local search produces a local optimal solution X, the PathRelinking procedure is called on line 7. For the CCPM, the elements in the symmetric difference are the agent paths which differ between the initial and guiding solutions. The value of m^* from Figure 4 is the path for an agent in the symmetric difference which results in the maximum increase in the total number of communications between the agents. In line 8, a function UpdateElite is called in which the elite pool is possibly updated. The solution returned from path-relinking is included in the elite pool if it is better than the best solution in $\mathcal{E}$ or if it is worse than the best but better than the worst and is sufficiently different from all elite solutions [31]. Finally, the optimal solution is updated in lines 12 to 14 if necessary.

4 Computational experiments

The proposed procedure was implemented in the C programming language and compiled using the Microsoft® Visual C++ 6.0. It was tested on a PC equipped with a 1800MHz Intel® Pentium® 4 processor and 256 megabytes of RAM operating under the Microsoft® Windows® 2000 Professional environment.

Both the pure GRASP and the GRASP with path-relinking were tested on a set of 60 random unit graphs with varying densities each having 50, 75, and 100 nodes. The radius of communication varies from 1 to 5 units (miles) in unit increments. We tested each case with three sets of mobile agents to achieve better comparisons and model real-world scenarios. Thus, in total 900

```
procedure GRASP+PR(MaxIter, RandomSeed)
1     X* ← ∅
2     E ← ∅
3     for i = 1 to MaxIter do
4         X ← ConstructionSolution(G, g, X)
5         X ← LocalSearch(X, MaxIterLS)
6         if |E| = MaxElite then
7             X ← PathRelinking(X, ℰ)
8             UpdateElite(X, ℰ)
9         else
10            ℰ ← ℰ ∪ {X}
11        end
12        if f(X) ≥ f(X*) then
13            X* ← X
14        end
15    end
16    return X*
end procedure GRASP+PR
```

Fig. 5. GRASP with path-relinking for maximization [15].

test cases were examined. The graphs were created by a generator used by Butenko et al. [11] on the TDMA MESSAGE SCHEDULING PROBLEM.

Since any instance of the CCPM is composed of several parameters, i.e. the number of mobile agents, their respective source and destination nodes, the radius of communication, and the maximum time horizon, each of which impacts the optimal solution for the instance, we will provide our numerical results in several sets of tables. First, we report solutions for several representative instances and provide all input parameters in order to establish an inference base for the overall experiment. Then we will summarize the overall results by providing the average solutions for each problem set[1].

In Table 1, we report solutions for three different instances on 50 node graphs. The **Source** vector and **Destination** vector provide the respective (s_i, d_i) pair for each agent respectively. The specific values of s_i were randomly selected from the first 20% of the nodes of the graph. Likewise, the d_i values were chosen randomly from the last 20% of nodes. This method of selection is preferred over a completely randomized design because in real-world situations such as a combat scenario, the available entry and exit points from a battle space are likely to be limited. However, using a random selection from the available subset of nodes allows for more thorough testing and helps avoid unintentional biases.

---

[1]Complete results for all experiments are at http://www.research.att.com/~mgcr/doc/gccpmdata.pdf.

Table 1. Three instances with different sets of agents on 50 node graphs are given. The value in the UBound column was found using Corollary 1.

Instance: 50r30i1 Nodes: 50 Agents: 10 MaxTime: 10			
Source: [6 10 10 3 5 7 4 2 10 6]			
Destination: [49 47 44 48 46 40 48 42 47 47]			
Radius	GRASP	GRASP+PR	UBound
1	291	303	450
2	365	373	450
3	412	423	450
4	443	443	450
5	449	449	450

(a)

Instance: 50r30i1 Nodes: 50 Agents: 15 MaxTime: 10			
Source: [10 9 8 9 6 8 1 7 2 5 5 5 2 1 1]			
Destination: [49 47 44 48 46 40 48 42 47 47]			
Radius	GRASP	GRASP+PR	UBound
1	756	757	1050
2	881	909	1050
3	963	972	1050
4	1029	1029	1050
5	1050	1050	1050

(b)

Instance: 50r40i4 Nodes: 50 Agents: 25 MaxTime: 10			
Source: [8 9 8 4 5 4 4 6 2 7 4 2 1 7 5 8 9 3 8 11 8 7 5 6 8]			
Destination: [49 48 44 48 46 42 49 40 48 49 45 46 49 45 48 44 42 41 48 43 40 49 45 49 43]			
Radius	GRASP	GRASP+PR	UBound
1	2613	2653	3000
2	2896	2918	3000
3	3000	3000	3000
4	3000	3000	3000
5	3000	3000	3000

(c)

The column MaxTime is the maximum time horizon T. Recall that all agents must reach their destination node by this time. The GRASP column provides the solution from GRASP after 1000 iterations and UBound is the upper bound on the solution value and was calculated by the equation in Corollary 1. Notice that as the radius value increases, the number of connections between the agents tends to converge to the value of the upper bound. Recall that the upper bound value from Corollary 1 is not an upper bound on the optimal solution for the given graph per se; it is an upper bound on the solution for the given time horizon and number of agents. Thus, the more dense the graph, the tighter the bound.

Table 2. Three instances with different sets of agents on 75 node graphs are given. The value in the UBound column was found using Corollary 1.

Instance: 75r30i2 Nodes: 75 Agents: 10 MaxTime: 15			
Source: [7 6 13 3 10 13 15 6 6 2]			
Destination: [68 68 71 73 68 68 73 70 74 62]			
Radius	GRASP	GRASP+PR	UBound
1	571	575	675
2	614	621	675
3	658	658	675
4	670	670	675
5	675	675	675

(a)

Instance: 75r40i4 Nodes: 75 Agents: 20 MaxTime: 15			
Source: [11 5 7 15 3 8 9 4 6 13 14 3 8 3 10 14 11 15 9 3]			
Destination: [63 66 69 61 62 68 62 67 68 66 62 60 61 66 63 73 72 64 71 71]			
Radius	GRASP	GRASP+PR	UBound
1	2535	2554	2850
2	2746	2758	2850
3	2842	2842	2850
4	2850	2850	2850
5	2850	2850	2850

(b)

Instance: 75r30i1 Nodes: 75 Agents: 30 MaxTime: 15			
Source: [14 15 2 8 10 4 13 3 4 5 12 6 4 9 2 3 15 8 5 12 9 3 7 7 11 3 4 1 15 4]			
Destination: [68 69 63 62 65 72 62 62 67 71 65 69 63 64 60 64 60 60 66 64 74 63 73 64 64 63 65 65 60 63]			
Radius	GRASP	GRASP+PR	UBound
1	4721	4870	6525
2	6002	6012	6525
3	6265	6285	6525
4	6497	6497	6525
5	6525	6525	6525

(c)

Table 2 presents the specific parameters and related solutions for three instances of the CCPM on 75 node graphs. On these networks, the number of agents varied from 10 to 30, and the maximum time horizon was 15. Again, we see that as the communication radius increases the solutions tend to the upper bound values. Similar results for three graphs having 100 nodes are provided in Table 3. For the 100 node instances, the number of agents varied from 15 to 35 and the maximum travel time was 20 units. The results for these instances also indicate that the heuristic is robust and able to provide excellent solutions for large instances.

Table 3. A 100 node instance with solutions with radius varying from 1 to 5 units. The value in UBound was found using Corollary 1.

Instance: 100r30i2 Nodes: 100 Agents: 15 MaxTime: 20			
Source: [9 19 10 18 13 18 12 18 15 8 6 6 20 18 1]			
Destination: [84 88 83 84 96 96 81 95 83 82 93 80 90 85 81]			
Radius	GRASP	GRASP+PR	UBound
1	1819	1821	2100
2	1960	1974	2100
3	2065	2067	2100
4	2100	2100	2100
5	2100	2100	2100

(a)

Instance: 100r30i1 Nodes: 100 Agents: 25 MaxTime: 20			
Source: [17 6 9 19 9 12 2 15 7 8 1 2 8 6 3 13 16 17 13 13 17 19 2 5 21]			
Destination: [81 89 84 82 88 99 93 89 93 97 84 96 96 91 90 86 98 86 81 89 82 89 81 80 99]			
Radius	GRASP	GRASP+PR	UBound
1	5183	5186	6000
2	5577	5647	6000
3	5898	5909	6000
4	5992	5992	6000
5	6000	6000	6000

(b)

Instance: 100r30i2 Nodes: 100 Agents: 35 MaxTime: 20			
Source: [3 5 11 2 14 4 4 5 10 12 14 13 17 4 17 8 16 7 15 7 15 13 12 9 5 6 19 18 3 16 18 19 10 5 2]			
Destination: [89 95 89 91 84 99 88 91 92 82 98 84 85 89 85 98 92 80 81 85 98 94 ...			
... 82 89 90 96 91 92 90 96 96 96 99 81 82]			
Radius	GRASP	GRASP+PR	UBound
1	10222	10255	11900
2	11108	11224	11900
3	11660	11704	11900
4	11842	11845	11900
5	11900	11900	11900

(c)

Tables 4, 5, and 6 show the evolution of the average solution values as the communication range increases for the 50, 75, and 100 node graphs, respectively. Notice once more that as the communication range increases, the average solution converges to the value of the upper bound given by Corol-

Table 4. Average solution values for GRASP and GRASP with path-relinking on 50 node graphs.

Nodes	Agents	Radius	GRASP	GRASP+PR	Bound
50	10	1	347	352.21	450
50	10	2	404.58	407.58	450
50	10	3	428.32	429.47	450
50	10	4	437.84	438.53	450
50	10	5	444.37	444.58	450
50	15	1	813.11	817.32	1050
50	15	2	937.74	945.47	1050
50	15	3	1001.11	1003.58	1050
50	15	4	1025.37	1026.21	1050
50	15	5	1037.16	1037.53	1050
50	25	1	2272.79	2315.58	3000
50	25	2	2686.26	2704.53	3000
50	25	3	2850.84	2861.95	3000
50	25	4	2924.05	2927.68	3000
50	25	5	2959	2959.26	3000
Average Comp Time (s)			2.89	4.29	−

Table 5. Average solution values for GRASP and GRASP with path-relinking on 75 node graphs.

Nodes	Agents	Radius	GRASP	GRASP+PR	Bound
75	10	1	574.95	577.42	675
75	10	2	629.42	631.37	675
75	10	3	653.53	654.63	675
75	10	4	665.42	665.89	675
75	10	5	669.47	669.84	675
75	20	1	2288	2319.63	2850
75	20	2	2639.37	2651.5	2850
75	20	3	2756.69	2762	2850
75	20	4	2805.53	2807.68	2850
75	20	5	2827.42	2828.42	2850
75	30	1	5349.84	5391.26	6525
75	30	2	6037.47	6064	6525
75	30	3	6310.90	6332.37	6525
75	30	4	6422.11	6430.80	6525
75	30	5	6472.42	6478.84	6525
Average Comp Time (s)			6.16	7.43	−

lary 1. In these tables we also report the average computing time required by both the pure GRASP and the GRASP+PR to find their best solutions within the specified number of iterations. For all of the experiments, the GRASP+PR found solutions at least as good as the pure GRASP, finding superior solutions for 45% of the instances tested.

Table 6. Average solution values for GRASP and GRASP with path-relinking on 100 node graphs.

Nodes	Agents	Radius	GRASP	GRASP+PR	Bound
100	15	1	1838.25	1840.45	2100
100	15	2	1996.75	2003.15	2100
100	15	3	2061.9	2064.7	2100
100	15	4	2083.1	2084.4	2100
100	15	5	2093.95	2094.05	2100
100	25	1	4979.1	5019.2	6000
100	25	2	5655.3	5674.35	6000
100	25	3	5869.35	5876.9	6000
100	25	4	5940.65	5944.7	6000
100	25	5	5978.2	5979.2	6000
100	35	1	9947.45	9997.15	11900
100	35	2	11254.55	11280	11900
100	35	3	11636.85	11664.5	11900
100	35	4	11787.9	11793	11900
100	35	5	11859.1	11860.35	11900
Average Comp Time (s)			5.17	8.05	−

In Figures 6, 7, and 8, we provide plots of the average objective function value versus communication range found using GRASP with path-relinking.

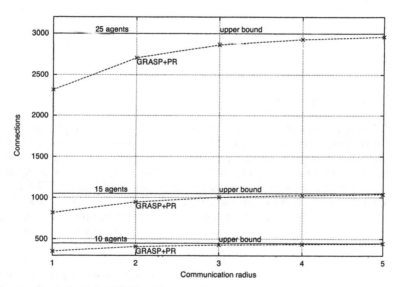

Fig. 6. Evolution of GRASP+PR solution values on 50 node graphs as the communication radius increases from 1 to 5 units.

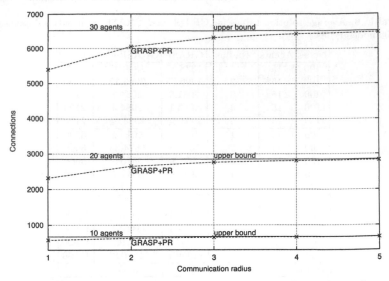

Fig. 7. Evolution of GRASP+PR solution values on 75 node graphs as the communication radius increases from 1 to 5 units.

The upper bound values for each case as computed by Corollary 1 are also plotted in the charts. These graphs indicate that on average, as the radius of communication increases, the objective function values tend to the upper bound values.

5 Concluding remarks

In this chapter, we extended the work of Commander et al. [13] by providing a greedy randomized adaptive search procedure (GRASP) for the COOPERA-TIVE COMMUNICATION PROBLEM ON MOBILE AD HOC NETWORKS. This is a new problem in cooperative control [21], which has many applications in path-planning for multiple autonomous agents. Since this problem is NP-hard, the need for efficient heuristics which quickly provide high-quality solutions arises. The proposed GRASP is one such heuristic. Extensive computational experiments indicate that this method is superior to the shortest path protocol and one-pass local search heuristic presented in [13]. Furthermore, the method finds near optimal solutions for the 900 cases tested and converges to the derived upper bound as the communication range increases. Future research efforts include developing an algorithm to find optimal solutions for small CCPM instances such as a branch and cut or a column generation method [22]. This will help evaluate the performance of heuristics whose solutions for

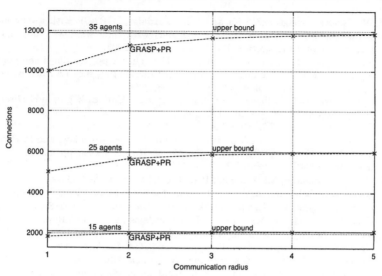

Fig. 8. Evolution of GRASP+PR solution values on 100 node graphs as the communication radius increases from 1 to 5 units.

certain instances are not verifiably optimal, i.e. the solutions that are not at the computed upper bound.

References

1. R.K. Ahuja, T.L. Magnanti, and J.B. Orlin, *Network Flows: Theory, Algorithms, and Applications*, Prentice-Hall, 1993.
2. R. Aiex, S. Binato, and M.G.C. Resende, Parallel GRASP with path-relinking for job shop scheduling, *Parallel Computing*, Vol. 29, pp. 393–430, 2003.
3. B.S. Baker, Approximation algorithms for np-complete problems on planar graphs, *Journal of the ACM*, Vol. 41(1), pp. 153–180, 1994.
4. S. Butenko, X. Cheng, D.-Z. Du and P.M. Pardalos, On the construction of virtual backbone for ad hoc wireless network, in S. Butenko, R. Murphey and P.M. Pardalos (eds), *Cooperative Control: Models, Applications and Algorithms*, pp. 43–54, Kluwer Academic Publishers, 2002.
5. S.I. Butenko, X. Cheng, C.A.S. Oliveira, and P.M. Pardalos, A new algorithm for connected dominating sets in ad hoc networks, in S. Butenko, R. Murphey and P. Pardalos (eds), *Recent Developments in Cooperative Control and Optimization*, pp. 61–73, Kluwer Academic Publishers, 2003.
6. S.I. Butenko, R.A. Murphey, and P.M. Pardalos (eds), *Cooperative Control: Models, Applications, and Algorithms*, Springer, 2003.
7. S.I. Butenko, R.A. Murphey and P.M. Pardalos (eds), *Recent Developments in Cooperative Control and Optimization*, Springer, 2004.

8. W. Chaovalitwongse, D. Kim and P.M. Pardalos, GRASP with a new local search scheme for vehicle routing problems with time windows, *Journal of Combinatorial Optimization*, Vol. 7, pp. 179–207, 2003.

9. X. Cheng, X. Huang, D. Li, W. Wu and D.Z. Du, A polynomial-time approximation scheme for the minimum-connected dominating set in ad hoc wireless networks, *Networks*, Vol. 42(4), pp. 202–208, 2003.

10. B.N. Clark, C.J. Colbourn, and D.S. Johnson, Unit disk graphs, *Discrete Mathematics*, Vol. 86, pp. 165–177, 1990.

11. C.W. Commander, S.I. Butenko and P.M. Pardalos, On the performance of heuristics for broadcast scheduling, in D. Grundel, R. Murphey, and P. Pardalos (eds), *Theory and Algorithms for Cooperative Systems*, pp. 63–80, World Scientific, 2004.

12. C.W. Commander, S.I. Butenko, P.M. Pardalos and C.A.S. Oliveira, Reactive grasp with path relinking for the broadcast scheduling problem, in *Proceedings of the 40th Annual International Telemetry Conference*, pp. 792–800, 2004.

13. C.W. Commander, C.A.S. Oliveira, P.M. Pardalos and M.G.C. Resende, A one-pass heuristic for cooperative communication in mobile ad hoc networks, in D.A. Grundel, R.A. Murphey, P.M. Prokopyev and P.M. Pardalos (eds), *Advances in Cooperative Control and Optimization*, World Scientific, 2006.

14. T.A. Feo and M.G.C. Resende, Greedy randomized adaptive search procedures, *Journal of Global Optimization*, Vol. 6, pp. 109–133, 1995.

15. P. Festa, P.M. Pardalos, L.S. Pitsoulis and M.G.C. Resende, GRASP with path-relinking for the weighted MAXSAT problem, *ACM Journal of Experimental Algorithmics*, accepted, 2006.

16. P. Festa, P.M. Pardalos, M.G.C. Resende and C.C. Ribeiro, Randomized heuristics for the MAX-CUT problem, *Optimization Methods and Software*, Vol. 7, pp. 1033–1058, 2002.

17. P. Festa and M.G.C. Resende, GRASP: An annotated bibliography, in C. Ribeiro and P.Hansen (eds), *Essays and surveys in metaheuristics*, pp. 325–367, Kluwer Academic Publishers, 2002.

18. M.R. Garey and D.S. Johnson, *Computers and Intractability: A Guide to the Theory of NP-Completeness*, W.H. Freeman and Company, 1979.

19. F. Glover, Tabu search and adaptive memory programming – advances, applications, and challenges, in R.S. Barr, R.V. Helgason, and J.L. Kenngington (eds), *Interfaces in Computer Science and Operations Research*, pp. 1–75, Kluwer Academic Publishers, 1996.

20. F. Glover, M. Laguna, and R. Martí, Fundamentals of scatter search and path-relinking, *Control Cybernetics*, Vol. 39, pp. 653–684, 2000.

21. D.A. Grundel, R.A. Murphey and P.M. Pardalos (eds), *Theory and Algorithms for Cooperative Systems*, World Scientific, 2004.

22. R. Horst, P.M. Pardalos and N.V. Thoai, *Introduction to Global Optimization, Nonconvex Optimization and its Applications*, Vol. 3, Kluwer Academic Publishers, 1995.

23. M. Laguna and R. Martí, Grasp with path relinking for 2-layer straight line crossing minimization, *INFORMS Journal on Computing*, Vol. 11, pp. 44–52, 1999.

24. Y. Li, P.M. Pardalos and M.G.C. Resende, A greedy randomized adaptive search procedure for the quadratic assignment problem, in P.M. Pardalos and

H. Wolkowicz (eds), *Quadratic Assignment and Related Problems, DIMACS Series on DIscrete Mathematics and Theoretical Computer Science*, Vol. 16, pp. 237–261, American Mathematical Society, 1994.

25. R.A. Murphey and P.M. Pardalos (eds), *Cooperative Control and Optimization*, Springer, 2002.

26. C.A.S. Oliveira and P.M. Pardalos, Ad hoc networks: Optimization problems and solution methods, in D.-Z. Du, M. Cheng, and Y. Li (eds), *Combinatorial Optimization in Communication Networks*, Kluwer, 2005.

27. C.A.S. Oliveira and P.M. Pardalos, An optimization approach for cooperative communication in ad hoc networks, submitted for publication, 2005.

28. C.A.S. Oliveira, P.M. Pardalos, and M.G.C. Resende, GRASP with path-relinking for the QAP, in *5th Metaheuristics International Conference*, pp. 57.1–57.6, 2003.

29. M.G.C. Resende and T.A. Feo, A GRASP for satisfiability, in D.S. Johnson and M.A. Trick (eds), *Cliques, Coloring, and Satisfiability: Second DIMACS Implementation Challenges*, Vol. 26, pp. 499–520, American Mathematical Society, 1996.

30. M.G.C. Resende and P.M. Pardalos, *Handbook of Optimization in Telecommunications*, Springer, 2006.

31. M.G.C. Resende and C.C. Ribeiro, GRASP and path-relinking: Recent advances and applications, in T. Ibaraki, K. Nonobe and M. Yagiura (eds), *Metaheuristics: Progress as Real Problem Solvers*, pp. 29–63, Springer, 2005.

11 Optimal Control of an ATR Module - Equipped MAV/Human Operator Team

Meir Pachter, Phillip R. Chandler and Dharba Swaroop

Summary. A distributed intelligence, surveillance and reconnaissance (ISR) scenario is considered where a team consisting of a flight of micro air vehicles (MAVs) and an operator is tasked with viewing and recognizing multiple targets in a battle space with many false targets. The entire search and classification chain, which includes a human operator for classification aiding, is modelled. The potential benefits accruing from mixed initiative operations, that is, cooperative automatic target recognition (ATR) and human target classification, are quantified. Analytical methods for the optimal control of an ISR team consisting of multiple distributed MAVs and an operator are developed.

1 Introduction

In support of the current COUNTER program [1], a distributed, micro air vehicles (MAVs) based, intelligence, surveillance and reconnaissance (ISR) scenario is considered. Previous work was reported in [2]. It is envisaged that a Small Air Vehicle (SAV) mother ship endowed with an automatic target recognition (ATR) module, and carrying m camera equipped Micro Air Vehicles (MAVs) is used. The high flying SAV mother ship is dispatched to the battle space, which is populated by a large number of clutter objects, only a small fraction of which, say p, $0 < p << 1$, are objects of interest. The latter are potential targets. The SAV surveys the battle space and designates a relatively small number of objects, e.g., vehicles, of interest, say mn, following which the m MAVs are released for a closer look at the designated objects. The SAV's ATR capability is somewhat limited: the SAV can geo-locate the designated vehicles, but cannot determine their orientation. Thus, each MAV released by the SAV is programmed to fly over n designated objects of interest so that all the SAV designated mn objects of interest will be inspected. Since the SAV could not determine the orientation of the designated objects of interest, the MAVs' viewing angles are random.

The performance of the SAV's ATR module is specified using a confusion matrix. The performance of the SAV's ATR module is limited because it has access to a top view only, whereas a target is characterized by a certain feature which is visible from a side view and from certain aspect angles only.

Hence, the diagonal elements of the confusion matrix which characterizes the performance of the ATR module on board the SAV are significantly less than 1. Therefore, only a fraction of the n vehicles of interest assigned to a MAV are a targets.

At the same time, the imagery recorded by a low flying MAV's camera is of good quality because the objects of interest are inspected at close range and low depression angles such that good side views are provided. Objects of interest are classified by the MAV's ATR module according to the presence of a certain feature. Targets carry this feature, which is only visible from the side and from a rather narrow, say $\bar{\theta}$, range of aspect angles - for example, $\bar{\theta} = 90$ degrees. When the object's pose is such that the said feature is visible, the MAV's ATR module will correctly report a target. Unfortunately, the MAV's viewing angle is random. Thus, in those cases where the characteristic feature is not directly in the MAV's field of view, the ATR module on board the MAV does not declare a target and instead, as the MAV flies over an object, the image of the inspected object/vehicle is transmitted to a human operator for classification. Even though the viewing angle was not ideal, a human uses subtle clues for discerning certain features of interest. Therefore, the classification performance is significantly enhanced by including a human operator in the classification loop. While the pattern recognition ability of the human operator tasked with analyzing the images sent by the MAVs as they sequentially inspect their assigned n objects of interest is not perfect and a confusion matrix which characterizes his performance is at work, its diagonal elements are close to 1. Hence, an *ATR aiding* scheme using a human operator is implemented: we then refer to a mixed initiative operation. This will enhance the performance of the mission critical classification task. However, even the human operator is not a perfect classifier, that is, his confusion matrix is not the unity matrix. Furthermore, the human operator introduces a delay, and the classification performance of the human operator deteriorates when his workload increases - for example, when more than four images per minute are received from the m MAVs.

An object of interest is first inspected by the MAV's ATR module. In those instances where, due to the object's pose angle, the MAV's ATR module was not in a position to discern the feature carried by a target and consequently report a target, there is still information gained about the inspected object. The information gained about the inspected object is determined by the targets' $\bar{\theta}$ parameter and by the MAV ATR module's confusion matrix parameters. Now, once the human operator's classification is in, and to further increase the information gained about the object, the MAV's schedule can be interrupted and instead of continuing on to the next object of interest, the MAV can be turned around and directed to revisit the object and obtain an additional view from a different aspect angle. There obviously is a cost associated with this course of action, in particular since an MAV's endurance, say k, is limited.

In analyzing the value of a second view, two alternative scenarios are investigated: a) From the imagery received the operator is able to estimate the object's pose and consequently the MAV is redirected to approach the object from a favorable viewing angle, and, b) the MAV is vectored to approach the object s.t. the second pass is flown on a heading which is orthogonal to the original ground track.

In scenario a), if the critical feature is there, the MAV's ATR module will classify the inspected object a target, which it is. If the feature is not visible, the object will be classified as a non target. In scenario b) and when the MAV's ATR module does not discern the said feature, the object's image will be forwarded to the human operator for further inspection and classification. In this chapter the entire search chain is modelled and the potential benefits accruing from cooperative target classification performed by the MAV/operator team are addressed. The following data are required.

1. Prior information on the ratio of Targets (Ts) to clutter objects, or False Targets (FTs): $\frac{p}{1-p}$.
2. The confusion matrix parameters of the SAV's ATR module.
3. The aspect angle range $\bar{\theta}$ wherefrom the feature associated with a target vehicle is clearly visible.
4. The operator's confusion matrix parameters. The latter might be specified as a function of his workload.
5. Operator delay statistics: The operator's delay $\tau \geq 0$ is a random variable and its p.d.f. is $f(\tau)$.

Each MAV must inspect n vehicles of interest, while his endurance is less than or equal to k.

The operator handles m data streams in parallel. Furthermore, the operator's delay, and, possibly, also his confusion matrix parameters, are a function of his work load.

Finally, one is interested in maximizing the average information [3] gained about the mn objects of interest. One would like to maximize a functional of the form

$$I(u_1, ..., u_n; k) = E(\sum_{i=1}^{mn} \log_2(\frac{P(O_i = T \mid Z_i)}{p}) + \log_2(\frac{P(O_i = FT \mid Z_i)}{1-p})) \ [bits]$$

where Z_i is the measurement taken on O_i and the probability $p = P(O_i = T)$ is the prior information on the object of interest O_i. Note: $\log_2(\cdot) = \frac{1}{\ln 2} \ln(\cdot) = \frac{1}{0.693} \ln(\cdot)$. When the basis of the log is e, the information is measured in *nats* (natural units of information).

A binary control variable $u \in \{0, 1\}$ is used. It entails the decision, upon receipt of the operator's classification, on when a MAV should be turned around to take a second look at an object of interest: when $u = 0$ the MAV continues on its path and when $u = 1$, the MAV is commanded to turn around and revisit the object. Whenever an MAV is turned around and directed to revisit an object of interest, the information gained increases, however a penalty is incurred. The penalty is proportional to the operator's delay.

The expected information gain I, given the MAV's available endurance, can be calculated ahead of time, before the inspection sequence commences. This affords the optimal allocation of MAV resources to geographically distributed objects of interest. This task will be performed on board the SAV. Furthermore, at the end of the day, at the completion of the sequential inspection of the n objects, one can calculate the posterior probability of each of the n inspected objects being a target, or, by all means not less important, the probability that the object of interest is a false target.

The objectives of our chapter are:

1. To develop a comprehensive model of distributed ISR operations using MAVs, and
2. To develop an analytical tool for evaluating the effectiveness of candidate Concepts of Operations (ConOps). Special attention is given to mixed intiative operations, namely operator-aided ATR.

The analytical approach is emphasized, with a view to capture and identify the critical MAV, SAV and target environment parameters that impact the effectiveness of the distributed ISR system.

2 Notation

$$
\begin{array}{rl}
d & \equiv \text{Threshold} \\
i & \equiv \text{SDP stage} \\
I & \equiv \text{Information } [bits] \\
k & \equiv \text{Endurance } [sec] \\
p & \equiv \text{Probability} \\
P & \equiv \text{Probability} \\
q & \equiv \text{Probability} \\
r_i & \equiv \text{Running cost } [bits] \\
t & \equiv \text{Time } [sec] \\
u & \equiv \text{Control/Decision variable} \\
Z & \equiv \text{Observation} \\
\alpha & \equiv \text{T/FT ratio} \\
\tau_i & \equiv \text{Operator delay } [sec] \\
f(\tau_i) & \equiv \text{P.D.F. of operator delay} \\
\tau & \equiv \text{Fixed delay } [sec]
\end{array}
$$

3 Confusion Matrix

The *confusion matrix* represents the probability of both correct and incorrect target reports. The concept of a confusion matrix has its roots in detection theory where the Receiver Operating Characteristic (ROC) plays a similar role.

The confusion matrix parameters are P_{TR}, the probability of target report when a target is encountered, and P_{FTR}, which is the probability of a false target report when a false target is encountered.

True/Rep.	T	FT
T	P_{TR}	$1 - P_{FTR}$
FT	$1 - P_{TR}$	P_{FTR}

Obviously, the sum of the entries in each column is 1. Ideally, one would like

$$P_{TR} = 1 \text{ and } P_{FTR} = 1 ,$$

i.e., one would like the confusion matrix to be the unity matrix, or, at least, one would like the confusion matrix to be diagonally dominant. Unfortunately, Autonomous Target Recognition (ATR) is far from achieving this lofty goal and the parameters $0 \leq P_{TR} \leq 1$ and $0 \leq P_{FTR} \leq 1$ of the confusion matrix play a crucial role in determining an autonomous system's effectiveness.

Example 1 (Target detection in ELINT): $P_{TR} = 0.95$ and $P_{FTR} = 1 - 10^{-4}$. Thus, the confusion matrix is

True/Rep.	T	FT
T	0.95	10^{-4}
FT	0.05	$1 - 10^{-4}$

Example 2 (Mammography screening): $P_{TR} = 0.8$ and $P_{FTR} = 0.1$.

True/Rep.	T	FT
T	0.8	0.9
FT	0.2	0.1

If different types of targets are of interest, the confusion matrix is of a higher dimension. Similarly, one could consider FTs and also Decoys (D). In this chapter we mainly confine our attention to targets and clutter vehicles/false targets and to relatively simple 2×2 confusion matrices, as illustrated above. However, one could/should also consider 3×2 confusion matrices where one allows for an inconclusive Ts/FTs classification. Then P_{TDK} is the probability of reporting "Don't Know" (DK) ($\equiv$ "can't tell") when a target is encountered, and P_{FTDK} is the probability of reporting "Don't Know" when a false target is encountered. The confusion matrix is

True/Rep.	T	FT
T	P_{TR}	$1 - P_{FTR} - P_{FTDK}$
FT	$1 - P_{TR} - P_{TDK}$	P_{FTR}
DK	P_{TDK}	P_{FTDK}

4 Sequential Inspection

The following sequential inspection sub-problem, which is interesting in its own right, and must be solved first, is fully addressed in this chapter. A MAV starting out at time $t = 0$ and tasked with sequentially inspecting n objects O_i, $i = 1, ..., n$ is considered.

Assume, without loss of generality, that it takes an MAV one unit of time to cover the distance between two adjacent objects of interest. Thus, the planned inspection time of object O_i is $t_i = i$, $i = 1, ..., n$. Also, the prior information $P(O_i = T) = p$ is available. Thus, from the outset, the number of Ts is governed by the binomial distribution:

$$P(\#T = l) = \binom{l}{n} p^l (1 - p)^{n-l} , \quad l = 0, 1, ..., n$$

The outcome of a classification performed by the human operator is governed by a specified confusion matrix parameterized by P_{TR_O} and P_{FTR_O}; the subscript O designates the presence of a human operator who *aids* the ATR module in performing the critical pattern recognition/classification task. The human operator - aided pattern recognition task causes the diagonal confusion matrix parameters to be quite close to 1, which is beneficial. At the same time, the human operator introduces a delay: the outcome of the inspection of O_i becomes available at time $i + \tau_i$. The operator's delay τ_i is a random variable with a known probability density distribution function $f(\tau_i)$, $0 \le \tau_i < 1$. The assumption $\tau_i < 1$ is not restrictive, considering an MAV's endurance of ca. 30' and the fact that the number of vehicles to be inspected $n \le 10$.

The result of the operator's classification of object i is a determination by the operator "$Z_i = T$" or "$Z_i = FT$". The inspection's outcome "$Z_i = T$" is taken to mean that the feature carried by a target has been discerned by the operator. When the result of the inspection of object i is "$Z_i = FT$", the feature characterizing a target was not discerned by the operator. In a more elaborate model one could allow for the contingency that the operator is not able to make a determination, that is, a measurement $Z_i = DK$ is possible, where DK stands for "Don't Know", and a 3×2 confusion matrix is then specified. In this chapter, the operator's performance is exclusively specified by a 2×2 confusion matrix.

If, after the first look, the MAV's ATR module has not declared the inspected object a target and the image has been forwarded to the human operator, one then waits for the human operator assisted ATR classification, following which a decision is taken. Specifically, the MAV's schedule can be altered and instead of continuing on to the next object of interest, O_{i+1}, the MAV can be turned around and directed to revisit object O_i and obtain an additional view of O_i, possibly from a more favorable aspect angle. The objective is to increase the information about the inspected object. Thus a binary decision variable is used: $u_i = 0$ stands for "continue to object O_{i+1}" and $u_i = 1$ signals a decision to turn the MAV around.

The endurance cost associated with the decision to turn the MAV around is

$$c_i = 2\tau_i + \tau$$

where $0 \leq \tau$ is a fixed cost. When the MAV is allowed to continue to object O_{i+1} the cost

$$c_i = 0$$

Let the remaining endurance of the MAV when object O_i is reached be k_i. Since it takes an MAV one unit of time to cover the distance between two adjacent objects of interest, the following must hold.

$$n - i \leq k_i \quad \forall\, i = 1, ..., n$$

At stage i, object O_i is being inspected, $i = 1, ..., n$. If the outcome of the inspection performed by the MAV's ATR module is a target classification, the MAV continues on its path to the next object of interest on its list. If the outcome of the inspection performed by the MAV's ATR module is not a target classification and the image is forwarded to the human operator, then upon the receipt of the operator's classification, a decision is called for. The binary control variable u can either be set to $u_i = 0$, meaning "continue to object O_{i+1}", or it is set to $u_i = 1$, meaning "turn around and revisit object O_i".

Hence, the stochastic optimal control problem is posed of maximizing the average information gain payoff I subject to the endurance constraint. We have obtained a discrete-time Stochastic Dynamic Programming (SDP) problem with n stages. At stage i the binary control variable is $u_i \in \{0, 1\}$ and it entails the decision on whether the MAV continues on its path ($u_i = 0$), or it turns around ($u_i = 1$) and revisits the last object. The scalar state variable is k_i and at stage i it must satisfy the condition

$$n - i \leq k_i \,, \quad i = 1, ..., n$$

The dynamics are

$$k_{i+1} = k_i - 1 - c_i(u_i)u_i \,, \quad i = 0, ..., n - 1 \,, \quad k_1 = k_0 - 1$$

and the control

$$u_0 \equiv 0$$

The sequential inspection optimization problem is solved on board the MAV. The role of the human operator is exclusively confined to ATR aiding.

5 ATR Action of the SAV

The original mix of Ts/FTs in the battle space is assumed known and is referred to as the prior information. For example, the ratio of targets to clutter targets is 0.001, or, in a high threat zone, the ratio could be 0.01. Thus, suppose the known prior probability that an object O is a T is $P^-(O = T) = p$; obviously $P^-(O = FT) = 1 - p$. We will use the notation $P^-(T) = p$ and $P^-(FT) = 1 - p$.

In this section, a simplified model of the performance of the SAV's ATR system is developed. Suppose the object O is subject to an inspection and the measurement Z is discrete, that is, the report of the ATR system is either $Z = T$ or $Z = FT$. Such a sensor is fully characterized by its confusion matrix.

True/Rep.	T	FT
T	P_{TR}	$1 - P_{FTR}$
FT	$1 - P_{TR}$	P_{FTR}

The objects in the battle space are sequentially inspected by the SAV and are classified as Ts or FTs. Evidently, the information on each examined object changes as a result of the measurement Z performed by the SAV's ATR system, Now, the ATR system's measurement performance is specified by a confusion matrix. This allows us to calculate the posterior probability that an examined object which has been designated by the SAV's ATR system a T, is indeed a T, and, conversely, an examined object which has been designated a FT, is indeed a FT: We are interested in the posterior probabilities $P^+(T \mid Z = T)$ and $P^+(FT \mid Z = FT)$; obviously, $P^+(FT \mid Z = T) = 1 - P^+(T \mid Z = T)$ and $P^+(FT \mid Z = FT) = 1 - P^+(T \mid Z = FT)$.

Applying Bayes' rule, one obtains

$$P^+(T \mid Z = T) = P(Z = T \mid T)\frac{P^-(T)}{P(Z = T)}$$

where

$$P^-(T) = p$$

Now, the SAV ATR system's report is T with probability $P(Z = T)$, calculated as follows.

$$P(Z = T) = P(Z = T \mid T)P^-(T) + P(Z = T \mid FT)P^-(FT)$$
$$= P_{TR} \cdot p + (1 - P_{FTR})(1 - p)$$

Hence

$$P^+(T \mid Z = T) = \frac{pP_{TR}}{pP_{TR} + (1 - p)(1 - P_{FTR})}$$

Similarly,

$$P^+(T \mid Z = FT) = \frac{p(1 - P_{TR})}{p(1 - P_{TR}) + (1 - p)P_{FTR}}$$

Define

$$\alpha = \frac{p}{1 - p} \cdot \frac{P_{TR}}{1 - P_{FTR}} \qquad (1)$$

Thus

$$P^+(T \mid Z = FT) = \frac{\alpha}{1 + \alpha}$$

α is the post inspection Ts/FTs ratio in the SAV designated set of objects of interest.

In summary, the SAV provided cueing designates a set of n objects of interest, where a more favorable mix α of Ts/FTs, $\alpha >> 0.001$ or, $\alpha >> 0.01$, is established. If, for example, $P_{TR} = P_{FTR} = 0.8$ then for $p = 0.001$ $\alpha = \frac{4}{999}$ and for $p = 0.01$, $\alpha = \frac{4}{99}$ so that in the set of designated objects of interest, the SAV's ATR action brings about a fourfold increase in the concentration of Ts. A set of designated objects of interest $\{O_1, ..., O_n\}$ is then handed over to a MAV.

6 ATR Action of a MAV

Also the MAVs are equipped with an ATR system. In this section, a simplified model of the performance of the MAV's ATR system is developed. The action of a human operator who provides aiding to the MAV's ATR system will be discussed in the sequel.

Consider first the time instant prior to the MAV's inspection. Denote by $P^-(T)$ the prior probability that object of interest O_i is a target, and denote by $P^-(FT)$ the prior probability that object of interest O_i is a clutter target. We know

$$P^-(T) = \frac{\alpha}{1 + \alpha} \text{ and } P^-(FT) = \frac{1}{1 + \alpha} ,$$

where α is given by eq. (1).

1. The environment: The MAVs move in the Euclidian plane. A finite set of objects of interest O_i, $i = 1, ..., n$, is inspected. The set of n objects of interest consists of Targets (Ts), and non-targets/clutter targets, also referred to as False Targets (FTs). The objects are modelled as rectangular pucks. Some pucks are indented: a rectangle with an indented side is a T. The angular extent $\bar{\theta}$ of the indentation is known. An object is a target (T) or a clutter target (FT), however Ts and FTs look alike, except for the indentation.

2. The measurement: The MAV is programmed to overfly each of the n objects of interest and obtain a side view l_i of puck O_i, $i = 1, ..., n$. Depending on the pose angle $0 \leq \theta < 2\pi$, the following three side views of a puck are possible: A, B and C. Thus, the measurement equation is

$$l_i = F(O_i, \theta)$$

where $l_i \in L$, $L = \{A, B, C\}$ and F is a measurement function. $l_i = A$ on a set of measure 0. Hence, in the sequel the reduced set of measurements $L = \{B, C\}$ is exclusively considered and a measurement l_i taken by the MAV's ATR module is a binary variable: $l_i = B$ or $l_i = C$. Evidently, when the measurement $l_i = C$ is recorded, $O_i = T$. The measurement $l_i = B$ does not allow the ATR module to resolve the ambiguity concerning the inspected object O_i and autonomously make a determination $O_i = T$ or $O_i = FT$.

The MAV approaches object O_i from a random direction and the aspect angle θ is a uniformly distributed random variable. The probabilities of recording a B or C measurement are easily calculated:

$$P(B \mid T) = 1 - \tfrac{\bar{\theta}}{2\pi} \quad P(B \mid FT) = 1,$$
$$P(C \mid T) \;\;= \tfrac{\bar{\theta}}{2\pi} \quad P(C \mid FT) = 0$$

and if, for example, $\bar{\theta} = \tfrac{\pi}{2}$ then

$$P(B \mid T) = \tfrac{3}{4} \quad P(B \mid FT) = 1$$
$$P(C \mid T) = \tfrac{1}{4} \quad P(C \mid FT) = 0$$

Hence, the classification action of the MAV's ATR system can conveniently be presented in the form of a confusion matrix

True/Rep.	O=T	O=FT
C	0.25	0
B	0.75	1

In the sequel, we'll assume that $\bar{\theta} = \tfrac{\pi}{2}$.

Consider now the outcome of an inspection of an object of interest. We are interested in the posterior conditional probabilities $P^+(T \mid B)$, $P^+(FT \mid B)$, $P^+(T \mid C)$ and $P^+(FT \mid C)$. Applying Bayes' rule, one obtains

$$P^+(T \mid B) = P(B \mid T)\frac{P^-(T)}{P(B)}$$

Similarly,

$$P^+(FT \mid B) = P(B \mid FT)\frac{P^-(FT)}{P(B)}$$

$$P^+(T \mid C) = P(C \mid T)\frac{P^-(T)}{P(C)}$$

$$P^+(FT \mid C) = P(C \mid FT)\frac{P^-(FT)}{P(C)}$$

Now, the MAV's ATR module report is B with probability

$$P(B) = P(B \mid T)P^-(T) + P(B \mid FT)P^-(FT)$$
$$= \frac{3}{4} \cdot \frac{\alpha}{1+\alpha} + 1 \cdot \frac{1}{1+\alpha} = \frac{4+3\alpha}{4(1+\alpha)}$$

Similarly, the MAV's ATR module report is C with probability

$$P(C) = P(C \mid T)P^-(T) + P(C \mid FT)P^-(FT)$$
$$= \frac{1}{4} \cdot \frac{\alpha}{1+\alpha} + 0 \cdot \frac{1}{1+\alpha} = \frac{\alpha}{4(1+\alpha)}$$

Hence, after the MAV ATR module's inspection of an object of interest whose pose angle is random and uniformly distributed, the following holds.

$$P^+(T \mid B) = \frac{3}{4} \cdot \frac{\frac{\alpha}{1+\alpha}}{\frac{4+3\alpha}{4(1+\alpha)}} = \frac{3\alpha}{4+3\alpha}$$

$$P^+(FT \mid B) = 1 \cdot \frac{\frac{1}{1+\alpha}}{\frac{4+3\alpha}{4(1+\alpha)}} = \frac{4}{4+3\alpha}$$

$$P^+(T \mid C) = \frac{1}{4} \cdot \frac{\frac{\alpha}{1+\alpha}}{\frac{\alpha}{4(1+\alpha)}} = 1 \quad P^+(FT \mid C) = 0$$

7 The Second Look

Two scenarios are investigated.

    a) First a CONOP is considered where, after the first look, a measurement of the object's pose becomes available and consequently, for the second measurement the MAV is vectored such that the ideal pose angle $\theta = \pi$ is attained. It is assumed that the navigation error in θ is negligible.

    We calculate

$$P(BB \mid FT) = 1$$
$$P(BC \mid FT) = P(CB \mid FT) = P(CC \mid FT) = 0$$
$$P(BB \mid T) = 0 \quad P(CB \mid T) = 0$$
$$P(BC \mid T) = P(B \mid T) = \frac{3}{4} \quad P(CC \mid T) = P(C \mid T) = \frac{1}{4}$$

and since

$$P^-(T) = \frac{\alpha}{1+\alpha} , \quad P^-(FT) = \frac{1}{1+\alpha}$$

we obtain

$$P(BB) = \tfrac{1}{1+\alpha} \quad P(BC) = \frac{3\alpha}{4(1+\alpha)}$$

$$P(CB) \ = 0 \quad P(CC) = \frac{\alpha}{4(1+\alpha)}$$

Finally, using Bayes' rule yields the information about object O_i

$$P(T \mid BB) \ = \ P(BB \mid T) \cdot \frac{P^-(T)}{P(BB)} = 0$$

$$P(FT \mid BB) = 1 \quad P(T \mid BC) = 1$$
$$P(FT \mid BC) = 0 \quad P(T \mid CC) = 1$$
$$P(FT \mid CC) = 0$$

Note that

- The prior information about the Ts/FTs ratio, α, plays no role here.
- The pose information is crucial: it is conducive to complete T/FT discrimination. After the second measurement the uncertainty is completely removed.

b) The alternative CONOP is now considered where, after the first look, the object's pose is not available and thus, for the second measurement the MAV is vectored such that the pose angle is increased by 90 degrees.
We calculate

$$P(BB \mid FT) \ = \ 1$$
$$P(BC \mid FT) \ = \ P(CB \mid FT) = P(CC \mid FT) = 0$$
$$P(BB \mid T) = \tfrac{1}{2} \quad P(BC \mid T) = \frac{1}{4}$$
$$P(CC \mid T) = 0 \quad P(CB \mid T) = \frac{1}{4}$$

and since

$$P^-(T) = \frac{\alpha}{1+\alpha} \ , \quad P^-(FT) = \frac{1}{1+\alpha}$$

we obtain

$$
\begin{aligned}
P(BB) \ &= \ P(BB \mid T)P^-(T) + P(BB \mid FT)P^-(FT) \\
&= \ \frac{1}{2} \cdot \frac{\alpha}{1+\alpha} + 1 \cdot \frac{1}{1+\alpha} = \frac{2+\alpha}{2(1+\alpha)} \\
P(BC) \ &= \ P(BC \mid T)P^-(T) + P(BC \mid FT)P^-(FT) \\
&= \ \frac{1}{4} \cdot \frac{\alpha}{1+\alpha} + 0 \cdot \frac{1}{1+\alpha} = \frac{\alpha}{4(1+\alpha)} \\
P(CC) &= 0 \quad P(CB) = \frac{\alpha}{4(1+\alpha)}
\end{aligned}
$$

We are interested in the probabilities $P(T \mid BB)$, $P(FT \mid BB)$, $P(T \mid BC)$ and $P(FT \mid BC)$. Applying Bayes' rule, the information on object O_i is obtained

$$P(T \mid BB) = P(BB \mid T) \cdot \frac{P^-(T)}{P(BB)} = \frac{1}{2} \cdot \frac{\frac{\alpha}{1+\alpha}}{\frac{2+\alpha}{2(1+\alpha)}} = \frac{\alpha}{2+\alpha}$$

$$P(FT \mid BB) = P(BB \mid FT) \cdot \frac{P^-(FT)}{P(BB)} = 1 \cdot \frac{\frac{1}{1+\alpha}}{\frac{2+\alpha}{2(1+\alpha)}} = \frac{2}{2+\alpha}$$

$$P(T \mid BC) = P(BC \mid T) \frac{P^-(T)}{P(BC)} = \frac{1}{4} \cdot \frac{\frac{\alpha}{1+\alpha}}{\frac{\alpha}{4(1+\alpha)}} = 1$$

$$P(FT \mid BC) = P(BC \mid FT) \cdot \frac{P^-(FT)}{P(BC)} = 0$$

$$P(T \mid CB) = 1 \quad P(FT \mid CB) = 0$$

8 ATR System Aiding

The MAV is scheduled to visit n objects of interest. Consider the MAV's "milk run". Suppose the MAV has arrived to object number i. At this point in time, the probabily that O_i is a T is $P^-(T) = \frac{\alpha}{1+\alpha}$ and the probability that O_i is not a FT is $P^-(FT) = \frac{1}{1+\alpha}$. At this time the MAV takes a measurement on object number i. The outcome of the measurement is discrete, that is, it is either B or C. The probabilities of the possible outcomes have been calculated in Section 6, namely

$$P(B) = \frac{4+3\alpha}{4(1+\alpha)} \quad P(C) = \frac{\alpha}{4(1+\alpha)}$$

When the MAV's ATR system returns a B, the probability that O_i is a target is

$$P(T \mid B) = \frac{3\alpha}{4+3\alpha}$$

We also calculated

$$P(T \mid C) = 1$$

Hence, a C return allows one to categorically state that O_i is a target, classify O_i as such, and proceed to inspect object O_{i+1}. However, when the MAV's ATR system returns a B, the confidence that object number i is a target decreases. Indeed,

$$\frac{3\alpha}{4+3\alpha} < \frac{\alpha}{1+\alpha} \quad \forall \, \alpha > 0$$

In light of the above, ATR system aiding is envisaged. This is achieved by including a human operator in the classification loop. Thus, when the outcome of the measurement performed by the MAV ATR system is C, O_i is classified a target and the MAV autonomously proceeds to object O_{i+1}. However, when the outcome of the measurement performed by the MAV ATR system is B, the image of the object is sent to the operator. The operator, upon being presented the image, takes on the classification task. The operator's classification performance is modelled by a known confusion matrix, as shown.

True/Rep.	O=T	O=FT
$Z_O =$T	P_{TR_O}	$1 - P_{FTR_O}$
$Z_O =$FT	$1 - P_{TR_O}$	P_{FTR_O}

Thus, the operator's output is T or FT. We use the subscript O to indicate that these are the operator's determinations.

After the operator's inspection of the image

$$P^+(T \mid Z_O = T) = \frac{P_{TR_O} P(T \mid B)}{P_{TR_O} P(T \mid B) + (1 - P_{FTR_O})[1 - P(T \mid B)]}$$

$$= \frac{P_{TR_O} \frac{3\alpha}{4+3\alpha}}{P_{TR_O} \frac{3\alpha}{4+3\alpha} + (1 - P_{FTR_O}) \frac{4}{4+3\alpha}} = \frac{3\alpha P_{TR_O}}{3\alpha P_{TR_O} + 4(1 - P_{FTR_O})}$$

Similarly

$$P^+(T \mid Z_O = FT) = \frac{(1 - P_{TR_O}) P(T \mid B)}{(1 - P_{TR_O}) P(T \mid B) + P_{FTR_O}[1 - P(T \mid B)]}$$

$$= \frac{(1 - P_{TR_O}) \frac{3\alpha}{4+3\alpha}}{(1 - P_{TR_O}) \frac{3\alpha}{4+3\alpha} + P_{FTR_O} \frac{4}{4+3\alpha}} = \frac{3\alpha(1 - P_{TR_O})}{3\alpha(1 - P_{TR_O}) + 4P_{FTR_O}}$$

In summary, once the operator's classification Z_O is in, that is, the discrete measurement T_O or FT_O is received, the information about object O_i is encapsulated in $P^+(T \mid Z_O = T)$ and $P^+(T \mid Z_O = FT)$. The above calculated probabilities are based on the prior information about the battlespace, p, the performance of the SAV's ATR system which is quantified by the parameter α, the return B of the ATR system onboard the MAV, and the operator's classification T_O or FT_O.

9 More Probability Calculations

The payoff functional is the average cummulative information gain resulting from the inspection of the set of objects of interest performed by the MAV. The information gain rendered by an inspection is determined by the prior information on the objects of interest and the inspection's outcome, that is, the recorded measurements. At each stage of the inspection round the decision must be taken whether to continue with the inspection sequence and move on to the next object of interest, and to the next stage, or, turn around and revisit the current object of interest for a second inspection. Thus, to evaluate the payoff functional the probabilities of all the possible inspection outcomes, which, in turn, are partially determined by the decision variable, must be derived. This will afford an analytic derivation of the optimal sequential search strategy and the solution of the attendant stochastic dynamic program.

The evaluation of the expected information gain resulting from an inspection of an object of interest necessitates the calculation of the following probabilities.

$$P(T,C) \quad = \frac{\alpha}{4(1+\alpha)} \qquad P(T,BT_O) = \frac{3\alpha}{4(1+\alpha)} P_{TR_O}$$

$$P(FT,BT_O) \quad = \frac{1-P_{FTR_O}}{1+\alpha} \qquad P(FT,BFT_O) = \frac{P_{FTR_O}}{1+\alpha}$$

$$P(T,BFT_O) = \frac{3\alpha}{4(1+\alpha)}(1 - P_{TR_O})$$

In addition, need $P(T, BT_O C)$:

$$P(T,BT_O C) = P(T,BC) \cdot P(Z_O = T \mid T) = P(T,BC) \cdot P_{TR_O}$$

Now

$$P(BC \mid T) = \frac{P(T,BC)}{P(T)} = \frac{1+\alpha}{\alpha} P(T,BC)$$

Hence,

$$P(T,BC) = \frac{\alpha}{1+\alpha} P(BC \mid T)$$

Next, recall that under CONOP a) from above

$$P_a(BC \mid T) = \frac{3}{4}$$

and under CONOP b)

$$P_b(BC \mid T) = \frac{1}{4}$$

Hence

$$P_a(T,BC) = \frac{3}{4} \cdot \frac{\alpha}{1+\alpha} \qquad P_b(T,BC) = \frac{1}{4} \cdot \frac{\alpha}{1+\alpha}$$

Hence, we finally obtain

$$P_a(T,BT_O C) = \frac{3\alpha}{4(1+\alpha)} P_{TR_O} \tag{2}$$

$$P_b(T,BT_O C) = \frac{\alpha}{4(1+\alpha)} P_{TR_O} \tag{3}$$

Also need $P(T, BT_O BT_O)$.

$$P(T, BT_O BT_O) = P(T, BB) \cdot P^2(Z_O = T \mid T) = P(T, BB) \cdot P_{TR_O}^2$$

Now

$$P(BB \mid T) = \frac{P(T, BB)}{P(T)} = \frac{1 + \alpha}{\alpha} P(T, BB)$$

Hence,

$$P(T, BB) = \frac{\alpha}{1 + \alpha} P(BB, T)$$

Next, recall that under CONOP a) from above

$$P_a(BB \mid T) = 0$$

and under CONOP b)

$$P_b(BB \mid T) = \frac{1}{2}$$

Hence

$$P_a(T, BB) = 0 \quad P_b(T, BB) = \frac{1}{2} \frac{\alpha}{1 + \alpha}$$

Hence, we finally obtain

$$P_a(T, BT_O BT_O) = 0 \tag{4}$$
$$P_b(T, BT_O BT_O) = \frac{\alpha}{2(1 + \alpha)} P_{TR_O}^2 \tag{5}$$

$P(T, BT_O BFT_O)$ is calculated as follows.

$$P(T, BT_O BFT_O) = P(T, BB) \cdot P(Z_O = T \mid T) \cdot P(Z_O = FT \mid T)$$
$$= P(T, BB) \cdot P_{TR_O} \cdot (1 - P_{TR_O})$$

Hence, we obtain

$$P_a(T, BT_O BFT_O) = 0 \tag{6}$$
$$P_b(T, BT_O BFT_O) = \frac{\alpha}{2(1 + \alpha)} P_{TR_O} (1 - P_{TR_O}) \tag{7}$$

$P(T, BFT_O C)$ is calculated as follows.

$$P(T, BFT_O C) = P(T, BC) \cdot P(Z_O = FT \mid T) = P(T, BC) \cdot (1 - P_{TR_O})$$

Hence, we obtain

$$P_a(T, BFT_OC) = \frac{3\alpha}{4(1+\alpha)}(1 - P_{TR_O}) \tag{8}$$

$$P_b(T, BFT_OC) = \frac{\alpha}{4(1+\alpha)}(1 - P_{TR_O}) \tag{9}$$

$P(T, BFT_OBT_O)$ is calculated as follows.

$$P(T, BFT_OBT_O) = P(T, BB) \cdot P(Z_O = FT \mid T) \cdot P(Z_O = T \mid T)$$
$$= P(T, BB) \cdot (1 - P_{TR_O})P_{TR_O}$$

Hence, we obtain

$$P_a(T, BFT_OBT_O) = 0 \tag{10}$$

$$P_b(T, BFT_OBT_O) = \frac{\alpha}{2(1+\alpha)}P_{TR_O}(1 - P_{TR_O}) \tag{11}$$

$P(T, BFT_OBFT_O)$ is calculated as follows.

$$P(T, BFT_OBFT_O) = P(T, BB) \cdot P^2(Z_O = FT \mid T)$$
$$= P(T, BB) \cdot (1 - P_{TR_O})^2$$

Hence, we obtain

$$P_a(T, BFT_OBFT_O) = 0 \tag{12}$$

$$P_b(T, BFT_OBFT_O) = \frac{\alpha}{2(1+\alpha)}(1 - P_{TR_O})^2 \tag{13}$$

$P(FT, BT_OBT_O)$ is calculated as follows.

$$P(FT, BT_OBT_O) = P(FT, BB) \cdot P^2(Z_O = T \mid FT)$$
$$= P(FT, BB) \cdot (1 - P_{FTR_O})^2$$

Now

$$P(BB \mid FT) = \frac{P(FT, BB)}{P(FT)} = (1 + \alpha) \cdot P(FT, BB)$$

Thus

$$P(FT, BB) = \frac{1}{1+\alpha}P(BB \mid FT)$$

Since

$$P_a(BB \mid FT) = 1 \quad P_b(BB \mid FT) = 1$$

we calculate

$$P_a(FT \mid BB) = \tfrac{1}{1+\alpha} \quad P_b(FT \mid BB) = \frac{1}{1+\alpha}$$

Hence, we obtain

$$P_a(FT, BT_O BT_O) = \frac{1}{1+\alpha}(1 - P_{FTR_O})^2 \tag{14}$$

$$P_b(FT, BT_O BT_O) = \frac{1}{1+\alpha}(1 - P_{FTR_O})^2 \tag{15}$$

$P(FT, BT_O BFT_O)$ is calculated as follows.

$$P(FT, BT_O BFT_O) = P(FT, BB) \cdot P(Z_O = T \mid FT) \cdot P(Z_O = FT \mid FT)$$
$$= P(FT, BB) \cdot (1 - P_{FTR_O}) \cdot P_{FTR_O}$$

Hence, we obtain

$$P_a(FT, BT_O BFT_O) = \frac{1}{1+\alpha} P_{FTR_O}(1 - P_{FTR_O}) \tag{16}$$

$$P_b(FT, BT_O BFT_O) = \frac{1}{1+\alpha} P_{FTR_O}(1 - P_{FTR_O}) \tag{17}$$

$P(FT, BFT_O BT_O)$ is calculated as follows.

$$P(FT, BFT_O BT_O) = P(FT, BB) \cdot P(Z_O = FT \mid FT) \cdot P(Z_O = T \mid FT)$$
$$= P(FT, BB) \cdot P_{FTR_O} \cdot (1 - P_{FTR_O})$$

Hence, we obtain

$$P_a(FT, BFT_O BT_O) = \frac{1}{1+\alpha} \cdot P_{FTR_O} \cdot (1 - P_{FTR_O}) \tag{18}$$

$$P_b(FT, BFT_O BT_O) = \frac{1}{1+\alpha} \cdot P_{FTR_O} \cdot (1 - P_{FTR_O}) \tag{19}$$

$P(FT, BFT_O BFT_O)$ is calculated as follows.

$$P(FT, BFT_O BFT_O) = P(FT, BB) \cdot P^2(Z_O = FT \mid FT)$$
$$= P(FT, BB) \cdot P_{FTR_O}^2$$

Hence, we obtain

$$P_a(FT, BFT_O BFT_O) = \frac{1}{1+\alpha} P_{FTR_O}^2 \tag{20}$$

$$P_b(FT, BFT_O BFT_O) = \frac{1}{1+\alpha} P_{FTR_O}^2 \tag{21}$$

Know:

$$P(T \mid C) = 1$$

$$P(T \mid BT_O) = \frac{3\alpha P_{TR_O}}{3\alpha P_{TR_O} + 4(1 - P_{FTR_O})}$$

$$P(FT \mid BT_O) = \frac{4(1 - P_{FTR_O})}{3\alpha P_{TR_O} + 4(1 - P_{FTR_O})}$$

$$P(T \mid BFT_O) = \frac{3\alpha(1 - P_{TR_O})}{3\alpha(1 - P_{TR_O}) + 4P_{FTR_O}}$$

$$P(FT \mid BFT_O) = \frac{4P_{FTR_O}}{3\alpha(1 - P_{TR_O}) + 4P_{FTR_O}}$$

Need $P(T \mid BT_O C)$.

$P(T \mid BT_O C)$ is calculated as follows.

$$P(T \mid BT_O C) = P(C \mid T, BT_O) \cdot \frac{P(T \mid BT_O)}{P(C \mid BT_O)}$$

Know

$$P(T \mid BT_O) = \frac{3\alpha P_{TR_O}}{3\alpha P_{TR_O} + 4(1 - P_{FTR_O})}$$

Calculate

$$P(C \mid BT_O) = P(C \mid T, BT_O) \cdot P(T \mid BT_O) + P(C \mid FT, BT_O) \cdot P(FT \mid BT_O)$$
$$= P(C \mid T, BT_O) \cdot P(T \mid BT_O)$$

whereupon it follows that

$$P(T \mid BT_O C) = 1 \ ,$$

as expected.

Next,

$$P(T \mid BT_O BT_O) = P(BT_O \mid T, BT_O) \cdot \frac{P(T \mid BT_O)}{P(BT_O \mid BT_O)}$$

Know

$$P(T \mid BT_O) = \frac{3\alpha P_{TR_O}}{3\alpha P_{TR_O} + 4(1 - P_{FTR_O})}$$

Calculate

$$P(BT_O \mid BT_O) = P(BT_O \mid T, BT_O) \cdot P(T \mid BT_O) + P(BT_O \mid FT, BT_O)$$
$$\cdot P(FT \mid BT_O) = P(BT_O \mid T, BT_O) \cdot P(T \mid BT_O)$$
$$+ P(BT_O \mid FT, BT_O) \cdot (1 - P(T \mid BT_O))$$

Now

$$P(BT_O \mid T, BT_O) = P(B \mid T, BT_O) \cdot P(Z_O = T \mid T, BT_O)$$
$$= P(B \mid T, B) \cdot P(Z_O = T \mid T)$$
$$= P(B \mid T, B) \cdot P_{TR_O}$$

Furthermore,

$$P_a(B \mid T, B) = 0$$
$$P_b(B \mid T, B) = \frac{P_b(BB \mid T)}{P(B \mid T)} = \frac{\frac{1}{2}}{\frac{3}{4}} = \frac{2}{3}$$

This yields

$$P_a(BT_O \mid T, BT_O) = 0$$
$$P_b(BT_O \mid T, BT_O) = \frac{2}{3} P_{TR_O}$$

Thus,

$$P_a(BT_O \mid BT_O) = P(BT_O \mid FT, BT_O) \cdot [1 - P(T \mid BT_O)]$$
$$P_b(BT_O \mid BT_O) = \frac{2}{3} P_{TR_O} \cdot P(T \mid BT_O) + P(BT_O \mid FT, BT_O) \cdot [1 - P(T \mid BT_O)]$$

Similar to the calculation from above,

$$P(BT_O \mid FT, BT_O) = P(B \mid FT, BT_O) \cdot P(Z_O = T \mid FT, BT_O)$$
$$= P(B \mid FT, B) \cdot P(Z_O = T \mid FT)$$
$$= P(B \mid FT, B) \cdot (1 - P_{FTR_O})$$

Furthermore,

$$P_a(B \mid FT, B) = 1 \text{ and } P_b(B \mid FT, B) = 1$$

which yields

$$P_a(BT_O \mid FT, BT_O) = 1 - P_{FTR_O}$$
$$P_b(BT_O \mid FT, BT_O) = 1 - P_{FTR_O}$$

and therefore

$$P_a(BT_O \mid BT_O) = (1 - P_{FTR_O}) \cdot [1 - P(T \mid BT_O)]$$

and

$$P_b(BT_O \mid BT_O) = \frac{2}{3} P_{TR_O} \cdot P(T \mid BT_O) + (1 - P_{FTR_O}) \cdot [1 - P(T \mid BT_O)]$$
$$= 1 - P_{FTR_O} + (\frac{2}{3} P_{TR_O} + P_{FTR_O} - 1) \cdot P(T \mid BT_O)$$

Hence

$$P_a(T \mid BT_O BT_O) = 0$$

and

$$P_b(T \mid BT_O BT_O) = \frac{2}{3} P_{TR_O} \cdot \frac{P(T \mid BT_O)}{1 - P_{FTR_O} + (\frac{2}{3} P_{TR_O} + P_{FTR_O} - 1) \cdot P(T \mid BT_O)}$$

Hence

$$P_a(T \mid BT_O BT_O) = 0 \tag{22}$$

$$P_b(T \mid BT_O BT_O) = \frac{\alpha P_{TR_O}^2}{\alpha P_{TR_O}^2 + 2(1 - P_{FTR_O})^2} \tag{23}$$

Next,

$$P(T \mid BT_O BFT_O) = P(BFT_O \mid T, BT_O) \cdot \frac{P(T \mid BT_O)}{P(BFT_O \mid BT_O)}$$

We know

$$P(T \mid BT_O) = \frac{3\alpha P_{TR_O}}{3\alpha P_{TR_O} + 4(1 - P_{FTR_O})}$$

We also calculate

$$P(BFT_O \mid BT_O) =$$

$$P(BFT_O \mid T, BT_O) \cdot P(T \mid BT_O) + P(BFT_O \mid FT, BT_O) \cdot P(FT \mid BT_O) =$$
$$P(BFT_O \mid T, BT_O) \cdot P(T \mid BT_O) + P(BFT_O \mid FT, BT_O) \cdot [1 - P(T \mid BT_O)]$$

Now

$$P(BFT_O \mid T, BT_O) = P(B \mid T, BT_O) \cdot P(Z_O = FT \mid T, BT_O)$$
$$= P(B \mid T, B) \cdot P(Z_O = FT \mid T) = P(B \mid T, B) \cdot (1 - P_{TR_O})$$

This yields

$$P_a(BFT_O \mid T, BT_O) = 0$$
$$P_b(BFT_O \mid T, BT_O) = \frac{2}{3}(1 - P_{TR_O})$$

and therefore

$$P_a(BFT_O \mid BT_O) = P(BFT_O \mid FT, BT_O) \cdot [1 - P(T \mid BT_O)]$$
$$P_b(BFT_O \mid BT_O) = \frac{2}{3}(1 - P_{TR_O}) \cdot P(T \mid BT_O)$$
$$+ P(BFT_O \mid FT, BT_O) \cdot [1 - P(T \mid BT_O)]$$

Similar to the calculation from above,

$$P(BFT_O \mid FT, BT_O) = P(B \mid FT, BT_O) \cdot P(Z_O = FT \mid FT, BT_O)$$
$$= P(B \mid FT, B) \cdot P(Z_O = FT \mid FT)$$
$$= P(B \mid FT, B) \cdot P_{FTR_O} = 1 \cdot P_{FTR_O}$$

This yields

$$P_a(BFT_O \mid FT, BT_O) = P_{FTR_O}$$
$$P_b(BFT_O \mid FT, BT_O) = P_{FTR_O}$$

and thus

$$P_a(BFT_O \mid BT_O) = P_{FTR_O}[1 - P(T \mid BT_O)]$$
$$P_b(BFT_O \mid BT_O) = \frac{2}{3}(1 - P_{TR_O}) \cdot P(T \mid BT_O) + P_{FTR_O}[1 - P(T \mid BT_O)]$$
$$= P_{FTR_O} + [\frac{2}{3}(1 - P_{TR_O}) - P_{FTR_O}] \cdot P(T \mid BT_O)$$

Hence

$$P_a(T \mid BT_O BFT_O) = 0$$
$$P_b(T \mid BT_O BFT_O) = \frac{2}{3}(1 - P_{TR_O}) \times$$
$$\frac{P(T \mid BT_O)}{\frac{2}{3}(1 - P_{TR_O})P(T \mid BT_O) + P_{FTR_O}[1 - P(T \mid BT_O)]}$$

which yields

$$P_a(T \mid BT_O BFT_O) = 0 \tag{24}$$
$$P_b(T \mid BT_O BFT_O) = \frac{\alpha P_{TR_O}(1 - P_{TR_O})}{\alpha P_{TR_O}(1 - P_{TR_O}) + 2P_{FTR_O}(1 - P_{FTR_O})} \tag{25}$$

Next,

$$P(T \mid BFT_O C) = P(C \mid T, BFT_O) \cdot \frac{P(T \mid BFT_O)}{P(C \mid BFT_O)}$$

and we calculate

$$P(C \mid BFT_O) = P(C \mid T, BFT_O) \cdot P(T \mid BFT_O) + P(C \mid FT, BFT_O) \cdot$$
$$P(FT \mid BFT_O) = P(C \mid T, BFT_O) \cdot P(T \mid BFT_O) + 0 \cdot P(FT \mid BFT_O)$$
$$= P(C \mid T, BFT_O) \cdot P(T \mid BFT_O)$$

This yields

$$P(T \mid BFT_O C) = 1 \, ,$$

as expected.

$$P(T \mid BFT_OBT_O) = P(BT_O \mid T, BFT_O) \cdot \frac{P(T \mid BFT_O)}{P(BT_O \mid BFT_O)}$$

First calculate $P(T \mid BFT_O)$. To this end, let

$$p \equiv P(T \mid B) = \frac{3\alpha}{4 + 3\alpha}$$

and let

$$P^+(T \mid Z_O = FT) \equiv P(T \mid BFT_O)$$

Apply Bayes' rule:

$$P^+(T \mid Z_O = FT) = P(Z_O = FT \mid T) \cdot \frac{P(T)}{P(Z_O = FT)}$$

that is,

$$P^+(T \mid Z_O = FT) = (1 - P_{TR_O}) \cdot \frac{p}{P(Z_O = FT)}$$

Now

$$P(Z_O = FT) = P(Z_O = FT \mid T) \cdot P(T) + P(Z_O = FT \mid FT) \cdot P(FT)$$

that is,

$$P(Z_O = FT) = (1 - P_{TR_O})p + P_{FTR_O}(1 - p)$$

Thus,

$$P^+(T \mid Z_O = FT) = (1 - P_{TR_O})\frac{p}{(1 - P_{TR_O})p + P_{FTR_O}(1 - p)}$$

and therefore :

$$P(T \mid BFT_O) = (1 - P_{TR_O})\frac{\frac{3\alpha}{4+3\alpha}}{(1 - P_{TR_O})\frac{3\alpha}{4+3\alpha} + P_{FTR_O}(1 - \frac{3\alpha}{4+3\alpha})} \; ,$$

that is,

$$P(T \mid BFT_O) = \frac{3\alpha(1 - P_{TR_O})}{3\alpha(1 - P_{TR_O}) + 4P_{FTR_o}}$$

We next calculate $P(BT_O \mid BFT_O)$.

$$P(BT_O \mid BFT_O) =$$

$$P(BT_O \mid T, BFT_O) \cdot P(T \mid BFT_O) + P(BT_O \mid FT, BFT_O) \cdot P(FT \mid BFT_O) =$$
$$P(BT_O \mid T, BFT_O) \cdot P(T \mid BFT_O) + P(BT_O \mid FT, BFT_O) \cdot [1 - P(T \mid BFT_O)]$$

Now

$$P(BT_O \mid T, BFT_O) = P(B \mid T, BFT_O) \cdot P(Z_O = T \mid T, BFT_O)$$
$$= P(B \mid T, B) \cdot P(Z_O = T \mid T) = P(B \mid T, B) \cdot P_{TR_O}$$

Hence, as in the calculation of $P(BT_O \mid T, BT_O)$,

$$P_a(BT_O \mid T, BFT_O) = 0 \text{ and } P_b(BT_O \mid T, BFT_O) = \frac{2}{3} P_{TR_O}$$

Hence

$$P_a(BT_O \mid BFT_O) = P(BT_O \mid FT, BFT_O) \cdot [1 - P(T \mid BFT_O)]$$

$$P_b(BT_O \mid BFT_O) = \frac{2}{3} P_{TR_O} \cdot P(T \mid BFT_O)$$
$$+ P(BT_O \mid FT, BFT_O) \cdot [1 - P(T \mid BFT_O)]$$
$$= P(BT_O \mid FT, BFT_O) + [\frac{2}{3} P_{TR_O}$$
$$- P(BT_O \mid FT, BFT_O)] \cdot P(T \mid BFT_O)$$

Similar to the calculation from above

$$P(BT_O \mid FT, BFT_O) = P(B \mid FT, BFT_O) \cdot P(Z_O = T \mid FT, BFT_O)$$
$$= P(B \mid FT, B) \cdot P(Z_O = T \mid FT)$$

Hence, as in the case of $P(BT_O \mid FT.BT_O)$,

$$P_a(BT_O \mid FT, BFT_O) = 1 - P_{FTR_O}$$
$$P_b(BT_O \mid FT, BFT_O) = 1 - P_{FTR_O}$$

which gives

$$P_a(BT_O \mid BFT_O) = (1 - P_{FTR_O})[1 - P(T \mid BFT_O)]$$
$$P_b(BT_O \mid BFT_O) = 1 - P_{FTR_O} + (\frac{2}{3} P_{TR_O} - 1 + P_{FTR_O}) \cdot P(T \mid BFT_O)$$

Hence

$$P_a(T \mid BFT_O BT_O) = 0$$
$$P_b(T \mid BFT_O BT_O) = \frac{2}{3} P_{TR_O}$$
$$\cdot \frac{P(T \mid BFT_O)}{1 - P_{FTR_O} + (\frac{2}{3} P_{TR_O} - 1 + P_{FTR_O}) \cdot P(T \mid BFT_O)}$$

Hence

$$P_a(T \mid BFT_O BT_O) = 0 \tag{26}$$

$$P_b(T \mid BFT_O BT_O) = \frac{\alpha P_{TR_O}(1 - P_{TR_O})}{\alpha P_{TR_O}(1 - P_{TR_O}) + 2P_{FTR_O}(1 - P_{FTR_O})} \tag{27}$$

Next

$$P(T \mid BFT_O BFT_O) = P(BFT_O \mid T, BFT_O) \cdot \frac{P(T \mid BFT_O)}{P(BFT_O \mid BFT_O)}$$

We know

$$P(T \mid BFT_O) = \frac{3\alpha(1 - P_{TR_O})}{3\alpha(1 - P_{TR_O}) + 4P_{FTR_O}}$$

We calculate

$$\begin{aligned}
P(BFT_O \mid BFT_O) &= P(BFT_O \mid T, BFT_O) \cdot P(T \mid BFT_O) \\
&+ P(BFT_O \mid FT, BFT_O) \cdot P(FT \mid BFT_O) \\
&= P(BFT_O \mid T, BFT_O) \cdot P(T \mid BFT_O) \\
&+ P(BFT_O \mid FT, BFT_O) \cdot [1 - P(T \mid BFT_O)]
\end{aligned}$$

Now

$$\begin{aligned}
P(BFT_O \mid T, BFT_O) &= P(B \mid T, BFT_O) \cdot P(Z_O = FT \mid T, BFT_O) \\
&= P(B \mid T, B) \cdot P(Z_O = FT \mid T) = P(B \mid T, B) \cdot (1 - P_{TR_O})
\end{aligned}$$

We know already

$$P_a(B \mid T, B) = 0 \quad \text{and} \quad P_b(B \mid T, B) = \frac{2}{3}$$

Hence

$$P_a(BFT_O \mid T, BFT_O) = 0$$

$$P_b(BFT_O \mid T, BFT_O) = \frac{2}{3}(1 - P_{TR_O})$$

which yields

$$P_a(BFT_O \mid BFT_O) = P(BFT_O \mid FT, BFT_O) \cdot [1 - P(T \mid BFT_O)]$$

$$\begin{aligned}
P_b(BFT_O \mid BFT_O) &= \frac{2}{3}(1 - P_{TR_O}) \cdot P(T \mid BFT_O) + \\
&\quad P(BFT_O \mid FT, BFT_O) \cdot [1 - P(T \mid BFT_O)]
\end{aligned}$$

Similar to the calculation from above,

$$P(BFT_O \mid FT, BFT_O) = P(B \mid FT, BFT_O) \cdot P(Z_O = FT \mid FT, BFT_O)$$
$$= P(B \mid FT, B) \cdot P(Z_O = FT \mid FT)$$
$$= P(B \mid FT, B) \cdot P_{FTR_O}$$

Furthermore,

$$P_a(B \mid FT, B) = 1 \quad \text{and} \quad P_b(B \mid FT, B) = 1$$

which yields

$$P_a(BFT_O \mid FT, BFT_O) = P_{FTR_O}$$
$$P_b(BFT_O \mid FT, BFT_O) = P_{FTR_O}$$

that is,

$$P_a(BFT_O \mid BFT_O) = P_{FTR_O} \cdot [1 - P(T \mid BFT_O)]$$
$$P_b(BFT_O \mid BFT_O) = \frac{2}{3}(1 - P_{TR_O}) \cdot P(T \mid BFT_O)$$
$$+ P_{FTR_O} \cdot [1 - P(T \mid BFT_O)]$$
$$= P_{FTR_O} + [\frac{2}{3}(1 - P_{TR_O}) - P_{FTR_O}] \cdot P(T \mid BFT_O)$$

Thus,

$$P_a(T \mid BFT_O BFT_O) = 0$$
$$P_b(T \mid BFT_O BFT_O) = \frac{2}{3}(1 - P_{TR_O}) \times$$
$$\frac{P(T \mid BFT_O)}{P_{FTR_O} + [\frac{2}{3}(1 - P_{TR_O}) - P_{FTR_O}]P(T \mid BFT_O)}$$

Hence

$$P_a(T \mid BFT_O BFT_O) = 0 \tag{28}$$
$$P_b(T \mid BFT_O BFT_O) = \frac{\alpha(1 - P_{TR_O})^2}{\alpha(1 - P_{TR_O})^2 + 2P_{FTR_O}^2} \tag{29}$$

Finally

$$P(FT \mid BT_O BT_O) = 1 - P(T \mid BT_O BT_O)$$

which yields

$$P_a(FT \mid BT_O BT_O) = 1 \tag{30}$$
$$P_b(FT \mid BT_O BT_O) = \frac{2(1 - P_{FTR_O})^2}{\alpha P_{TR_O}^2 + 2(1 - P_{FTR_O})^2} \tag{31}$$

Similarly,

$$P_a(FT \mid BT_O BFT_O) = 1 \tag{32}$$

$$P_b(FT \mid BT_O BFT_O) = \frac{2P_{FTR_O}(1 - P_{FTR_O})}{\alpha P_{TR_O}(1 - P_{TR_O}) + 2P_{FTR_O}(1 - P_{FTR_O})} \tag{33}$$

$$P_a(FT \mid BFT_O BT_O) = 1 \tag{34}$$

$$P_b(FT \mid BFT_O BT_O) = \frac{2P_{FTR_O}(1 - P_{FTR_O})}{\alpha P_{TR_O}(1 - P_{TR_O}) + 2P_{FTR_O}(1 - P_{FTR_O})} \tag{35}$$

$$P_a(FT \mid BFT_O BFT_O) = 1 \tag{36}$$

$$P_b(FT \mid BFT_O BFT_O) = \frac{2P_{FTR_O}^2}{\alpha(1 - P_{TR_O})^2 + 2P_{FTR_O}^2} \tag{37}$$

10 Optimal Control

The MAV is tasked to inspect n objects of interest. Thus, a Dynamic Program with n stages must be solved. The scalar state variable k is the remaining endurance of the MAV: at stage i, that is, when object O_i is reached, the MAV's remaining endurance is k_i. The endurance of the MAV at the time instant of its release from the SAV at time $t = 0$ is k_0. We assume, without loss of generality, that the time required to travel from object i to object $i + 1$ is fixed and is one unit. Thus, the following must hold:

$$k_i \geq n - i , \quad i = 0, 1, ..., n$$

The control signal at stage i is the binary variable $u_i \in \{0, 1\}$. When the control variable $u_i = 0$, the MAV proceeds to inspect object O_{i+1} and when $u_i = 1$, the MAV will revisit object O_i. Initially, $u_0 \equiv 0$ and thus, the state at stage 1 is deterministically set to $k_1 = k_0 - 1$. The control u_i at stage i is decided on once the measurement $Z_i \in \{C, BT_O, BFT_O\}$ on object O_i, obtained from the first look, is in. This occurs τ_i time units after the MAV has reached stage i; τ_i is the operator's delay at stage i. The operator delay $0 \leq \tau_i < 1$ is a random number which is specified by a known probability density function $f(\tau_i)$; the assumption $\tau_i < 1$ is not particulary restrictive. Thus, the optimal feedback control law assumes the form

$$u_i = u_i(k_i, Z_i, \tau_i) , \quad i = 1, ..., n$$

We set

$$u_i^*(k_i, Z_i, 0) = 0 \quad \forall \, Z_i = C , \quad i = 1, ..., n$$

Note that the control law uses the current state, the received measurement, and the operator delay information. The state is a deterministic variable, whereas the measurement and operator delay are random variables which are communicated to the controller at decision time i.

Concerning the dynamics: when the decision variable $u_i = 0$, the state propagates according to

$$k_{i+1} = k_i - 1$$

If however $Z_i \neq C$ and $u_i = 1$, the state propagates according to

$$k_{i+1} = k_i - 1 - 2\tau_i - \tau$$

where τ is a fixed maneuver delay. Now, the control $u_i = 1$ is a viable option, provided that $k_{i+1} \geq n - (i+1)$, that is, $k_i \geq n - i + 2\tau_i + \tau$. In other words, it is required that

$$\tau_i \leq \frac{1}{2}(k_i + i - n - \tau)$$

If however the random variable τ_i is s.t. the above inequality does not hold, then

$$u_i^*(k_i, Z_i, \tau_i) = 0 \quad \forall\, Z_i \ , \quad i = 1, ..., n \ ,$$

and, as before, the state propagates according to

$$k_{i+1} = k_i - 1$$

In summary, the (nonlinear) dynamics are

$$k_{i+1} = k_i - 1 - (2\tau_i + \tau)u_i \ , \quad i = 0, 1, ..., n - 1$$

Once the MAV arrives at O_i, the i'th stage of the DP is reached. At this i'th stage, the control variable $u_i \in \{0, 1\}$ is decided on based on the information available at decision time i. At stage i the physical decision time is $t = k_0 - k_i + \tau_i$. The information available at decision time at stage i is k_i, τ_i, and Z_i, where $k_0 - 1 \geq k_i \geq n - i$, $0 \leq \tau_i < 1$, and $Z_i \in \{C, BT_O, BFT_O\}$. Thus, the optimal feedback control law is

$$u_i^*(k_i, Z_i, \tau_i) = 0$$

provided that $Z_i = C$ or $Z_i \in \{BT_O, BFT_O\}$ _and_ $\tau_i > \frac{1}{2}(k_i + i - n - \tau])$. If however $Z_i \in \{BT_O, BFT_O\}$ _and_ $\tau_i \leq \frac{1}{2}(k_i + i - n - \tau)$ then the optimal control u_i^* is obtained by maximizing the expected (average) information to be gained by choosing one of the two alternative courses of action $u_i = 0$, or, $u_i = 1$, as discussed in the sequel.

11 Stochastic Dynamic Programming

11.1 The Optimal Control at the n'th Stage

According to the dynamic programming paradigm, the $n'th$ stage is considered first and the optimal feedback control u_n^* is derived.

1. Suppose $Z_n = BT_O$ $\quad and \quad$ $\tau_n \leq \frac{1}{2}(k_n - \tau)$. Then

$$u_n^* = 1 \quad if \quad d_n \geq 0$$
$$u_n^* = 0 \quad if \quad d_n < 0$$

where the threshold

$$d_n = P(T, BT_O C) \log(\frac{P(T \mid BT_O C)}{P(T)}) + P(T, BT_O BT_O) \log(\frac{P(T \mid BT_O BT_O)}{P(T)})$$

$$+ P(T, BT_O BFT_O) \log(\frac{P(T \mid BT_O BFT_O)}{P(T)}) + P(FT, BT_O C)$$

$$\cdot \log(\frac{P(FT \mid BT_O C)}{P(FT)}) + P(FT, BT_O BT_O) \log(\frac{P(FT \mid BT_O BT_O)}{P(FT)})$$

$$+ P(FT, BT_O BFT_O) \log(\frac{P(FT \mid BT_O BFT_O)}{P(FT)})$$

$$- P(T, BT_O) \log \frac{P(T \mid BT_O)}{P(T)} - P(FT, BT_O) \log \frac{P(FT \mid BT_O)}{P(FT)}$$

2. Suppose $Z_n = BFT_O$ $\quad and \quad$ $\tau_n \leq \frac{1}{2}(k_n - \tau)$. Then

$$u_n^* = 1 \quad if \quad d_n \geq 0$$
$$u_i^* = 0 \quad if \quad d_n < 0$$

where the threshold

$$d_n = P(T, BFT_O C) \log(\frac{P(T \mid BFT_O C)}{P(T)}) + P(T, BFT_O BT_O)$$

$$\cdot \log(\frac{P(T \mid BFT_O BT_O)}{P(T)}) + P(T, BFT_O BFT_O) \log(\frac{P(T \mid BFT_O BFT_O)}{P(T)})$$

$$+ P(FT, BFT_O C) \log(\frac{P(FT \mid BFT_O C)}{P(FT)}) + P(FT, BFT_O BT_O)$$

$$\cdot \log(\frac{P(FT \mid BFT_O BT_O)}{P(FT)})$$

$$+ P(FT, BFT_O BFT_O) \log(\frac{P(FT \mid BFT_O BFT_O)}{P(FT)})$$

$$- P(T, BFT_O) \log \frac{P(T \mid BFT_O)}{P(T)} - P(FT, BFT_O) \log \frac{P(FT \mid BFT_O)}{P(FT)}$$

3. Suppose $Z_n = C$. Then

$$u_n^* = 0$$

4. Suppose $Z_n = BT_O$ and $\tau_n > \frac{1}{2}(k_n - \tau)$. Then

$$u_n^* = 0$$

5. Suppose $Z_n = BFT_O$ and $\tau_n > \frac{1}{2}(k_n - \tau)$. Then

$$u_n^* = 0$$

Evidently, the following probabilities, which were calculated in Section 8, are needed.

$$P(T) = \frac{\alpha}{1+\alpha}, \qquad P(FT) = \frac{1}{1+\alpha}$$

$$P(T, BT_O) = \frac{3\alpha}{4(1+\alpha)} P_{TR_O}, \quad P(FT, BT_O) = \frac{1}{1+\alpha}(1 - P_{FTR_O})$$

$$P(T, BFT_O) = \frac{3\alpha}{4(1+\alpha)}(1 - P_{TR_O}), \; P(FT, BFT_O) = \frac{1}{1+\alpha} P_{FTR_O}$$

In addition,

$$P(T \mid BT_O) = \frac{3\alpha P_{TR_O}}{3\alpha P_{TR_O} + 4(1 - P_{FTR_O})}$$

$$P(FT \mid BT_O) = \frac{4(1 - P_{FTR_O})}{3\alpha P_{TR_O} + 4(1 - P_{FTR_O})}$$

$$P(T \mid BFT_O) = \frac{3\alpha(1 - P_{TR_O})}{3\alpha(1 - P_{TR_O}) + 4P_{FTR_O}}$$

$$P(FT \mid BFT_O) = \frac{4P_{FTR_O}}{3\alpha(1 - P_{TR_O}) + 4P_{FTR_O}}$$

Next, we need to differentiate between the two second look scenarios, a) and b):

$$P_a(FT, BT_OC) = 0, \; P_a(T, BT_OC) = \frac{3\alpha}{4(1+\alpha)} P_{TR_O},$$

$$P_a(FT, BFT_OC) = 0, \; P_a(T, BFT_OC) = \frac{3\alpha}{4(1+\alpha)}(1 - P_{TR_O}),$$

$$P_a(T, BT_OBT_O) = 0, \; P_a(FT, BT_OBT_O) = \frac{1}{1+\alpha}(1 - P_{FTR_O})^2$$

$$P_a(T, BFT_OBT_O) = 0, \; P_a(FT, BFT_OBT_O) = \frac{1}{1+\alpha} P_{FTR_O}(1 - P_{FTR_O})$$

$$P_a(T, BT_OBFT_O) = 0, \; P_a(FT, BT_OBFT_O) = \frac{1}{1+\alpha} P_{FTR_O}(1 - P_{FTR_O})$$

$$P_a(T, BFT_OBFT_O) = 0, \; P_a(FT, BFT_OBFT_O) = \frac{1}{1+\alpha} P_{FTR_O}^2$$

and

$$P_b(T, BT_OC) = \frac{\alpha}{4(1+\alpha)}P_{TR_O}, \ P_b(FT, BT_OC) = 0$$

$$P_b(T, BFT_OC) = \frac{\alpha}{4(1+\alpha)}(1 - P_{TR_O}), \ P_b(FT, BFT_OC) = 0$$

$$P_b(T, BT_OBT_O) = \frac{\alpha}{2(1+\alpha)}P_{TR_O}^2, \ P_b(FT, BT_OBT_O) = \frac{1}{1+\alpha}(1 - P_{FTR_O})^2$$

$$P_b(T, BFT_OBT_O) = \frac{\alpha}{2(1+\alpha)}P_{TR_O}(1 - P_{TR_O})$$

$$P_b(FT, BFT_OBT_O) = \frac{1}{1+\alpha}P_{FTR_O}(1 - P_{FTR_O})$$

$$P_b(T, BT_OBFT_O) = \frac{\alpha}{2(1+\alpha)}P_{TR_O}(1 - P_{TR_O})$$

$$P_b(FT, BT_OBFT_O) = \frac{1}{1+\alpha}P_{FTR_O}(1 - P_{FTR_O})$$

$$P_b(T, BFT_OBFT_O) = \frac{\alpha}{2(1+\alpha)}(1 - P_{TR_O})^2$$

$$P_b(FT, BFT_OBFT_O) = \frac{1}{1+\alpha}P_{FTR_O}^2$$

In addition,

$$P_a(T \mid BT_OC) = 1, \ P_a(FT \mid BT_OC) = 0$$
$$P_a(T \mid BFT_OC) = 1, \ P_a(FT \mid BFT_OC) = 0$$
$$P_a(T \mid BT_OBT_O) = 0, \ P_a(FT \mid BT_OBT_O) = 1$$
$$P_a(T \mid BT_OBFT_O) = 0, \ P_a(FT \mid BT_OBFT_O) = 1$$
$$P_a(T \mid BFT_OBT_O) = 0, \ P_a(FT \mid BFT_OBT_O) = 1$$
$$P_a(T \mid BFT_OBFT_O) = 0, \ P_a(FT \mid BFT_OBFT_O) = 1$$

and

$$P_b(T \mid BT_OC) = 1, \ P_b(FT \mid BT_OC) = 0$$
$$P_b(T \mid BFT_OC) = 1, \ P_b(FT \mid BFT_OC) = 0$$

$$P_b(T \mid BT_OBT_O) = \frac{\alpha P_{TR_O}^2}{\alpha P_{TR_O}^2 + 2(1 - P_{FTR_O})^2}$$

$$P_b(FT \mid BT_OBT_O) = \frac{2(1 - P_{FTR_O})^2}{\alpha P_{TR_O}^2 + 2(1 - P_{FTR_O})^2}$$

$$P_b(T \mid BT_O BFT_O) = \frac{\alpha P_{TR_O}(1 - P_{TR_O})}{\alpha P_{TR_O}(1 - P_{TR_O}) + 2P_{FTR_O}(1 - P_{FTR_O})}$$

$$P_b(FT \mid BT_O BFT_O) = \frac{2P_{FTR_O}(1 - P_{FTR_O})}{\alpha P_{TR_O}(1 - P_{TR_O}) + 2P_{FTR_O}(1 - P_{FTR_O})}$$

$$P_b(T \mid BFT_O BT_O) = \frac{\alpha P_{TR_O}(1 - P_{TR_O})}{\alpha P_{TR_O}(1 - P_{TR_O}) + 2P_{FTR_O}(1 - P_{FTR_O})}$$

$$P_b(FT \mid BFT_O BT_O) = \frac{2P_{FTR_O}(1 - P_{FTR_O})}{\alpha P_{TR_O}(1 - P_{TR_O}) + 2P_{FTR_O}(1 - P_{FTR_O})}$$

$$P_b(T \mid BFT_O BFT_O) = \frac{\alpha(1 - P_{TR_O})^2}{\alpha(1 - P_{TR_O})^2 + 2P_{FTR_O}^2}$$

$$P_b(FT \mid BFT_O BFT_O) = \frac{2P_{FTR_O}^2}{\alpha(1 - P_{TR_O})^2 + 2P_{FTR_O}^2}$$

This allows us to express the decision thresholds d_n as a function of the problem parameters α, P_{TR_O}, and P_{FTR_O}. For $Z_n = BT_O$ and $\tau_n \leq \frac{1}{2}(k_n - \tau)$ we calculate the fixed threshold

$$d_{n_a} = \frac{1}{4(1 + \alpha)}[3\alpha P_{TR_O} \log(\frac{3\alpha P_{TR_O} + 4(1 - P_{FTR_O})}{3\alpha P_{TR_O}}) +$$

$$4(1 - P_{FTR_O}) \log \frac{3\alpha P_{TR_O} + 4(1 - P_{FTR_O})}{4(1 - P_{FTR_O})}]$$

$$d_{n_b} =$$

The d_{n_a} threshold is an expression of the form

$$d_{n_a} = \frac{1}{4(1 + \alpha)}[a \log(\frac{a + b}{a}) + b \log(\frac{a + b}{b})]$$

$$= \frac{1}{4(1 + \alpha)}[\log((1 + \frac{b}{a})^a \cdot (1 + \frac{a}{b})^b)]$$

and $\forall\ a > 0, b > 0$,

$$(1 + \frac{b}{a})^a \cdot (1 + \frac{a}{b})^b > 1$$

Hence the threshold

$$d_{n_a} > 0$$

and in case a)

$$u_n^* = 1$$

For $Z_n = BFT_O$ and $\tau_n \leq \frac{1}{2}(k_n - \tau)$ we calculate the fixed threshold

$$d_{n_a} = \frac{1}{4(1+\alpha)}[3\alpha(1-P_{TR_O})\log(\frac{3\alpha(1-P_{TR_O})+4P_{FTR_O}}{3\alpha(1-P_{TR_O})})+$$

$$4P_{FTR_O}\log\frac{3\alpha(1-P_{TR_O})+4P_{FTR_O}}{4P_{FTR_O}}]$$

$$d_{n_b} =$$

Similar to the argument from above, the threshold

$$d_{n_a} > 0$$

and in case a)

$$u_n^* = 1$$

Remark: If $\tau_n > k_n$ then by the time the operator's classification of object O_n is received, the MAV will have fallen out of the sky.

11.2 The Optimal Control in the First $n-1$ Stages

The optimality principle of dynamic programming is applied and the optimal feedback control $u_i^*(k_i, Z_i, \tau_i)$ for $i = 1, ..., n-1$, is derived.

1. Suppose $Z_i = BT_O$ and $\tau_i \leq \frac{1}{2}(k_i + i - n - \tau)$. Then

$$u_i^* = 1 \quad if \quad d_i \geq 0$$
$$u_i^* = 0 \quad if \quad d_i < 0$$

where the threshold

$$d_i = P(T, BT_OC)\log(\frac{P(T \mid BT_OC)}{P(T)}) + P(T, BT_OBT_O)\log(\frac{P(T \mid BT_OBT_O)}{P(T)})$$

$$+ P(T, BT_OBFT_O)\log(\frac{P(T \mid BT_OBFT_O)}{P(T)}) + P(FT, BT_OC)$$

$$\cdot \log(\frac{P(FT \mid BT_UC)}{P(FT)}) + P(FT, BT_OBT_O)\log(\frac{P(FT \mid BT_OBT_O)}{P(FT)})$$

$$+ P(FT, BT_OBFT_O)\log(\frac{P(FT \mid BT_OBFT_O)}{P(FT)}) + [P(T, BT_OC)$$

$$+ P(T, BT_OBT_O) + P(T, BT_OBFT_O) + P(FT, BT_OBT_O)$$

$$+ P(FT, BT_OBFT_O)] \cdot I_{i+1}(k_i - 1 - 2\tau_i - \tau) - P(T, BT_O)\log\frac{P(T \mid BT_O)}{P(T)}$$

$$- P(FT, BT_O)\log\frac{P(FT \mid BT_O)}{P(FT)} - [P(T, BT_O) + P(FT, BT_O)] \cdot I_{i+1}(k_i - 1)$$

Thus, for $i = n - 1, ..., 1$,

$$d_i = d_n + [P(T, BT_OC) + P(T, BT_OBT_O) + P(T, BT_OBFT_O)$$
$$+ P(FT, BT_OBT_O) + P(FT, BT_OBFT_O)] \cdot I_{i+1}(k_i - 1 - 2\tau_i - \tau)$$
$$- [P(T, BT_O) + P(FT, BT_O)] \cdot I_{i+1}(k_i - 1)$$

We calculate

$$d_{i_a} = d_{n_a} + \frac{1}{4(1+\alpha)}[3\alpha P_{TR_O} + 4(1 - P_{FTR_O})]$$
$$\cdot [I_{i+1}(k_i - 1 - 2\tau_i - \tau) - I_{i+1}(k_i - 1)]$$

and

$$d_{i_b} = d_{n_b} + \frac{1}{4(1+\alpha)}[3\alpha P_{TR_O} + 4(1 - P_{FTR_O})]$$
$$\cdot [I_{i+1}(k_i - 1 - 2\tau_i - \tau) - I_{i+1}(k_i - 1)]$$

2. Suppose $Z_i = BFT_O$ and $\tau_i \leq \frac{1}{2}(k_i + i - n - \tau)$. Then

$$u_i^* = 1 \quad if \quad d_i \geq 0$$
$$u_i^* = 0 \quad if \quad d_i < 0$$

where the threshold

$$d_i = P(T, BFT_OC)\log(\frac{P(T \mid BFT_OC)}{P(T)}) + P(T, BFT_OBT_O)$$

$$\cdot \log(\frac{P(T \mid BFT_OBT_O)}{P(T)}) + P(T, BFT_OBFT_O)\log(\frac{P(T \mid BFT_OBFT_O)}{P(T)})$$

$$+ P(FT, BFT_OBT_O)\log(\frac{P(FT \mid BFT_OBT_O)}{P(FT)}) + P(FT, BFT_OBFT_O)$$

$$\cdot \log(\frac{P(FT \mid BFT_OBFT_O)}{P(FT)}) + [P(T, BFT_OC) + P(T, BFT_OBT_O)$$

$$+ P(T, BFT_OBFT_O) + P(FT, BFT_OBT_O) + P(FT, BFT_OBFT_O)]$$

$$\cdot I_{i+1}(k_i - 1 - 2\tau_i - \tau) - P(T, BFT_O)\log\frac{P(T \mid BFT_O)}{P(T)}$$

$$- P(FT, BFT_O)\log\frac{P(FT \mid BFT_O)}{P(FT)}$$

$$- [P(T, BFT_O) + P(FT, BFT_O)] \cdot I_{i+1}(k_i - 1)$$

Thus, for $i = n - 1, ..., 1$,

$$d_i = d_n + [P(T, BFT_OC) + P(T, BFT_OBT_O) + P(T, BFT_OBFT_O)$$
$$+ P(FT, BFT_OBT_O) + P(FT, BFT_OBFT_O)] \cdot I_{i+1}(k_i - 1 - 2\tau_i - \tau)$$
$$- [P(T, BFT_O) + P(FT, BFT_O)] \cdot I_{i+1}(k_i - 1)$$

We calculate

$$d_{i_a} = d_{n_a} + \frac{1}{4(1+\alpha)}[3\alpha(1 - P_{TR_O}) + 4P_{FTR_O}]$$
$$\cdot [I_{i+1}(k_i - 1 - 2\tau_i - \tau) - I_{i+1}(k_i - 1)]$$

and

$$d_{i_b} = d_{n_b} + \frac{1}{4(1+\alpha)}[3\alpha(1 - P_{TR_O}) + 4P_{FTR_O}]$$
$$\cdot [I_{i+1}(k_i - 1 - 2\tau_i - \tau) - I_{i+1}(k_i - 1)]$$

3. Suppose $Z_i = C$. Then

$$u_i^* = 0$$

4. Suppose $Z_i = BT_O$ and $\tau_i > \frac{1}{2}(k_i + i - n - \tau)$. Then

$$u_i^* = 0$$

5. Suppose $Z_i = BFT_O$ and $\tau_i > \frac{1}{2}(k_i + i - n - \tau)$. Then

$$u_i^* = 0$$

Note: To calculate the optimal control u_i^*, $i = n-1, ..., 1$, the value function $I_{i+1}(k_{i+1})$, $i = n, ..., 2$, is needed. When the threshold d_i at stage i is calculated, according to the dynamic programming paradigm, the value function $I_{i+1}(k_{i+1})$ is already available.

11.3 Running Cost

The running cost at stage i,

$$r_i = r_i(Z_i, u_i)$$

is the average information gained at stage i. The function $r_i(Z_i, u_i)$, $i = 1, ..., n$, is specified as follows.

1. For $Z_i = C$

$$r_i = P(T, C)\log(\frac{P(T \mid C)}{P(T)}) + P(FT, C)\log(\frac{P(FT \mid C)}{P(FT)})$$
$$= P(T, C)\log(\frac{P(T \mid C)}{P(T)})$$

We calculate the probability

$$P(T, C) = \frac{\alpha}{4(1+\alpha)}$$

Hence

$$r_i = \frac{\alpha}{4(1+\alpha)}\log(\frac{1+\alpha}{\alpha})$$

2. For $Z_i = BT_O$, $u_i = 0$

$$r_i = P(T, BT_O) \log(\frac{P(T \mid BT_O)}{P(T)}) + P(FT, BT_O) \log(\frac{P(FT \mid BT_O)}{P(FT)})$$

Hence

$$r_i = \frac{1}{4(1+\alpha)}[3\alpha P_{TR_O} \log(\frac{3(1+\alpha)P_{TR_O}}{3\alpha P_{TR_O} + 4(1 - P_{FTR_O})})$$
$$+ 4(1 - P_{FTR_O}) \log(\frac{4(1+\alpha)(1 - P_{FTR_O})}{3\alpha P_{TR_O} + 4(1 - P_{FTR_O})})]$$

3. For $Z_i = BFT_O$, $u_i = 0$

$$r_i = P(T, BFT_O) \log(\frac{P(T \mid BFT_O)}{P(T)}) + P(FT, BFT_O) \log(\frac{P(FT \mid BFT_O)}{P(FT)})$$

Hence

$$r_i = \frac{1}{4(1+\alpha)}[3\alpha(1 - P_{TR_O}) \log(\frac{3(1+\alpha)(1 - P_{TR_O})}{3\alpha(1 - P_{TR_O}) + 4P_{FTR_O}})$$
$$+ 4P_{FTR_O} \log(\frac{4(1+\alpha)P_{FTR_O}}{3\alpha(1 - P_{TR_O}) + 4P_{FTR_O}})]$$

4. For $Z_i = BT_O$, $u_i = 1$

$$r_i = P(T, BT_OC) \log(\frac{P(T \mid BT_OC)}{P(T)}) + P(T, BT_OBT_O) \log(\frac{P(T \mid BT_OBT_O)}{P(T)})$$
$$+ P(T, BT_OBFT_O)(\log \frac{P(T \mid BT_OBFT_O)}{P(T)}) + P(FT, BT_OC)$$
$$\cdot \log(\frac{P(FT \mid BT_OC)}{P(FT)}) + P(FT, BT_OBT_O) \log(\frac{P(FT \mid BT_OBT_O)}{P(FT)})$$
$$+ P(FT, BT_OBFT_O) \log(\frac{P(FT \mid BT_OBFT_O)}{P(FT)})$$

Hence

$$r_{i_a} = \frac{1}{4(1+\alpha)}[3\alpha \log(\frac{1+\alpha}{\alpha}) + 4(1 - P_{FTR_O}) \log(1+\alpha)]$$

and

$$r_{i_b} = \frac{1}{4(1+\alpha)}[\alpha P_{TR_O} \log(\frac{1+\alpha}{\alpha}) + 2\alpha P^2_{TR_O} \log(\frac{(1+\alpha)P^2_{TR_O}}{\alpha P^2_{TR_O} + 2(1 - P_{FTR_O})^2})$$
$$+ 2\alpha P_{TR_O}(1 - P_{TR_O}) \log(\frac{(1+\alpha)P_{TR_O}(1 - P_{TR_O})}{\alpha P_{TR_O}(1 - P_{TR_O}) + 2P_{FTR_O}(1 - P_{FTR_O})})$$
$$+ 4(1 - P_{FTR_O})^2 \log(\frac{2(1+\alpha)(1 - P_{FTR_O})^2}{\alpha P^2_{TR_O} + 2(1 - P_{FTR_O})^2})$$
$$+ 4P_{FTR_O}(1 - P_{FTR_O}) \log(\frac{2(1+\alpha)P_{FTR_O}(1 - P_{FTR_O})}{\alpha P_{TR_O}(1 - P_{TR_O}) + 2P_{FTR_O}(1 - P_{FTR_O})})]$$

5. For $Z_i = BFT_O$, $u_i = 1$

$$r_i = P(T, BFT_O C) \log(\frac{P(T \mid BFT_O C)}{P(T)}) + P(T, BFT_O BT_O)$$

$$\cdot \log(\frac{P(T \mid BFT_O BT_O)}{P(T)}) + P(T, BFT_O BFT_O)$$

$$\cdot \log(\frac{P(T \mid BFT_O BFT_O)}{P(T)}) + P(FT, BFT_O C) \log(\frac{P(FT \mid BFT_O C)}{P(FT)})$$

$$+ P(FT, BFT_O BT_O) \log(\frac{P(FT \mid BFT_O BT_O)}{P(FT)})$$

$$+ P(FT, BFT_O BFT_O) \log(\frac{P(FT \mid BFT_O BFT_O)}{P(FT)})$$

Hence

$$r_{i_a} = \frac{1}{4(1+\alpha)}[3\alpha(1 - P_{TR_O}) \log(\frac{1+\alpha}{\alpha}) + 4P_{FTR_O}(1 - P_{FTR_O}) \log(1 + \alpha)]$$

and

$$r_{i_b} = \frac{1}{4(1+\alpha)}[\alpha(1 - P_{TR_O}) \log(\frac{1+\alpha}{\alpha})$$

$$+ 2\alpha P_{TR_O}(1 - P_{TR_O}) \log(\frac{(1+\alpha)P_{TR_O}(1 - P_{TR_O})}{\alpha P_{TR_O}(1 - P_{TR_O}) + 2P_{FTR_O}(1 - P_{FTR_O})})$$

$$+ 2\alpha(1 - P_{TR_O})^2 \log(\frac{(1+\alpha)(1 - P_{TR_O})^2}{\alpha(1 - P_{TR_O})^2 + 2P_{FTR_O}^2})$$

$$+ 4P_{FTR_O}(1 - P_{FTR_O}) \log(\frac{2(1+\alpha)P_{FTR_O}(1 - P_{FTR_O})}{\alpha P_{TR_O}(1 - P_{TR_O}) + 2P_{FTR_O}(1 - P_{FTR_O})})$$

$$+ 4P_{FTR_O}^2 \log(\frac{2(1+\alpha)P_{FTR_O}^2}{\alpha(1 - P_{TR_O})^2 + 2P_{FTR_O}^2})]$$

Note that the running cost function is not i - dependent, and

$$r_i(\cdot, \cdot) = r_n(\cdot, \cdot) \quad \forall \quad i = 1, ..., n-1$$

11.4 Terminal Value Function

In this section the terminal value function, that is, the value function at the terminal stage n, $I_n(k_n)$, is calculated. The starting point: the knowledge of the optimal control at stage n, namely the function $u_n^*(k_n, Z_n, \tau_n)$; the latter is needed to evaluate r_n.

$$I_n(k_n) = \mathbf{E}_{Z_n, \tau_n} (r_n(Z_n, u_n^*(k_n, Z_n, \tau_n)))$$

$$= \mathbf{E}_{\tau_n} (r_n(C, u_n^*(k_n, C)) \cdot P(C) + r_n(BT_0, u_n^*(k_n, BT_O, \tau_n)) \cdot P(BT_O)$$

$$+ r_n(BFT_O, u_n^*(k_n, BFT_O, \tau_n)) \cdot P(BFT_O))$$

Recall

$$P(C) = \frac{\alpha}{4(1+\alpha)}$$

and we calculate

$$\begin{aligned}
P(BT_O) &= P(BT_O \mid T)P(T) + P(BT_O \mid FT)P(FT) \\
&= \frac{P(BT_O, T)}{P(T)}P(T) + \frac{P(BT_O, FT)}{P(FT)}P(FT) \\
&= P(T, BT_O) + P(FT, BT_O) \\
&= \frac{3\alpha P_{TR_O} + 4(1 - P_{FTR_O})}{4(1+\alpha)}
\end{aligned}$$

Similarly,

$$P(BFT_O) = \frac{3\alpha(1 - P_{TR_O}) + 4P_{FTR_O}}{4(1+\alpha)}$$

Consequently,

$$\begin{aligned}
I_n(k_n) = &\frac{\alpha}{4(1+\alpha)} r_n(C, u_n^*(k_n, C)) \\
&+ \frac{1}{4(1+\alpha)} \int_0^1 [(3\alpha P_{TR_O} + 4(1 - P_{FTR_O})) \cdot r_n(BT_O, u_n^*(k_n, BT_O, \tau_n)) \\
&+ (3\alpha(1 - P_{TR_O}) + 4P_{FTR_O}) \cdot r_n(BFT_O, u_n^*(k_n, BFT_O, \tau_n))] \cdot f(\tau_n) d\tau_n ,
\end{aligned}$$

that is,

$$\begin{aligned}
I_n(k_n) = &\frac{1}{4(1+\alpha)} \{ \alpha r_n(C, u_n^*(k_n, C)) + \int_0^1 [(3\alpha P_{TR_O} + 4(1 - P_{FTR_O})) \\
&\cdot r_n(BT_O, u_n^*(k_n, BT_O, x)) \\
&+ (3\alpha(1 - P_{TR_O}) + 4P_{FTR_O}) \cdot r_n(BFT_O, u_n^*(k_n, BFT_O, x))] \cdot f(x) dx \}
\end{aligned}$$

Now, for $Z_n = C$, $u_n^* = 0$ and

$$r_n(C, 0) = \frac{\alpha}{4(1+\alpha)} \log(\frac{1+\alpha}{\alpha})$$

Hence, the terminal value function

$$\begin{aligned}
I_n(k_n) = &\frac{1}{4(1+\alpha)} \{ \frac{\alpha^2}{4(1+\alpha)} \log(\frac{1+\alpha}{\alpha}) + \int_0^1 [(3\alpha P_{TR_O} + 4(1 - P_{FTR_O})) \\
&\cdot r_n(BT_O, u_n^*(k_n, BT_O, x)) + (3\alpha(1 - P_{TR_O}) + 4P_{FTR_O}) \\
&\cdot r_n(BFT_O, u_n^*(k_n, BFT_O, x))] \cdot f(x) dx \}
\end{aligned} \tag{38}$$

If

$$k_n < 2 + \tau$$

the integral

$$I \equiv \int_0^1 [(3\alpha P_{TR_O} + 4(1 - P_{FTR_O})) \cdot r_n(BT_O, u_n^*(k_n, BT_O, x))$$
$$+ (3\alpha(1 - P_{TR_O}) + 4P_{FTR_O}) \cdot r_n(BFT_O, u_n^*(k_n, BFT_O, x))] \cdot f(x)dx$$
$$= \int_0^{\frac{1}{2}(k_n-\tau)} [(3\alpha P_{TR_O} + 4(1 - P_{FTR_O})) \cdot r_n(BT_O, u_n^*(k_n, BT_O, x))$$
$$+ (3\alpha(1 - P_{TR_O}) + 4P_{FTR_O}) \cdot r_n(BFT_O, u_n^*(k_n, BFT_O, x))] \cdot f(x)dx$$
$$+ \int_{\frac{1}{2}(k_n-\tau)}^1 [(3\alpha P_{TR_O} + 4(1 - P_{FTR_O})) \cdot r_n(BT_O, u_n^*(k_n, BT_O, x))$$
$$+ (3\alpha(1 - P_{TR_O}) + 4P_{FTR_O}) \cdot r_n(BFT_O, u_n^*(k_n, BFT_O, x))] \cdot f(x)dx$$
$$= \int_0^{\frac{1}{2}(k_n-\tau)} [(3\alpha P_{TR_O} + 4(1 - P_{FTR_O})) \cdot r_n(BT_O, u_n^*(k_n, BT_O, x))$$
$$+ (3\alpha(1 - P_{TR_O}) + 4P_{FTR_O}) \cdot r_n(BFT_O, u_n^*(k_n, BFT_O, x))] \cdot f(x)dx$$
$$+ \int_{\frac{1}{2}(k_n-\tau)}^1 [(3\alpha P_{TR_O} + 4(1 - P_{FTR_O})) \cdot r_n(BT_O, 0)$$
$$+ (3\alpha(1 - P_{TR_O}) + 4P_{FTR_O}) \cdot r_n(BFT_O, 0)] \cdot f(x)dx$$
$$= \int_0^{\frac{1}{2}(k_n-\tau)} [(3\alpha P_{TR_O} + 4(1 - P_{FTR_O})) \cdot r_n(BT_O, u_n^*(k_n, BT_O, x))$$
$$+ (3\alpha(1 - P_{TR_O}) + 4P_{FTR_O}) \cdot r_n(BFT_O, u_n^*(k_n, BFT_O, x))] \cdot f(x)dx$$
$$+ \frac{1}{4(1+\alpha)}[(3\alpha P_{TR_O} + 4(1 - P_{FTR_O}))[3\alpha P_{TR_O} \log(\frac{3(1+\alpha)P_{TR_O}}{3\alpha P_{TR_O} + 4(1 - P_{FTR_O})})$$
$$+ 4(1 - P_{FTR_O}) \log(\frac{4(1+\alpha)(1 - P_{FTR_O})}{3\alpha P_{TR_O} + 4(1 - P_{FTR_O})})] + (3\alpha(1 - P_{TR_O})$$
$$+ 4P_{FTR_O}) \cdot [3\alpha(1 - P_{TR_O}) \log(\frac{3(1+\alpha)(1 - P_{TR_O})}{3\alpha(1 - P_{TR_O}) + 4P_{FTR_O}})$$
$$+ 4P_{FTR_O} \log(\frac{4(1+\alpha)P_{FTR_O}}{3\alpha(1 - P_{TR_O}) + 4P_{FTR_O}})]]q_n(k_n)$$

where

$$q_n(k_n) \equiv \int_{\frac{1}{2}(k_n-\tau)}^1 f(x)dx$$

Thus,

$$I_n(k_n) = \frac{1}{4(1+\alpha)} \{ \frac{\alpha^2}{4(1+\alpha)} \log(\frac{1+\alpha}{\alpha}) + \frac{1}{4(1+\alpha)} q_n(k_n)$$

$$\cdot [(3\alpha P_{TR_O} + 4(1 - P_{FTR_O})) \cdot [3\alpha P_{TR_O} \log(\frac{3(1+\alpha)P_{TR_O}}{3\alpha P_{TR_O} + 4(1 - P_{FTR_O})})$$

$$+ 4(1 - P_{FTR_O}) \log(\frac{4(1+\alpha)(1 - P_{FTR_O})}{3\alpha P_{TR_O} + 4(1 - P_{FTR_O})})] + (3\alpha(1 - P_{TR_O})$$

$$+ 4P_{FTR_O}) \cdot [3\alpha(1 - P_{TR_O}) \log(\frac{3(1+\alpha)(1 - P_{TR_O})}{3\alpha(1 - P_{TR_O}) + 4P_{FTR_O}})$$

$$+ 4P_{FTR_O} \log(\frac{4(1+\alpha)P_{FTR_O}}{3\alpha(1 - P_{TR_O}) + 4P_{FTR_O}})]]$$

$$+ \int_0^{\frac{1}{2}(k_n - \tau)} [(3\alpha P_{TR_O} + 4(1 - P_{FTR_O})) \cdot r_n(BT_O, u_n^*(k_n, BT_O, x))$$

$$+ (3\alpha(1 - P_{TR_O}) + 4P_{FTR_O}) \cdot r_n(BFT_O, u_n^*(k_n, BFT_O, x))] \cdot f(x)dx \}$$

Hence, if $k_n < 2 + \tau$, the terminal value function

$$I_n(k_n) = \frac{1}{16(1+\alpha)^2} \{ \frac{\alpha^2}{4(1+\alpha)} \log(\frac{1+\alpha}{\alpha}) + q_n(k_n)$$

$$\cdot [(3\alpha P_{TR_O} + 4(1 - P_{FTR_O})) \cdot [3\alpha P_{TR_O} \log(\frac{3(1+\alpha)P_{TR_O}}{3\alpha P_{TR_O} + 4(1 - P_{FTR_O})})$$

$$+ 4(1 - P_{FTR_O}) \log(\frac{4(1+\alpha)(1 - P_{FTR_O})}{3\alpha P_{TR_O} + 4(1 - P_{FTR_O})})] + (3\alpha(1 - P_{TR_O})$$

$$+ 4P_{FTR_O}) \cdot [3\alpha(1 - P_{TR_O}) \log(\frac{3(1+\alpha)(1 - P_{TR_O})}{3\alpha(1 - P_{TR_O}) + 4P_{FTR_O}})$$

$$+ 4P_{FTR_O} \log(\frac{4(1+\alpha)P_{FTR_O}}{3\alpha(1 - P_{TR_O}) + 4P_{FTR_O}})]] + 4(1+\alpha)$$

$$\cdot \int_0^{\frac{1}{2}(k_n - \tau)} [(3\alpha P_{TR_O} + 4(1 - P_{FTR_O})) \cdot r_n(BT_O, u_n^*(k_n, BT_O, x))$$

$$+ (3\alpha(1 - P_{TR_O}) + 4P_{FTR_O}) \cdot r_n(BFT_O, u_n^*(k_n, BFT_O, x))]$$

$$\cdot f(x)dx \} \tag{39}$$

11.5 Recursion

In this section the recursion for the value function is derived. The starting point: the knowledge of the optimal control at stage i, namely the function $u_i^*(k_i, Z_i, \tau_i)$. The principle of optimality yields

$$I_i(k_i) = \mathbf{E}_{Z_i, \tau_i} \ (r_i(Z_i, u_i^*(k_i, Z_i, \tau_i)) + I_{i+1}(k_i - 1 - (2\tau_i + \tau)u_i^*(k_i, Z_i, \tau_i)) \ )$$

We calculate

$$I_i(k_i) = \mathbf{E}_{\tau_i} \left(\, [r_i(C, u_i^*(k_i, C) + I_{i+1}(k_i - 1 - (2\tau_i + \tau)u_i^*(k_i, C))]P(C) \right.$$
$$+ [r_i(BT_O, u_i^*(k_i, BT_O, \tau_i)) + I_{i+1}(k_i - 1 - (2\tau_i + \tau)u_i^*(k_i, BT_O, \tau_i))]P(BT_O)$$
$$+ [r_i(BFT_O, u_i^*(k_i, BFT_O, \tau_i))$$
$$\left. + I_{i+1}(k_i - 1 - (2\tau_i + \tau)u_i^*(k_i, BFT_O, \tau_i))]P(BFT_O) \, \right),$$

that is,

$$I_i(k_i) = [r_i(C) + I_{i+1}(k_i - 1)]P(C)$$
$$+ \mathbf{E}_{\tau_i} \left(\, [r_i(BT_O, u_i^*(k_i, BT_O, \tau_i)) + I_{i+1}(k_i - 1 - (2\tau_i + \tau)u_i^*(k_i, BT_O, \tau_i))] \right.$$
$$\cdot P(BT_O) + [r_i(BFT_O, u_i^*(k_i, BFT_O, \tau_i))$$
$$\left. + I_{i+1}(k_i - 1 - (2\tau_i + \tau)u_i^*(k_i, BFT_O, \tau_i))]P(BFT_O) \, \right)$$

Hence

$$I_i(k_i) = [P(T, C)\log(\frac{P(T \mid C)}{P(T)}) + I_{i+1}(k_i - 1)]P(C)$$

$$+ \int_0^1 \{[r_i(BT_O, u_i^*(k_i, BT_O, \tau_i)) + I_{i+1}(k_i - 1 - (2\tau_i + \tau)u_i^*(k_i, BT_O, \tau_i))]P(BT_O)$$
$$+ [r_i(BFT_O, u_i^*(k_i, BFT_O, \tau_i)) + I_{i+1}(k_i - 1 - (2\tau_i + \tau)u_i^*(k_i, BFT_O, \tau_i))]$$
$$\cdot P(BFT_O)\}f(\tau_i)d\tau_i$$

Thus,

$$I_i(k_i) = \frac{\alpha}{4(1 + \alpha)}[\frac{\alpha}{4(1 + \alpha)}\log(\frac{1 + \alpha}{\alpha}) + I_{i+1}(k_i - 1)] + \int_0^1 \{[r_i(BT_O, u_i^*(k_i, BT_O, \tau_i))$$
$$+ I_{i+1}(k_i - 1 - (2\tau_i + \tau)u_i^*(k_i, BT_O, \tau_i))]\frac{3\alpha P_{TR_O} + 4(1 - P_{FTR_O})}{4(1 + \alpha)}$$
$$+ [r_i(BFT_O, u_i^*(k_i, BFT_O, \tau_i)) + I_{i+1}(k_i - 1 - (2\tau_i + \tau)u_i^*(k_i, BFT_O, \tau_i))]$$
$$\cdot \frac{3\alpha(1 - P_{TR_O}) + 4P_{FTR_O}}{4(1 + \alpha)}\}f(\tau_i)d\tau_i$$

which yields the recursion

$$I_i(k_i) = \frac{1}{4(1 + \alpha)}[\frac{\alpha^2}{4(1 + \alpha)}\log(\frac{1 + \alpha}{\alpha}) + \alpha I_{i+1}(k_i - 1) + \int_0^1 \{[r_i(BT_O, u_i^*(k_i, BT_O, x))$$
$$+ I_{i+1}(k_i - 1 - (2x + \tau)u_i^*(k_i, BT_O, x))](3\alpha P_{TR_O} + 4(1 - P_{FTR_O}))$$
$$+ [r_i(BFT_O, u_i^*(k_i, BFT_O, x)) + I_{i+1}(k_i - 1 - (2x + \tau)u_i^*(k_i, BFT_O, x))]$$
$$\cdot (3\alpha(1 - P_{TR_O}) + 4P_{FTR_O})\}f(x)dx \,], \quad i = n - 1, ..., 1 \tag{40}$$

If

$$k_i < 2 + \tau + n - i \, ,$$

the integral

$$
\begin{aligned}
I \equiv &\int_0^1 \{[r_i(BT_O, u_i^*(k_i, BT_O, x)) + I_{i+1}(k_i - 1 - (2x + \tau)u_i^*(k_i, BT_O, x))] \\
&\cdot (3\alpha P_{TR_O} + 4(1 - P_{FTR_O})) + [r_i(BFT_O, u_i^*(k_i, BFT_O, x)) \\
&+ I_{i+1}(k_i - 1 - (2x + \tau)u_i^*(k_i, BFT_O, x))] \cdot (3\alpha(1 - P_{TR_O}) + 4P_{FTR_O})\} f(x) dx \\
= &\int_0^{\frac{1}{2}(k_i + i - n - \tau)} \{[r_i(BT_O, u_i^*(k_i, BT_O, x)) + I_{i+1}(k_i - 1 - (2x + \tau)u_i^*(k_i, BT_O, x))] \\
&\cdot (3\alpha P_{TR_O} + 4(1 - P_{FTR_O})) + [r_i(BFT_O, u_i^*(k_i, BFT_O, x)) \\
&+ I_{i+1}(k_i - 1 - (2x + \tau)u_i^*(k_i, BFT_O, x))] \cdot (3\alpha(1 - P_{TR_O}) + 4P_{FTR_O})\} f(x) dx \\
+ &\int_{\frac{1}{2}(k_i + i - n - \tau)}^1 \{[r_i(BT_O, u_i^*(k_i, BT_O, x)) + I_{i+1}(k_i - 1 - (2x + \tau)u_i^*(k_i, BT_O, x))] \\
&\cdot (3\alpha P_{TR_O} + 4(1 - P_{FTR_O})) + [r_i(BFT_O, u_i^*(k_i, BFT_O, x)) \\
&+ I_{i+1}(k_i - 1 - (2x + \tau)u_i^*(k_i, BFT_O, x))] \cdot (3\alpha(1 - P_{TR_O}) + 4P_{FTR_O})\} f(x) dx \\
= &\int_0^{\frac{1}{2}(k_i + i - n - \tau)} \{[r_i(BT_O, u_i^*(k_i, BT_O, x)) + I_{i+1}(k_i - 1 - (2x + \tau)u_i^*(k_i, BT_O, x))] \\
&\cdot (3\alpha P_{TR_O} + 4(1 - P_{FTR_O})) + [r_i(BFT_O, u_i^*(k_i, BFT_O, x)) \\
&+ I_{i+1}(k_i - 1 - (2x + \tau)u_i^*(k_i, BFT_O, x))] \cdot (3\alpha(1 - P_{TR_O}) + 4P_{FTR_O})\} f(x) dx \\
+ &\int_{\frac{1}{2}(k_i + i - n - \tau)}^1 \{[r_i(BT_O, 0) + I_{i+1}(k_i - 1)](3\alpha P_{TR_O} + 4(1 - P_{FTR_O})) \\
&+ [r_i(BFT_O, 0) + I_{i+1}(k_i - 1)] \cdot (3\alpha(1 - P_{TR_O}) + 4P_{FTR_O})\} f(x) dx
\end{aligned}
$$

Now

$$
\begin{aligned}
I_1 \equiv &\int_{\frac{1}{2}(k_i + i - n - \tau)}^1 \{[r_i(BT_O, 0) + I_{i+1}(k_i - 1)](3\alpha P_{TR_O} + 4(1 - P_{FTR_O})) \\
&+ [r_i(BFT_O, 0) + I_{i+1}(k_i - 1)] \cdot (3\alpha(1 - P_{TR_O}) + 4P_{FTR_O})\} f(x) dx \\
= &\{ [\frac{1}{4(1 + \alpha)}[3\alpha P_{TR_O} \log(\frac{3(1 + \alpha)P_{TR_O}}{3\alpha P_{TR_O} + 4(1 - P_{FTR_O})}) + 4(1 - P_{FTR_O}) \\
&\cdot \log(\frac{4(1 + \alpha)(1 - P_{FTR_O})}{3\alpha P_{TR_O} + 4(1 - P_{FTR_O})})] + I_{i+1}(k_i - 1)](3\alpha P_{TR_O} + 4(1 - P_{FTR_O})) \\
+ &[\frac{1}{4(1 + \alpha)}[3\alpha(1 - P_{TR_O}) \log(\frac{3(1 + \alpha)(1 - P_{TR_O})}{3\alpha(1 - P_{TR_O}) + 4P_{FTR_O}}) + 4P_{FTR_O} \\
&\cdot \log(\frac{4(1 + \alpha)P_{FTR_O}}{3\alpha(1 - P_{TR_O}) + 4P_{FTR_O}})] \\
+ &I_{i+1}(k_i - 1)] \cdot (3\alpha(1 - P_{TR_O}) + 4P_{FTR_O}) \} q_i(k_i)
\end{aligned}
$$

where

$$q_i(k_i) \equiv \int_{\frac{1}{2}(k_i+i-n-\tau)}^{1} f(x)dx$$

Hence, if $k_i < 2 + \tau + n - i$, the recursion is

$$
\begin{aligned}
I_i(k_i) = {} & \frac{1}{4(1+\alpha)} \Big[\frac{\alpha^2}{4(1+\alpha)} \log(\frac{1+\alpha}{\alpha}) + \alpha I_{i+1}(k_i - 1) \\
& + \{ \, [\, \frac{1}{4(1+\alpha)} [3\alpha P_{TR_O} \log(\frac{3(1+\alpha)P_{TR_O}}{3\alpha P_{TR_O} + 4(1 - P_{FTR_O})}) \\
& + 4(1 - P_{FTR_O}) \log(\frac{4(1+\alpha)(1 - P_{FTR_O})}{3\alpha P_{TR_O} + 4(1 - P_{FTR_O})})] \\
& + I_{i+1}(k_i - 1)](3\alpha P_{TR_O} + 4(1 - P_{FTR_O})) \\
& + [\, \frac{1}{4(1+\alpha)} [3\alpha(1 - P_{TR_O}) \log(\frac{3(1+\alpha)(1 - P_{TR_O})}{3\alpha(1 - P_{TR_O}) + 4P_{FTR_O}}) \\
& + 4P_{FTR_O} \log(\frac{4(1+\alpha)P_{FTR_O}}{3\alpha(1 - P_{TR_O}) + 4P_{FTR_O}})] + I_{i+1}(k_i - 1) \,] \\
& \cdot (3\alpha(1 - P_{TR_O}) + 4P_{FTR_O}) \, \} q_i(k_i) \\
& + \int_0^{\frac{1}{2}(k_i+i-n-\tau)} \{ [r_i(BT_O, u_i^*(k_i, BT_O, x)) \\
& + I_{i+1}(k_i - 1 - (2x + \tau)u_i^*(k_i, BT_O, x))](3\alpha P_{TR_O} + 4(1 - P_{FTR_O})) \\
& + [r_i(BFT_O, u_i^*(k_i, BFT_O, x)) + I_{i+1}(k_i - 1 - (2x + \tau)u_i^*(k_i, BFT_O, x))] \\
& \cdot (3\alpha(1 - P_{TR_O}) + 4P_{FTR_O}) \} f(x)dx \,], \qquad i = n - 1, ..., 1 \qquad (41)
\end{aligned}
$$

Remark: If the operator delay τ_i is a constant - is not random - then an integration is not needed.

12 Stochastic Dynamic Programming Algorithm

Calculate all required probabilities.
The following algorithm is executed.

1. Calculate the binary valued function $u_n(k_n, Z_n, \tau_n)$ for $0 \leq k_n \leq k_0 - n$, $Z_n \in \{C, BT_O, BFT_O\}$, $0 \leq \tau_n \leq 1$.
2. Calculate the terminal value function $I_n(k_n)$ for $0 \leq k_n \leq k_0 - n$.
3. For $i = n - 1, ..., 1$
   a) Calculate the binary valued optimal control function $u_i^*(k_i, Z_i, \tau_i)$ for $n - i \leq k_i \leq k_0 - i$, $Z_i \in \{C, BT_O, BFT_O\}$, $0 \leq \tau_i \leq 1$.
   b) Calculate the value functions $I_i(k_i)$ for $n - i \leq k_i \leq k_0 - i$.

The optimal feedback control law and the value function are obtained. The maximal expected information gain provided by an optimal inspection strategy, given the MAV's endurance k_0, is $I_1(k_0 - 1)$.

13 Conclusion

The effectiveness of a distributed ISR team consisting of autonomous MAVs and an operator for target classification aiding is analyzed. The presence of false targets is acknowledged and the decision chain, including the human operator, is modelled. The data required by our model is readily derivable from physical considerations and controlled human effectiveness experiments. Thus, a realistic stochastic model of the surveillance mission and decision process has been developed. A sequential inspection problem is formulated and elementary probability theory [4] and the closed form solution of a stochastic dynamic program are used to obtain an optimal inspection strategy. The analytical approach is promulgated, with a view to capture and identify the critical target environment, SAV, MAV, and operator classification performance parameters that impact the effectiveness of the distributed ISR system and mixed initiative operations. We believe that good modeling and analytic methods will produce results which are useful to the war fighter.

References

1. Cooperative Operations in UrbaN TERrain (COUNTER), AFMC 04-197, Approved for public release and obtainable upon request from AFRL/VACA.
2. A.R. Girard, M. Pachter and P. R. Chandler, Decision Making Under Uncertainty and Human Operator Model for Small UAV Operations, submitted to the 2006 ACC.
3. R. J. Gallager, Information Theory and Reliable Communication, Wiley 1968, pp. 18.
4. E. Parzen, *Modern Probability Theory and its Applications*, Wiley 1960.

12 An Investigation of a Dynamic Sensor Motion Strategy

Nathan P. Yerrick, Abhishek Tiwari and David E. Jeffcoat

Summary. This chapter considers a dynamic sensor coverage problem in which a single mobile sensor attempts to monitor multiple sites. Sensor motion is modeled using a discrete time, discrete state Markov process. State dynamics at each site are modeled as a linear system. A stochastic simulation is used to demonstrate previously derived theoretical conditions under which a single sensor is or is not sufficient to maintain a bounded estimate of the state of every site. Observations are made about the relationship of sensor motion to system dynamics. A strategy is presented to find a good sensor motion model based upon the system dynamics and to determine the convexity of the solution set.

1 Introduction

"Sensor coverage is the problem of deploying multiple sensors in an unknown environment for the purpose of automatic surveillance, cooperative exploration or target detection [5]." Sensors can either cover an area of interest statically (fixed sensors) or dynamically (mobile sensors). Static coverage can be used when the area or objects of interest can be completely covered by the associated sensors or the attribute of interest does not change with time. However, dynamic coverage becomes necessary when the available sensors cannot adequately cover the area or objects of interest from fixed positions.

2 Problem Description

This chapter focuses on the dynamic sensor coverage problem when there is only one sensor available to cover N sites. This research is an extension of the work accomplished by Tiwari, *et al.* in [5]. Each site is defined to be a discrete time linear system located at a unique point in space. The sensor maintains discrete time Kalman filter estimates of an attribute of interest at each site, but is constrained to only measuring the attribute of the site where it is physically located at a given instant in time. An indicator function, described by Sinopoli in [4], ensures that if the sensor is over site i at time k, both the measurement and time updates of the Kalman filter are executed for site i.

However, if the sensor is not over site i at time k, then only the time update is executed for site i.

The physical realization of this problem could range from estimating the changing temperature in several buildings to estimating the pollution emission levels of several manufacturing plants in an area. In these examples, the attributes would be temperature and pollution emissions, respectively. The changing attributes of each site are modeled in the following manner. Consider N independently evolving linear time invariant (LTI) systems, whose dynamics are given by

$$\begin{aligned} x_{k+1} &= Ax_k + w_k \\ y_k &= Cx_k + v_k \end{aligned} \tag{1}$$

where x_k, x_{k+1}, $w_k \in \mathbb{R}^{n\times 1}$, and y_k, $v_k \in \mathbb{R}^{m\times 1}$, w and v are Gaussian random vectors with zero mean and covariance matrices Q and R respectively. In this chapter, we track only one attribute per site. Since all sites' attributes are independent, the information pertaining to the attribute is found on the diagonals of A, C, Q, and R. For instance, all information regarding the attribute at site two would be contained at A_{22}, C_{22}, Q_{22}, and R_{22}.

The goal of the Kalman filter is to keep a bounded estimate of the *a priori* error covariance between the random vectors x_k and $\hat{x}_k$; that is, the true state of the system and the estimate of the system state at time k, respectively. The *a priori* error covariance, i.e. the error covariance on the 'time update' side of the Kalman filter, at time k and location i is denoted $P_{i,k}^-$ and is defined as

$$P_{i,k}^- = E[(x_k - \hat{x}_k^-)(x_k - \hat{x}_k^-)^T] \tag{2}$$

In general, the dynamic coverage problem is said to have been solved, if for N sites, the limit of the expected value of the error covariance as k approaches infinity is finite for all sites, if $P_{i,0}^- \geq 0$. Similarly, if the limit of the expected value of the error covariance as k approaches infinity is unbounded for some $P_{i,0}^- \geq 0$, then the dynamic coverage problem is not solved. If the error covariance is unbounded at any site, the system is referred to as unstable; if the system is bounded for all sites, it is referred to as stable [6].

Tiwari, et al. [5] model the sensor's motion as an independently and identically distributed (IID) random process or by a discrete time discrete state (DTDS) ergodic Markov chain. This chapter only considers the Markov chain model. The concept of steady state ergodic Markov chains is described in depth in [2] and [3]. For simplicity, we assume that the sensor can move instantaneously from one site to another at any time k. The DTDS Markov chain has a transition probability matrix T, where T_{ij} is the probability that the sensor will be at site j at time $k+1$ given that the sensor was at site i at k. T_{ii}, the ith diagonal entry of the transition probability matrix, denotes the probability that the sensor remains at site i at time $k+1$ given that the sensor was at site i at k. Also, let π_i be the steady state probability of finding the sensor at site i.

The conditions under which the coverage problem is or is not solved were already discussed in a qualitative sense; now those conditions will be explicitly defined for sensor motion described by an ergodic Markov chain.

(a) Let (A, C) be detectable and $(A, \sqrt{Q})$ be observable, then the sensor fails to solve the coverage problem if at least one of the following conditions holds:

$$\frac{1 - \pi_i(2 - T_{ii})}{1 - \pi_i} > \frac{1}{\alpha_i^2}, \ i \in 1, 2, \ldots, N \tag{3}$$

where α is the eigenvalue of A associated with site i. Note: in this chapter, since every site is independent, all off-diagonal elements will be zero, so α_i will simply be the value of A_{ii} [5].

(b) If matrix C is invertible, then the sensor solves the coverage problem if all of the following conditions hold [5]:

$$\frac{1 - \pi_i(2 - T_{ii})}{1 - \pi_i} < \frac{1}{\alpha_i^2}, \ i \in 1, 2, \ldots, N \tag{4}$$

If the coverage problem is solved, then a lower and upper bound of the expected value of the error covariance, $E[P_{i,k}^-]$, can be obtained based on π_i, T_{ii}, A, C, Q, R and $P_{i,0}$. Reference [5] provides additional detail on these bounds. If the bounds diverge for at least one site, then the coverage problem is not solved. However, if the bounds converge for all sites, then the coverage problem is solved. Equations (3) and (4) will later be referred to as feasibility inequalities. Figures 1(a) and 1(b) illustrate bounds for both unstable and stable sites.

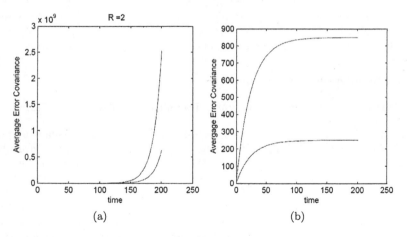

(a)   (b)

Fig. 1. Figure (a) shows a typical unstable site: both bounds diverge. Figure (b) shows a stable site: both bounds converge.

3 Simulation: Checking the Theory

3.1 Simulation Approach

Intuition suggests that the faster the system dynamics of a site relative to other sites, the more time the sensor must spend over this site. Also, it makes sense that the sensor must divide its attention among the sites according to the dynamics of the attribute of each site. The theory in [5] confirms this intuition. We now provide a simulation both to demonstrate the theory and to provide additional insights into the relationship between sensor motion and the dynamics of each site.

A Monte Carlo simulation was built in MATLAB ® using a DTDS Kalman filter with an indicator function and a DTDS Markov chain. The error covariance was computed at each site for all time iterations over multiple replications. The average error covariance over the multiple replications was calculated for all time iterations and plotted with the appropriate error covariance bounds for each site. The values used for equation 1 in the simulation were $C_{ii} = 0.2$, $Q_{ii} = 10$, and $R_{ii} = 2$ for all sites.

The feasibility inequalities in equations (3) and (4) not only convey whether or not the coverage problem is solved, but also indicate, given the dynamics of each site, whether or not a given transition probability matrix provides a feasible solution to the problem. Refer to Figure 2 for the following discussion. As discussed in section 2 if at least one site's bounds are divergent; i.e., if the feasibility inequality in (4) is not satisfied for at least one site, then the transition probability matrix does not provide a feasible solution to the coverage problem. Note that the bounds in Figure 2(c) are divergent and that the inequality in Figure 2(f) is not satisfied. However, in Figures 2(d) and 2(e), the bounds are convergent, indicating that both inequalities in Figures 2(g) and 2(h) are satisfied. In this case, the overall coverage problem is not solved because the dynamics at site one cannot be adequately monitored.

3.2 Simulation Observations

Case I: System Dynamics Equal, Transition Probabilities Varied

Next, we examine a case in which the system dynamics are the same across all sites. The transition probability matrix is constructed so that the coverage problem is solved, even though each site receives a different amount of "sensor attention." In this case, site one will be visited by the sensor most often, and site three will be visited least often [Figure 3]. Each dot in Fig. 3 represents the average error covariance over 10,000 Monte Carlo replications. There are 200 dots at each site, one for each time step. As a site gets less sensor attention, we notice several effects: the bounds converge more slowly, the scale of the bounds is increased, and more dots are outside of the bounds. As a site nears the point of instability, the bounds begin to look more like a line than a curve.

$$A = \begin{bmatrix} 1.3 & 0 & 0 \\ 0 & 1.2 & 0 \\ 0 & 0 & 1.1 \end{bmatrix}$$

(a) System Dynamics

$$T = \begin{bmatrix} .36111 & .33333 & .30556 \\ .36111 & .33333 & .30556 \\ .36111 & .33333 & .30556 \end{bmatrix}$$

$$\pi = \begin{bmatrix} .36111 & .33333 & .30556 \end{bmatrix}$$

(b) Sensor Motion

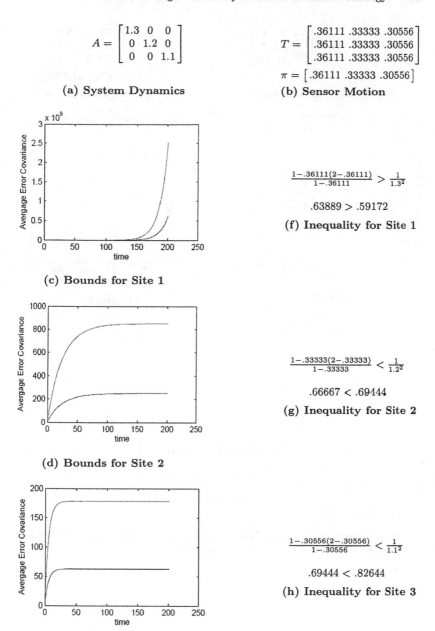

(c) Bounds for Site 1

$$\frac{1-.36111(2-.36111)}{1-.36111} > \frac{1}{1.3^2}$$

$$.63889 > .59172$$

(f) Inequality for Site 1

(d) Bounds for Site 2

$$\frac{1-.33333(2-.33333)}{1-.33333} < \frac{1}{1.2^2}$$

$$.66667 < .69444$$

(g) Inequality for Site 2

(e) Bounds for Site 3

$$\frac{1-.30556(2-.30556)}{1-.30556} < \frac{1}{1.1^2}$$

$$.69444 < .82644$$

(h) Inequality for Site 3

Fig. 2. This figure shows that the bounds computed in simulation match the feasibility inequalities. The feasibility inequality at site 1 is not satisfied, and the bounds at site 1 are divergent. The feasibility inequalities at site 2 and 3 are satisfied, indicating that the bounds at sites 2 and 3 converge.

$$A = \begin{bmatrix} 1.2 & 0 & 0 \\ 0 & 1.2 & 0 \\ 0 & 0 & 1.2 \end{bmatrix}$$

(a) System Dynamics

$$T = \begin{bmatrix} .372 & .32 & .308 \\ .372 & .32 & .308 \\ .372 & .32 & .308 \end{bmatrix}$$

$$\pi = \begin{bmatrix} 0.372 & 0.32 & 0.308 \end{bmatrix}$$

(b) Sensor Motion

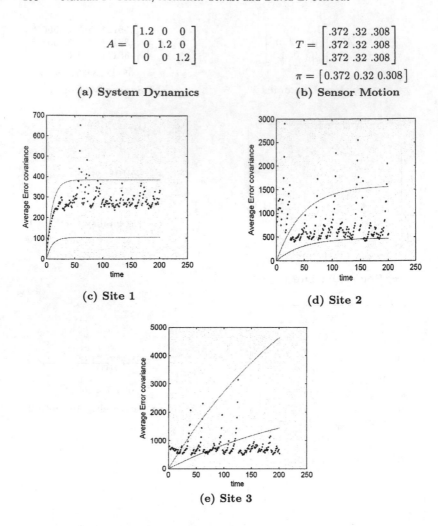

(c) Site 1

(d) Site 2

(e) Site 3

Fig. 3. This figure shows Monte Carlo simulation trends when the system dynamics are equal at each site and the sensor motion is varied.

Case II: System Dynamics Varied, Transition Probabilities Equal

In this case, the transition probability matrix is approximately equal for all sites, meaning that every site will be visited approximately the same number of times. The system dynamics vary across the three sites: site one has the fastest dynamics and site three has the slowest dynamics [Figure 4]. The varied dynamics cause results similar to the varied sensor attention in the last scenario. As a site's dynamics get slower, the bounds' scales decrease, the number of dots outside the bounds decrease, and the less time is required for the bounds to converge. Also, as can be seen in Figure 4(e), when a site's dynamics are slow, and it receives sufficient attention from the sensor, the sample error covariances tend to form a horizontal line well within the bounds.

Summary of Observations

The Monte Carlo simulations confirm both theory and intuition. It is clear in the simulation results that as a site gets more sensor attention, the average error covariance is more likely to cluster within the bounds. We also note that the slower the dynamics of a site, the less sensor attention is needed to ensure that the error covariance remains bounded and stable. Finally, when a slowly evolving site has sufficient sensor attention, the average error covariance remains well within the bounds.

In the next section, we introduce a method that exploits the system dynamics to find a good, although not necessarily optimal, sensor motion model.

4 Optimization: Finding a good sensor motion model given the system dynamics

4.1 Optimization Approach

In this section, we present an approach to determine a good transition probability matrix given a model of site dynamics. Note that the existence of a method that guarantees an optimal solution remains an open question.

The approach presented in this section is a genetic-based heuristic algorithm called scatter search implemented in MATLAB ®. The goal of this meta-heuristic is to provide both intensification and diversification of the solution: intensification meaning that the algorithm uses randomly created feasible solutions to progressively find better results with respect to the objective function, and diversification meaning that the algorithm uses differences between randomly created feasible solutions to increase the chances of finding the best solution available.

The objective function is based on the feasibility inequalities discussed earlier.

$$A = \begin{bmatrix} 1.2 & 0 & 0 \\ 0 & 1.15 & 0 \\ 0 & 0 & 1.1 \end{bmatrix} \qquad T = \begin{bmatrix} .3334 & .3333 & .3333 \\ .3333 & .3334 & .3333 \\ .3333 & .3333 & .3334 \end{bmatrix}$$

$$\pi = \begin{bmatrix} 0.3333 & 0.3333 & 0.3333 \end{bmatrix}$$

(a) System Dynamics   (b) Sensor Motion

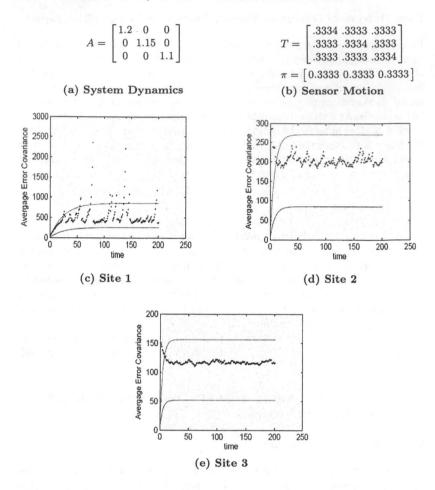

(c) Site 1   (d) Site 2

(e) Site 3

Fig. 4. This figure shows Monte Carlo simulation trends when the system dynamics are varied at each site and the sensor motion is equal for all sites.

The objective is to minimize the maximum ratio over all N sites, as shown in (5), subject to typical probability constraints on π and on the rows of T.

$$\min \left\{ max_{i \in [1,2,...,N]} \left(\frac{\frac{1-\pi_i(2-T_{ii})}{1-\pi_i}}{\frac{1}{\alpha_i^2}} \right) \right\} \tag{5}$$

This min/max objective will be referred to as the performance indicator; the smaller the performance indicator, the better the solution.

In general, the algorithm performs the following process. First, initial solutions for T are randomly generated. A solution is retained if it is determined to be feasible based on equation (4). Next, linear combinations of feasible solutions are constructed to form new solutions. Each solution is assigned a score based upon the objective function (5), and each solution is ranked according to its score. After a set number of iterations in which no better score is obtained, the algorithm is terminated. The solution with the lowest performance indicator is presented as the best solution found.

The simulation was also used to investigate whether the set of feasible solutions is convex. First, solutions were constructed so that the solutions would be feasible, but very close to the infeasibility boundary. Next, linear combinations of these close-to-the-boundary solutions were formed, using equation (6) with $\lambda \in [0, 1]$. If any linear combination of two initial solutions was found to be infeasible, then the solution set cannot be convex. In fact, more than one infeasible solution was found, so the solution set of all feasible solutions is not convex by demonstration. The matrices shown in Figure 5 show one example where the linear combination of two feasible solutions formed an infeasible solution. In this example, $\lambda = 0.5$, the system dynamics are shown in Figure 5(a) , and T_1 and T_2 represent the two feasible solutions in Figures 5(b) and 5(c).

$$T_3 = \lambda T_1 + (1 - \lambda)T_2 \tag{6}$$

$$A = \begin{bmatrix} 1.3 & 0 & 0 \\ 0 & 1.2 & 0 \\ 0 & 0 & 1.1 \end{bmatrix}$$

(a)

$$T_1 = \begin{bmatrix} 0.63105 & 0.3185 & 0.050452 \\ 0.37456 & 0.20869 & 0.41675 \\ 0.46796 & 0.27469 & 0.25734 \end{bmatrix}$$

(b)

$$T_2 = \begin{bmatrix} 0.32893 & 0.27175 & 0.39932 \\ 0.46596 & 0.53137 & 0.0026781 \\ 0.29887 & 0.41671 & 0.28442 \end{bmatrix}$$

(c)

Fig. 5.

4.2 Optimization Results

This section presents results of the optimization algorithm. The effectiveness of the heuristic is demonstrated by comparing solutions generated by the algorithm with randomly generated feasible solutions. Feasible solutions are generated by creating random transition probability matrices constrained so that each row sum is one. The equilibrium probability vector, π_i, is computed for each matrix, then the feasibility inequalities are checked. Those transition matrices that provide a feasible solution are saved. Heuristic solutions are compared to random feasible solutions under two cases that differ in the underlying site dynamics.

Case I: Faster Dynamics

For the first case, one hundred feasible solutions were randomly created with site dynamics modeled by matrix A in Figure 6(a). The mean of the performance indicators from the 100 random solutions was 0.9729, the sample standard deviation was 0.0246, and the minimum and maximum values were 0.9010 and .9997, respectively. The heuristic solution had a performance indicator of 0.8901, a 9% improvement. Figures 6 and 7 show a graphical representation of a randomly created solution and a heuristic solution, respectively. The main difference between the figures is the scale of the bounds. Also, there are more dots outside the bounds in Figure 6 than in Figure 7.

Case II: Slower Dynamics

In the second comparison, one hundred feasible solutions were randomly created with slower dynamics, modeled by Figure 8(a). The mean of the performance indicators from the 100 random solutions was 0.9398, the sample standard deviation was 0.0469, and the minimum and maximum values were 0.7624 and 0.9991, respectively. The heuristic solution had a performance indicator of 0.7027, a 25.2% improvement. From the two comparison scenarios representing faster and slower dynamics, we note that the heuristic's performance is better when the dynamics are slower. Figures 8 and 9 depict a randomly created solution and a heuristic solution, respectively, for the case of slower dynamics. In both figures, the error covariances tend to form a straight line within the bounds. The main difference between the figures is in the mean of the average error covariances. Also, the magnitude of the bounds in Figures 9(c), 9(d), and 9(e) is consistently less than in Figures 8(c), 8(d), and 8(e) .

 One observation from Figure 9(b) is counter-intuitive. We would expect a site with faster dynamics to warrant more sensor attention than sites with slower dynamics, but the heuristic solution in this case does not bear this out. We note that π_3 is greater than π_1 and π_2, even though site three has the slowest dynamics [Figure 9(b)]. Intuition would suggest that π_3 should be the smallest steady-state probability.

5 Summary and Future Work

The goal of this chapter was to demonstrate the theory of the dynamic sensor coverage problem in the case of a single sensor whose motion is modeled by

$$A = \begin{bmatrix} 1.35 & 0 & 0 \\ 0 & 1.3 & 0 \\ 0 & 0 & 1.25 \end{bmatrix}$$

(a) System Dynamics

$$T = \begin{bmatrix} .22925 & .41638 & .35437 \\ .53443 & .060045 & .405525 \\ .49143 & .44609 & .06248 \end{bmatrix}$$

$$\pi = \begin{bmatrix} 0.40002 & 0.31327 & 0.28671 \end{bmatrix}$$

(b) Sensor Motion

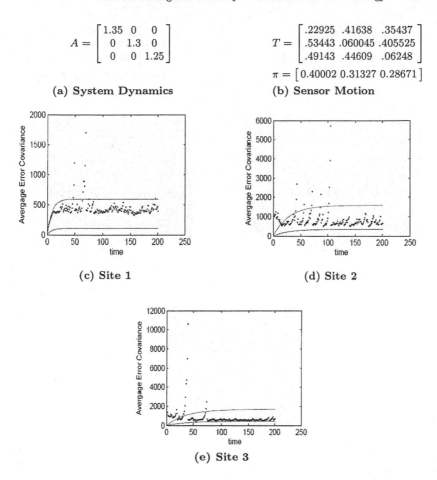

(c) Site 1   **(d) Site 2**

(e) Site 3

Fig. 6. This figure shows the graphical depiction of a representative randomly created solution with a performance indicator of 0.9737.

a DTDS Markov chain. The Monte Carlo simulation demonstrated that the theory developed in [5] matched both intuition and empirical results. Another goal was to show that an improved transition probability matrix can be found using a heuristic. An improvement over random feasible solutions was found using scatter search, though this method does not guarantee a global optimum. The search did provide solutions that were sometimes counter-intuitive. The heuristic performed best in cases with relatively slower system dynamics. The algorithm run time depends on the exact conditions specified, but for the conditions specified in this research, the optimization algorithm required

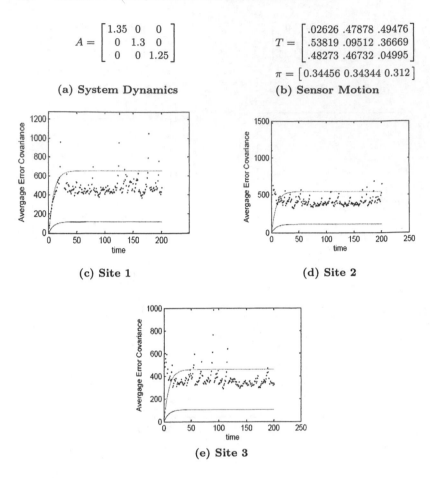

$$A = \begin{bmatrix} 1.35 & 0 & 0 \\ 0 & 1.3 & 0 \\ 0 & 0 & 1.25 \end{bmatrix}$$

(a) System Dynamics

$$T = \begin{bmatrix} .02626 & .47878 & .49476 \\ .53819 & .09512 & .36669 \\ .48273 & .46732 & .04995 \end{bmatrix}$$

$$\pi = \begin{bmatrix} 0.34456 & 0.34344 & 0.312 \end{bmatrix}$$

(b) Sensor Motion

(c) Site 1

(d) Site 2

(e) Site 3

Fig. 7. This figure shows the heuristic solution in graphical form for the case of faster dynamics. The performance indicator in this case is 0.89007.

from two to four minutes of run time on a desktop PC. We also showed by demonstration that the set of feasible solutions is not convex.

Future work will include refined optimization approaches for the single sensor case and extensions to multiple sensors. For the single sensor, an effort will be made to understand why the heuristic produces results that are counter-intuitive with respect to the division of sensor attention, and to further explore under what conditions the scatter search heuristic performs best. The heuristic will also be used to provide insights that might lead to a theoretical solution to the problem of determining an optimal transition matrix.

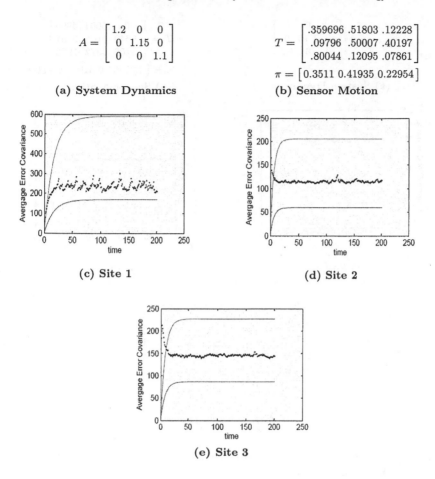

(a) System Dynamics

(b) Sensor Motion

(c) Site 1

(d) Site 2

(e) Site 3

Fig. 8. This figure shows the graphical depiction of a representative randomly created solution that has a performance indicator of 0.9411.

Acknowledgments

The research presented in this chapter was supported in part by the U.S. Air Force Office of Scientific Research under grant number LRIR 99MN01COR. The content of this chapter does not necessarily reflect the position or policy of the U.S. Government. The authors wish to thank Dr. Mesut Yavuz of the University of Florida for his contribution to section 4 on the topic of the scatter search meta-heuristic.

$$A = \begin{bmatrix} 1.2 & 0 & 0 \\ 0 & 1.15 & 0 \\ 0 & 0 & 1.1 \end{bmatrix}$$

(a) System Dynamics

$$T = \begin{bmatrix} 0 & .5004 & .4996 \\ .6649 & 0 & .3351 \\ .3706 & .4372 & .1922 \end{bmatrix}$$

$$\pi = \begin{bmatrix} 0.33891 & 0.31911 & 0.34198 \end{bmatrix}$$

(b) Sensor Motion

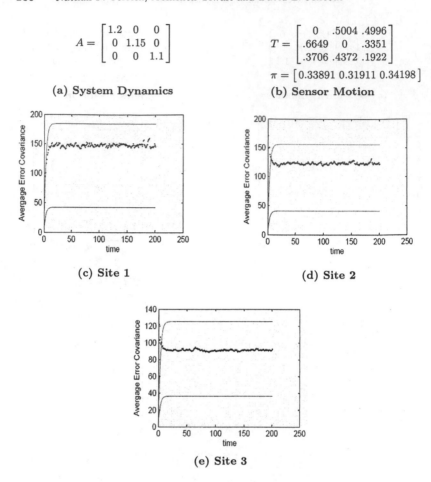

(c) Site 1

(d) Site 2

(e) Site 3

Fig. 9. This figure shows the heuristic solution in graphical form for the case of slower dynamics. The performance indicator was 0.7027.

References

1. R.G. Brown and P. Hwang, *Introduction to Random Signals and Applied Kalman Filtering*, 2nd edn, John Wiley & Sons, New York, 1992.
2. M. Fisz, *Probability Theory and Mathematical Statistics*, 3rd edn, John Wiley & Sons, New York, 1963..
3. L. Kleinrock, *Queueing Systems*, Vol. 1, John Wiley & Sons, New York, 1975.
4. B. Sinopoli, L. Schenato, M. Franceschetti, K. Poolla, M.I. Jordan and S.S. Sastry, Kalman Filtering with Intermittent Observations, *IEEE Transactions on Automatic Control*, Vol. 49/9, pp. 1453–1464, 2004.

5. A. Tiwari, M. Jun, D. Jeffcoat and R. Murray, The Dynamic Sensor Coverage Problem, *16th International Federation of Automatic Control (IFAC) World Congress*, Prague, July 2005.
6. G. Welch and G. Bishop, *An Introduction to the Kalman Filter*, University of North Carolina at Chapel Hill, 2004.

13 Market Based Adaptive Task Allocations for Autonomous Agents

Jared W. Patterson and Kendall E. Nygard

Summary. Allocation of tasks among multiple cooperating agents is fundamentally important in many applications. The application that motivates our work is mission control for teams of unmanned aircraft systems (UAS), where the tasks to allocate are conducting surveillance, dispatching weapons, classifying targets, and assessing battle damage. Our approach is distributed, to avoid the communication overhead and computation time associated with centralized schemes. An additional design goal for the method is to readily support adaptation to new information in near-real time. We utilize a Contract Net Auction Protocol that is extended to two-levels. By including secondary auctions that can potentially transfer previously allocated tasks among agents, the approach partially addresses the inherent non-linearity of a task allocation process. The method consistently produces better solutions than can be provided by a single-level auction. The method is well-suited for dynamic problems if available a priori information is incomplete.

1 Introduction

The work is motivated by a military application in which a team of unmanned aircraft systems (UAS) is dispatched to visit a set of target locations and carry out a single task at each one. The tasks can include such things as conducting surveillance, dispatching weapons, classifying a target, or assessing battle damage. We assume that the aircraft are all launched from known locations, travel at a known speed, and return to specified destination points for recovery. Each aircraft has a task list that specifies a sequence in which multiple targets are potentially visited, and defines a complete tour. Fuel capacity constrains the distance that each aircraft can travel, and all targets must be visited. The objective is to complete the mission in minimum time, which corresponds to the time required for the longest tour. A significant complication is that the problem is dynamic in the sense that some targets become known only after the tours have begun, and must be assigned to an aircraft and sequenced into the task list in near-real time. The core mathematical problem is np-complete because it is a substantial extension of the vehicle routing and traveling salesman problems [1], [2]. Thus, an optimal solution methodology would require computation time that increases exponentially with the number of tasks. Further, the need to dynamically adapt the tours when new targets are discovered mandates the use of a fast heuristic. Although heuristics rarely produce optimal solutions, when well-designed they can produce high quality solutions while still meeting computational time constraints.

We refer to our method as a two-level auction heuristic. Tasks are treated as items available for bid, and agents are treated as both auctioneers and bidders. Bids are based not only upon an agent's current state, but also incorporate secondary bids that model the agent's potential state after auctioning and transferring already allocated tasks. The method is distributed, adaptive, and scales well in terms of the number of tasks and agents involved. Adaptation is accomplished by utilizing bids that modify an incumbent solution in real-time in response to dynamically changing problem data. The method is fundamentally based upon the Contract Net Protocol [3], [4], [5], but extends that paradigm via the secondary bids. Distribution of the computation allows workload sharing and removes the risk of a single point of failure. No single external processing entity is omniscient and thus the removal of a single agent from the team does not hinder the method.

2 Related Work

Although inspired by the application to military unmanned aircraft, the methodology applies more generally to other problem domains. Viewing the aircraft as agents and the targets as tasks reveals that the method applies to many task allocation problems. Examples include distributed control for network routing [6], load balancing [7], electronic marketplaces [8] and distributed robotics [9]. The related vehicle routing problem has a rich literature. Surveys are available in Golden and Assad [1] and Toth and Vigo [2]. Many VRP solution methods are based on simulated annealing [10], tabu search [11], genetic algorithms [12], and branch and cut [13]. Some of the literature on the traveling salesmen problem (TSP) also applies [14]. However, these methods are either fully centralized solutions that require complete communication among all of the agents, or are not suited for the adaptation required for dynamic problems. One of the most popular VRP solution methods is the Clarke-Wright Savings Heuristic [15]. Because of its popularity and ability to quickly generate solutions, the method is used as a benchmark for the method developed in this chapter. A complete exposition on implementation of the Clarke-Wright method and its variants is available in Nelson [16]. Solutions generated by any of the heuristic methods can potentially be improved through post-processing using the n-opt or Or-opt techniques of Lin [17], Lin and Kernighan [18] and Or [19], these are basically heuristics that seek route improving exchanges of arcs in a tour for arcs currently not in the tour. We utilize the 3-opt heuristic in our work. Market-based task allocation methods based on the Contract Net Protocol of Smith [3], are well-suited for distributed problems that require adaptation. Contract Nets were further developed by

Sandholm and Leser [4] and Aknine et al. [5]. In Contract Nets, agents are both bidders and auctioneers and a one-level auction is utilized. Combinatorial auctions [20] allow for bidding on bundles of tasks. Our two-level method, although not fully combinatorial, does capture inter-task dependencies through the secondary auction. The approach of applying market-based techniques to task allocation among autonomous agents is described by Wellman and Wurman [21]. Additional related research in market-based agent interaction is found in Berhault et al. [22], Golfarelli et al. [23], Gerekey and Matarić [24], Zlot et al. [25], and Dias et al. [26]. Berhault in particular presents a combinatorial auction approach to robot routing. Golfarelli et al. present a variation of the Contract Net Protocol where task swapping is an allowable method of task transfer. The Dias work provides a method in which sub-team leaders are responsible for efficiently reallocating tasks within their sub-team. Lemaire [27] describes a one-level market-based approach to a UAS application similar to the one that we address, but is focused on load-balancing among tours.

3 The Two-level Market-based Method

In the procedure, all agents perform the roles of both auctioneer and bidder. Prior to agent movement to conduct the mission, there can be a set of initially known tasks to auction. These pre-movement auctions build the initial routes and provide the initial tours. After the initial tours are generated, agents begin flying their tours. When a new task appears, it is auctioned in the same manner as the previous tasks. The mission continues until all tasks have been accomplished and all agents have returned to their designated end locations. Start and end locations need not coincide. An agent that begins with an empty task list will proceed directly from its start to end location.

When a new task becomes available for auction, each agent that is informed that the task is available formulates a two-level bid based upon two calculations. The agent serving as auctioneer is also a bidder. The first calculation is the minimum tour length incurred by the bidder if it assumes the new task and transfers exactly one of the tasks already on its task list to another agent. Calculating this minimum requires that each task on the list be hypothetically considered for transferring in turn, and that the new task is hypothetically inserted into the best possible location in the tour (this is the cheapest insertion heuristic [14]). The null case, in which no task is transferred, is included. The second calculation is the result of the agent hypothetically offering each task available for transferring to other agents in a simulated auction. Each agent that bids in the simulated auction formulates their bid by calculating the minimum tour length that would result from their inserting the newly transferred task in the best possible location. The maximum of the two calculations is the bid. Thus, the bid is the maximum tour length over all agents that would result from the bidding agent taking on the new task and possibly transferring a previously assigned task. Once all of the bids are received, the best bid is accepted, the new task is allocated to the best bidder, who in turn will carry out the previously calculated best transfer of a task to another agent (possibly null). We note that as new tasks are auctioned in turn, some previously assigned tasks could potentially be transferred multiple times among the agents.

To formalize the procedure, we adopt the notation that T_{new} is the unallocated task, L is the task list for the bidder of interest, T_i is the ith task contained in L, and n is the total number of tasks in L. The tasks in L are indexed starting with 1 and ending with n. The case in which $i = 0$ is the null-task case. In that case, T_i refers to no task in L and therefore steps 2, 5, and most of 4 below are skipped. The bid calculation for a single agent is given below.

1. Let $i = 0$.
2. Run a simulated one-level auction for T_i. Remember the best bid as B_{sim} and then temporarily remove T_i from L.
3. With T_i removed from L, use cheapest insertion to determine the minimum route length possible with T_{new} inserted into L. Let this route length be B_{new}.
4. Compare the best bid from the simulated auction, B_{sim}, with the minimum route length from Step 3, B_{new}, and remember the maximum of the two. This value, denoted $B[i]$, is the new maximum route length if this bidder is awarded T_{new} and auctions task T_i to another agent. Formally, $B[i] = \max(B_{new}, B_{sim})$.
5. Remove T_{new} from L and insert T_i back into L in its original location.
6. Increment i by 1 so that T_i is the next task in L.
7. Repeat steps 2 through 7 until $i = n$.
8. Return as a bid $B_{final} = \min(B[0], B[1], \ldots, B[n])$. Remember task T_i so that it can be auctioned if T_{new} is awarded to this agent and also remember the cheapest insertion point for T_{new} into L.

If a bidder wins the two-level auction and $B[i]$ (where $0 < i \leq n$) was its bid, it generated its bid by accepting T_{new} and simulating an auction of T_i. This simulated auction is now instantiated, and the winning bidder also inserts T_{new} into its task list. The null case happens if $B[0]$ is the winning bid, in which case the winning agent simply accepts and inserts T_{new} without transferring any of its tasks. As a final tour improvement step, a 3-opt procedure is applied to the tours that were changed by a task auction. It would be feasible to employ 3-opt in bid generation as well, but tests reveal little improvement in tour quality by so doing.

The two-level auction scheme is designed to award tasks so that the maximum route length over all of the agents is kept small, which results in a load-leveling aspect of the procedure. The task transfers capture some of the non-linearity in the task allocation solution.

4 Benchmarking Methods

To evaluate the new method, two competing heuristics based on previous research were implemented and used in the empirical testing. These competing methods are given below.

4.1 Sorted Cheapest Insertion with Post 3-opt

In this method, all tasks are first sorted in spatial order with reference to the location of the launch point. Starting with the furthest from the launch point, each task is assigned to a tour using cheapest insertion. Tasks are inserted in order, beginning with the task furthest from the launch point (the literature generally supports this convention). After all tasks are assigned, a 3-opt procedure is performed on each tour to further improve the solution.

4.2 Iterative Clarke-Wright with Post 3-opt

The classical Clarke-Wright method is designed to minimize the sum of all route lengths and must be modified to incorporate the objective of minimizing the maximum tour length. The Clarke-Wright method is based on a "savings" calculation, which is defined as the distance saved when two tours are combined into one. The savings are processed and incorporated into the solution in descending order, ultimately producing a set of tours, one for each vehicle. We implement an Iterative Clarke-Wright (ICW) procedure by setting a small upper bound on tour length and increasing the bound iteratively until the number of tours matches that generated by the new auction procedure. This establishes a point of comparison which is the best that the Clarke-Wright can produce. After the tasks have been assigned to tours and sequenced, the 3-opt procedure is applied to each tour to further improve the solution.

4.3 Test Results

Empirical testing was conducted on suites of test problems for both the static and the dynamic cases. The tests on static problems provide direct comparisons with the benchmark methods. No complete benchmark methods are available for the dynamic case. The dynamic cases validate the scalability of the method and the stability and predictability of results under conditions in which the method was forced to adapt solutions to new information. The dynamic problem test cases were evaluated for varying time lags prior to the information becoming available. Tasks were made available at intervals of 5, 10, 15 and 20 time units of lag. The geographical region is modeled on a 200 X 200 grid with the Euclidean metric used for all distance calculations. Alternatively, a distance matrix or another distance metric could be used. Vehicle start points were set to the center of the grid. Speed was set to one Euclidean distance-unit per time-unit.

For the static problem tests, mean maximum tour length values were generated over 100 replications for multiple combinations of number of tasks and number of agents. The number of agents was varied from 2 to 10 agents and tasks were varied from 3 to 25 tasks in increments of 1 task and 30 to 100 tasks in increments of 10 tasks.

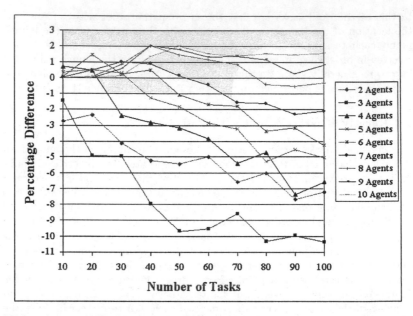

Fig. 1. Average difference between maximum tour lengths for the two-level auction method as a percentage of the Iterative Clarke-Wright value.

The tests reveal that the new two-level auction method and the Iterative Clarke-Wright (ICW) procedure are uniformly superior to the cheapest insertion method. Thus, we present comparisons here only with the ICW. At worst, the ICW method generated tour lengths that were 2.04% longer than the ICW method. At best, the two-level auction produced tours 10.36% shorter than the ICW. Figure 1 provides a more complete percentage based comparison of the two-level auction and ICW. Results indicate that as the number of tasks and agents increase, the two-level auction gains advantage. The explanation is that when there are small numbers of tasks and agents, the two-level method is limited by a lack of task transfer possibilities. For larger numbers of tasks and agents, the richer set of available transfers results in superior solutions, thus validating the value of the method as problems scale. However, even with small numbers of agents and tasks, the method is competitive. The method has special advantage when the task-to-agent ratio is high. Figure 1 and Table 1 illustrate crossover points that indicate the number of tasks necessary for the two-level auction and the ICW to produce equivalent results. Graphically the crossover points are where the percentage difference line crosses the zero percent line. These results provide evidence that the task-to-agent ratio is a predictor of solution quality for the two-level auction method. This clearly establishes that the method is able to take advantage of task transfers, particularly when there are numerous agents submitting bids for tasks.

Table 1. Crossover Points as Illustrated in Figure 1.

Number of Agents	Number of Tasks at the Crossover Point	Task-to-Agent Ratio at the Crossover Point
2	n/a	n/a
3	n/a	n/a
4	22	5.5
5	32	6.4
6	43	7.17
7	53	7.57
8	77	9.63
9	n/a	n/a
10	n/a	n/a

Test problems in which the existence of some tasks is revealed only after some delay were also evaluated. In the UAS application, these delayed tasks are referred to as "pop-up" targets. Problems with pop-up tasks are expected to have solutions that are uniformly inferior to corresponding static problems, because the delay in the appearance of some tasks limits the task allocation possibilities. In these tests, average tour lengths were generated for 100 replications over combinations of from 2 to 10 agents and 10, 20, 50, and 100 tasks. Half of the tasks were auctioned prior to agent movement and the other half were auctioned at the specified intervals.

Table 2 shows the effects of pop-up tasks on the two-level auction method. Table 2 is for 50 tasks with 25 delayed auctions, but the results for 10, 20 and 100 task cases are similar. The numbers are the difference between each method's average results and the results of the two-level auction method with no delayed auctions, expressed in percentages. The ICW method is also included for comparison purposes. Although greater delay in providing tasks lowers solution quality, the degradation is consistent and graceful.

We also note that as the number of agents increases, a lower bound on the average maximum route length is induced. For example, a 20 unit delay in a 50 task problem with 25 delayed auctions implies that the final task will be auctioned 500 time units after the mission has begun, forcing the maximum route length to be at least 500 units. This motivated testing for alternative

Table 2. Comparison of Methods with 50 total tasks and 25 delayed tasks

Agents	ICW	2L w/ 5 Unit Delay	2L w/ 10 Unit Delay	2L w/ 15 Unit Delay	2L w/ 20 Unit Delay
2	6.21%	4.70%	13.43%	25.45%	37.43%
3	9.42%	6.41%	19.87%	35.76%	52.56%
4	3.83%	7.37%	24.41%	47.49%	73.30%
5	2.66%	7.04%	28.40%	58.65%	89.57%
6	0.11%	6.51%	29.81%	65.83%	100.60%
7	-0.90%	5.00%	34.36%	75.38%	110.20%
8	-1.40%	5.58%	40.50%	83.47%	120.20%
9	-1.30%	6.04%	45.71%	87.62%	125.70%
10	-1.70%	5.85%	53.02%	95.63%	134.20%

numbers of delayed tasks. In Table 3, the relative performance of the two-level method is shown for 50 tasks in which 10 tasks are auctioned at delayed intervals.

Table 3. Comparison of Methods with 50 total tasks and 10 delayed tasks

Agents	ICW	2L w/ 5 Unit Delay	2L w/ 10 Unit Delay	2L w/ 15 Unit Delay	2L w/ 20 Unit Delay
2	5.44%	0.67%	1.91%	3.33%	5.13%
3	8.94%	1.03%	2.34%	4.68%	7.59%
4	3.52%	1.05%	2.69%	5.42%	9.88%
5	1.69%	1.33%	3.00%	5.75%	10.65%
6	0.51%	0.71%	2.40%	6.18%	12.69%
7	-0.20%	0.13%	1.38%	5.38%	14.44%
8	-1.65%	0.20%	1.59%	4.78%	18.04%
9	-1.47%	0.36%	1.45%	5.53%	20.77%
10	-1.60%	0.50%	1.35%	4.69%	22.22%

In the two and three agent test cases, the two-level auction method out-performed the ICW method regardless of the delay length. All but the 20 unit delay results averaged within 5% of the results provided by the two-level method without delayed auctions. This is especially interesting in the 10 agent test case since many of the 10 additional tasks were auctioned after agents had already accomplished one or more of the other 40 tasks. This demonstrates the strength of the two-level auction method, but is also expected given that there are a greater number of agents within the same area and thus a greater likelihood that a new task is within a relatively short distance of one of those agents. The two-level auction method easily identifies and allocates each new task to the appropriate agent thereby minimizing the negative effects of introducing tasks late in the scenario.

5 Conclusions

The two-level auction method performs best in test cases with a large task-to-agent ratio, although it is competitive for other problems as well. We note that the method lends itself well to using other tour performance criteria such as task priorities, mobile tasks, task hostility, and task list size limitations. The market-based approach is distributed, scales well, and readily adapts solutions to new information with little computational time required. Complete communication among the agents is not required. The two-level auction method is a viable approach for mission planning for UAS and related problems involving task allocation.

References

1. B. L. Golden and A. A. Assad (eds), *Vehicle Routing: Methods and Studies*, Amsterdam, Netherlands: North-Holland, 1988.
2. P. Toth and D. Vigo (eds), *The Vehicle Routing Problem*, Philadelphia, SIAM, 2002.
3. R. G. Smith, The Contract Net Protocol: high-level communication and control in a distributed problem solver, *IEEE Transaction on Computers*, Vol. C–29, no. 12, December 1980.
4. T. Sandholm and V. Leser, Issues in automated negotiation and electronic commerce: extending the Contract Net framework, in *Proceedings of First International Conference on Multiagent Systems (ICMAS-95)*, pp. 328–335, 1995.
5. S. Aknine, S. Pinson, and M. F. Shakun, An extended multi-agent negotiation protocol, *Autonomous Agents and Multi-Agent Systems*, Vol. 8, pp. 5–45, 2004.
6. R. Schoonderwoerd, O. Holland, and J. Bruten. Ant-like agents for load balancing in telecommunications networks. In W. Lewis Johnson and Barbara Hayes-Roth, editors, *Proceedings of the 1st International Conference on Autonomous Agents*, pp. 209–216, New York, February 5–8 1997. ACM Press.
7. A. Schaerf, Y. Shoham, and M. Tennenholtz. Adaptive load balancing: a study in co-learning. In Sandip Sen, editor, *IJCAI-95 Workshop on Adaptation and Learning in Multiagent Systems*, pp. 78–83, 1995.
8. K. Lerman and O. Shehory. Coalition Formation for Large-Scale Electronic Markets. *Proceedings of ICMAS' 2000*, 2000.
9. D. Goldberg and M. J. Mataric. Coordinating mobile robot group behavior using a model of interaction dynamics, in O. Ozioni, J. Muller, and J. Bradshaw, editors, *Proceedings of the Third Annual Conference on Autonomous Agents (AGENTS-99)*, pp. 100–107, New York, May 1–5 1999. ACM Press
10. W. C. Chiang and R. A. Russell, Simulated annealing metaheuristics for the vehicle routing problem with time windows, *Annals of Operations Research*, Vol. 63, pp. 3–27, 1996.

11. O. Brysy and M. Gendreau, Tabu search heuristics for the vehicle routing problem with time windows, SINTEF Applied Mathematics, Department of Optimization, Oslo, Norway, Tech. Rep. STF42–A01022, 2001.
12. D. Goldberg, *Genetic Algorithms in Search, Optimization, and Machine Learning*. Reading, MA: Addison–Wesley, 1989.
13. D. Applegate, W. Cook, S. Dash, and A. Rohe, Solution of a min-max vehicle routing problem, *INFORMS Journal on Computing*, Vol. 14, no. 2, pp. 132–143, 2002.
14. W. R. Stewart, Jr., A Computational comparison of five heuristic algorithms for the Euclidean traveling salesman problem, in *Evaluating Mathematical Programming Techniques*, no 199., *Lecture Notes in Economics and Mathematical Systems*, J. M. Mulvey (ed.), Springer-Verlag, New York, pp. 104–116, 1982.
15. G. Clarke and J.W. Wright, Scheduling of vehicles from a central depot to a number of delivery points, *Operations Research*, Vol. 12, pp. 568–581, 1964.
16. M. D. Nelson, Implementation techniques for vehicle routing problem, M.S. thesis, North Dakota State University, Fargo, ND, USA, 1983.
17. S. Lin, Computer solution of the traveling salesman problem, *Bell System Computer Journal*, Vol. 44, pp. 2245–2269, 1965.
18. S. Lin and B. W. Kernighan, An effective heuristic algorithm for the TSP, *Operations Research*, Vol. 21, pp. 498–516, 1973.
19. I. Or, Traveling salesman-type combinatorial problems and their relations to the logistics of regional blood banking, Ph.D. dissertation, Northwestern University, Evanston, IL, 1976.
20. S. de Vries and R. V. Vohra, Combinatorial auctions: a survey, *INFORMS Journal on Computing*, Vol. 15. no. 3, pp. 284–309, summer 2003.
21. M. P. Wellman and P.R. Wurman, Market–aware agents for a multiagent world, *Robotics and Autonomous Systems*, Vol. 24, pp. 115–125, 1998.
22. M. Berhault, H. Huang, P. Keskinocak, S. Koenig, W. Elmaghraby, P. Griffin and A. Kleywegt, Robot exploration with combinatorial auctions, in *Proceedings of the International Conference on Intelligent Robots and Systems*, 2003.
23. M. Golfarelli, D. Maio, and D. Rizzi, A task-swap negotiation protocol based on the Contract Net paradigm, Research Centre for Informatics and Telecommunication Systems (CSITE), University of Bologna, Tech. Rep. 005–97, 1997.
24. B. P. Gerekey and M. J. Matarić, Sold!: auction methods for multirobot coordination, *IEEE Transactions on Robotics and Automation*, Vol. 18, no. 5, October 2002.
25. R. Zlot, A. Stentz, M. B. Dias, and S. Thayer, Multi-Robot exploration controlled by a market economy, in *Proceedings of the IEEE International Conference on Robotics and Automation*, pp. 3016–3023, 2002.
26. M. B. Dias and A. Stentz, Opportunistic optimization for market-based Multirobot Control, in *Proceedings of IEEE/RSJ International Conference on Intlligent Robots and Systems (IROS)*, September 2002.
27. T. Lemaire, R. Alami, and S. Lacroix, A distributed tasks allocation scheme in multi-UAV context, in *Proceedings of IEEE International Conference on Robotics and Automation (ICRA '04)*, Vol. 4, pp. 3622–3627, 2004.

14 Cooperative Persistent Surveillance Search Algorithms using Multiple Unmanned Aerial Vehicles

Daniel J. Pack and George W. P. York

Summary. Developing a robust and effective distributed control architecture for multiple platform is one of the key tasks necessary to create a multiple cooperative Unmanned Aerial Vehicles (UAVs) system. The work reported in [1] presented our efforts toward developing such an architecture using behavior-based state machines. In this chapter, we present an in-depth comparative study on one aspect of the control architecture: cooperative persistent search over a large area using multiple UAVs. We compare the performance of four different search algorithms and show the validity of the adaptive search algorithm of the proposed distributed control architecture in allowing multiple UAVs to (1) maximize the coverage of the search area and (2) increase the variability of search patterns, decreasing the flight pattern predictability. The second criterion is a unique aspect of search important to our particular search application of interest involving intelligent mobile ground targets. Simulation results are used to support our findings.

1 Introduction

One of the important applications of multiple UAVs is to survey and monitor a large area to detect missing persons as in a search and rescue mission or a variety of targets in military missions. For convenience, we will refer search objects as targets regardless of its nature for the remainder of this chapter. If some a priori information about a target such as its last known location or its direction of the last movement in the search space is known, one can use a technique based on probability theories to narrow the search to find targets [2]. Our interest, however, is for a more complicated scenario where no target information is available prior to the start of the search.

Our 'smart' mobile targets are capable of moving in any direction and emitting radio frequency (RF) signals. To make the matters worse, the duration of signal emission is random and targets can go silent for some indefinite period of time. For our application scenario, a conventional way of searching a space will no longer suffice. With no a priori knowledge of targets, the optimal strategy is not only to maximize the coverage of a search area but, at the same time, also to create unpredictable search patters. In this chapter, we compare four different cooperative search algorithms for multiple Unmanned

Aerial Vehicles (UAVs) to provide persistent surveillance over a search area. We assume that we have limited UAV resources that cannot achieve 100% coverage of the search area at one time. We also assume that only one UAV can be launched at a time and, before running out of fuel, each UAV must return to the launch pad to be re-fueled and re-launched. We further assume we have perfect communication among all the UAVs and the ground station throughout the search area, and information regarding searched locations is continually broadcasted to all UAVs and the ground station. Figure 1 shows the search environment and a sample search pattern in a two dimensional world. The small rectangle on the left represents the launch pad and the large rectangle on the right is the search area. The particular figure shows four UAVs searching the area with some generic onboard sensors. Each UAV is represented as a single white point and the circular area surrounding the point represents the sensor area one UAV can search at a time. The figure also shows the trace of the areas searched by each UAV over time. Such areas collectively make up the cooperative searched area for a group of UAVs. A careful reader will notice that the intensities used to depict the search trace diminishes over time. To better show the effect of time for a search area a three dimensional graph is useful. Figure 2 shows such a graph showing the search area. The two dimensional axes represent the area and the third axis shows the time elapsed from the last time each location was visited by any UAV. As the time elapses the entire area in the last axis goes down unless UAVs continually search the area.

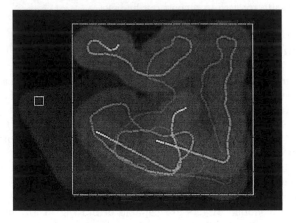

Fig. 1. Four UAVs searching a rectangular search area using an adaptive search algorithm.

We evaluate the search algorithms using a performance metric which considers the coverage frequency of the entire search area, the overall time to

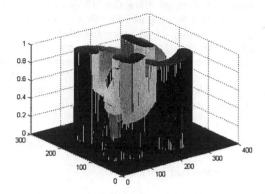

Fig. 2. A search area showing each location with the UAV visit history.

cover the search area, and the number of sorties among UAVs required for a fixed mission time. The four algorithms we compare are (1) creeping line search [3], (2) a zone allocation based search with lawnmower patterns, (3) a random search, and (4) a cooperative adaptive search presented in [1].

For the comparative study, we vary the size of the search areas and the number of UAVs for each of the four algorithms. The chapter is organized as follows. In the next section, we present the adaptive search algorithm in detail within the context of the proposed distributed control architecture. In section III, we describe the other three search algorithms that we will use for the comparative study. Section IV discusses the experimental results. We conclude the chapter with few remarks in Section V.

2 Cooperative Control Search Algorithm

Our proposed distributed control architecture[1] allows each UAV to operate in one of four states: *(1) global search for targets, (2) approach detected target, (3) orbit and locate target, and (4) target re-acquisition.* Each UAV determines its own operating state based on the sensor data and the information it receives from neighboring UAVs. Only the first state of operation, global search for targets, will be discussed next.

The goal for each UAV operating in this state is to balance the following three subtasks: (a) explore locations that has not recently visited, (b) fly away from other UAVs to maximize the search efforts, and (c) fly straight if possible to reduce the fuel consumption. To do so, each UAV constantly updates the flight path history of all neighboring UAVs including its own. Since our application is searching for mobile targets that do not continually emit and can evade detection, we allow the UAVs to re-visit areas that were previously searched. This is accomplished by flying to locations with the most outdated history or no history at all using the flight path history of UAVs. When a location is visited by a UAV, we set the history value at the location to a maximum value and then decrease this value incrementally over time. Figure 2 depicts the search area with visit history. The UAVs seek out locations with no information (not previously visited) or aged information. A UAV uses the following rules [4] to determine the next search point.

1. Fly to a point with the minimum explored history
2. Fly to a point farthest from other neighboring UAVs and the search boundaries
3. Fly straight (maximize fuel efficiency)

These rules are incorporated in the following cost function:

$$search = H(\sum \frac{1}{D_i} + \sum \frac{1}{D_j})(\sqrt{\frac{|\phi|}{\frac{\pi}{p}}} + 1) \tag{1}$$

where H corresponds to a numerical value representing the explored history of a location. D_i represents the distance from the location to each known UAV i and D_j represents the shortest distance from the location to each search area boundaries j. Symbol ϕ is the turn angle required in radians and symbol p represents the number of discrete points used to determine a turning angle. Instead of exhaustively calculating this cost function for all potential points, only a finite number of points equally spaced on a 180 degree arc in front of the UAV are evaluated.

We have found the cost function in the above equation successfully implements the above three rules, encouraging the UAVs to seek out unexplored locations but allowing them to return to previously searched areas as our targets can temporarily disappear. The D_i and D_j terms force UAVs to spread over the search area. Finally the ϕ term prevents excessive turning (dampens oscillations) resulting in less fuel consumption or greater flight times, which is a critical concern for the smaller UAVs. Figure 1 shows a typical pattern of a set of UAVs search an area using the proposed search technique.

3 Three Other Search Algorithms

3.1 Creeping Line Search

This well-known search algorithm is what one uses to completely cover an area from one end to another as shown in Figure 3.

For our application, we expand the number of UAVs involved in search from one to many and assign each UAV a starting point equi-distance apart from others along the creeping line, accounting for flight time from air field.

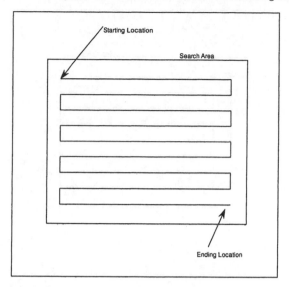

Fig. 3. A creeping line search pattern for a single UAV.

Figure 4 shows a case with three UAVs involved. The figure shows segments allocated for each UAV to cover the entire search area collectively.

Note that for us to execute the search task, some mission plan must be done prior to the beginning of the search based on the distance one UAV can fly without re-fueling.

3.2 Zone-allocation based search algorithm

This particular search algorithm also requires a degree of mission planning to segment the search area into the number of UAVs available and assignment of each UAV with one of the segmented zones prior to the search mission. Once a UAV reaches to its designated zone, it then performs the local search using the lawn-mower search technique similar to the creeping line search to cover its area as shown in Figure 5.

3.3 Random search algorithm

In this method, each UAV makes its own decision to randomly visit one of its neighboring locations without any coordination with other UAVs. The random search algorithm is the most unpredictable search algorithm out of the ones we consider. The unpredictability, however, has the significant consequence in reducing the overall coverage of the search space, since no coordination among UAVs is applied.

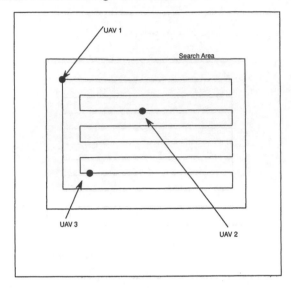

Fig. 4. A creeping line search for multiple UAVs.

4 Results

We evaluated the four search algorithms based on two criteria: (1) search area coverage and (2) predictability of search patterns. For our experiment, we varied the search area size and the number of UAVs involved in the search efforts. According to the selected criteria one can generally rank order the performance of the four search algorithms. For the coverage criterion, the creeping line search algorithm outperforms the other three search algorithms. The zone allocation based algorithm, our adaptive search algorithm, and the random algorithm follow the creeping line search algorithm in covering search area in that order on the coverage performance. On the other hand, for the unpredictability of search pattern criterion, the performance order is the opposite of the one obtained when the search coverage criterion is used: the random search algorithm, our adaptive search algorithm, the zone allocation based algorithm, and the creeping line search algorithm, respectively.

The metric used to score the search performance (surveillance metric) is discussed next. The search space is represented as a set of points placed equal distance from its neighboring points as shown in Figure 6.

To compute the overall search score, we execute the following steps. For each location (i, j), we initialize score zero (not searched) and give a high score if it is recently visited. We normalize a high score with numerical value one. For each time unit, each location's score is decremented. We represent the process with the following equation.

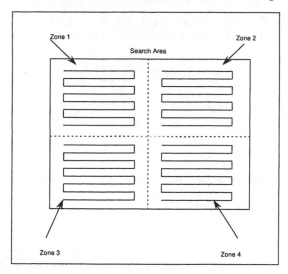

Fig. 5. A search pattern with pre-assigned zone search.

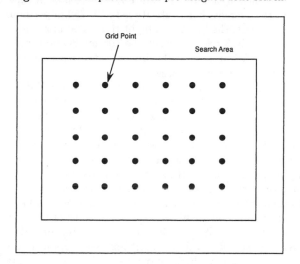

Fig. 6. A search area represented by a grid of points.

$$H_{i,j}(0) = 0.0 \text{ if a locaiton is not visited} \tag{2}$$

$$H_{i,j}(t) = 1.0 \text{ if a location is visited} \tag{3}$$

$$H_{i,j}(t) = H_{i,j}(t-1) - \frac{\delta t}{T} \tag{4}$$

where t is the discrete time variable and δt is the time period for one time unit, and T is the total search time. The score at a particular time is defined as

$$SearchScore(t) = \sum_{i=1}^{I}\sum_{j=1}^{J} \frac{H_{i,j}(t)}{I \cdot J} \qquad (5)$$

The overall score for the search is

$$OverallScore = \frac{1}{T}\sum_{t=1}^{T} SearchScore(t) \qquad (6)$$

The overall search score reflects both the search coverage and the frequency of visits for the overall search area. Note that visiting a portion of the search space repeatedly without covering other portions of the search area will significantly lower the overall search score.

For our experiment, we varied the number of UAVs involved in the search mission from four, six to nine. The search area is predetermined and rectangular. We used three separate search areas: small (15 km by 11 km), medium (25 km by 18 km), and large (34 km by 25 km). Only a single UAV can launch or land on the launch area and a 5 minute re-fuel/re-launch time is required. For each UAV, the average flying velocity is set to 75 km per hour, the sensor range is 1.5 km radius, the maximum flight time is 60 minutes, and the total search mission time is 3 hours. We set the communication interval every 10 seconds and each UAV broadcasts its time stamped position to all UAVs and the ground station in the launch area.

We found that for a small search area, the comparison did not render significant difference. Figure 7 shows a typical result of a run for the small search space using the four different search algorithms with six UAVs. The Pack-York method represents to the adaptive search method. The x-axis represents time in seconds and the y-axis shows the overall search score.

As we increased the size of the search area, we observed noticeable difference among the performance of the search methods. Figure 8 frame (a) shows the performance of the four algorithms in the medium search area with six UAVs and frame (b) shows the same frame (a) zoomed in 1000 seconds after the search began. One can clearly see the superiority of the adaptive search algorithm over the other three algorithms. Frames (c) and (d) of the figure show the performance of the four algorithms when applied to search tasks in the large search space with six and nine UAVs. One can draw a similar conclusion that the adaptive search algorithm came out on top with the highest overall search score. The sinusoidal progression of the performance for all methods reflect the time required for UAVs to abandon search for re-fueling purposes.

Note that when the search area was small, all four algorithms performed similarly, including the random path generation algorithm, which indicates that the random path generation algorithm is a good candidate to meet our

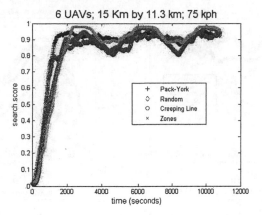

Fig. 7. A graph over time depicting the overall search score for the four search algorithms.

requirement when we are dealing with a small space. As the size of the search area grows, the effectiveness of the random path generation method diminishes.

The overall average search results for the large area are shown in Table 1. When the number of UAVs available for the search is four, the zone allocation-based search method came slightly ahead of the adaptive and creeping line search methods, but as the number of UAVs increases the adaptive search algorithm outperformed all the rest of the search schemes. As predicted, the random research method performed poorly compared to the other three search methods.

Table 1. Search Results

Large Area	4 UAVs	6 UAVs	9 UAVs
Creeping	0.4318	0.5284	0.5803
Zone	0.4514	0.5434	0.5919
Adaptive	0.4497	0.5491	0.6245
Random	0.3788	0.4495	0.5148

5 Conclusion

In this chapter, we presented a comparative study of four different search algorithms: (1) a creeping line search algorithm, (2) a zone allocation based

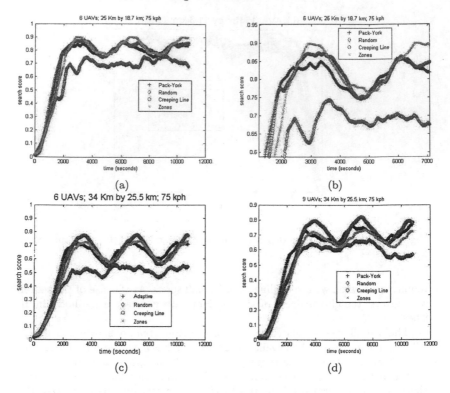

Fig. 8. Search performance on the medium and the large search areas using the four search algorithms.

search algorithm, (3) a random search algorithm, and (4) the adaptive search algorithm. The two criteria we used to evaluate each search algorithm are (1) maximum search area coverage and (2) the unpredictability of search patterns. The second criterion is important for the particular application of our interest: finding RF targets with some intelligence to avoid detection. Our results show that our proposed adaptive search algorithm performed on par or better with the creeping line search algorithm, the best search coverage algorithm, when covering a large search area and outperformed the creeping line search algorithm and the zone allocation based search algorithm when the unpredictability criterion is used. The proposed algorithm finished at the top when we considered both criteria, illustrating the effectiveness of the proposed search algorithm for our particular application. One other note of significance is that the proposed search algorithm does not require pre-planning while the creeping line search algorithm and the zone allocation search algorithm do.

References

1. G.W.P. York, D.J. Pack and J. Harder, Comparison of Cooperative Search Algorithms for Mobile RF Targets using multiple Unmanned Aerial Vehicles, D. Grundel, R. Murphey, P. Pardalos, O. Prokopyev (eds), *Cooperative Systems: COntrol and Optimization*, Springer, Berlin, 2007.
2. T. Furukawa, F. Bourgault, B. Lavis and H.F. Durrant-Whyte, Bayesian Search-and-Tracking Using Coordinated UAVs for Lost Targets, *IEEE International Conference on Robotics and Automation*, pp. 2521–2526, Orlando, May 2006.
3. H. Wollan, Incorporating Heuristically Generated Search Patterns in Search and Rescue, *University of Edinburgh Press*, 2004.
4. D. Pack and B. Mullins, Toward Finding an Universal Search Algorithm for Swarm Robots, *Proceedings of the 2003 IEEE/RJS Conference on Intelligent Robotic Systems (IROS)*, pp. 1945–1950, 2003.

15 Social Networks for Effective Teams

Paul Scerri and Katia Sycara

Summary. In fields as diverse as sociology and physics researchers have been investigating the rich networks that exist in nature. More recently, a small number of multi-agent researchers have shown that the performance of a group can be significantly impacted by the nature of the network that connects them. In this chapter, we build on these initial efforts, performing systematic experiments in an attempt to understand how and why social networks affect group performance. Our key conclusion is that performance of a team can sometimes be improved by imposing a social network with relatively few connections even if it were feasible to connect the agents with a complete network.

1 Introduction

In a variety of important domains hundreds or thousands of heterogeneous agents are required to work together to achieve very complex goals. For example, for a large scale disaster response police, firefighters, paramedics and many others need to work together, albeit loosely, to mitigate the effects of the disaster. With current and future high speed network infrastructure any two members of the team could potentially communicate directly with one another. However, some recent work by Gaston and des Jardinss [5] has hinted that fully connected networks may not necessarily facilitate the most effective coordinated behavior. In this chapter, we explore this question in detail and attempt to quantify and explain the benefits of *not* having complete connectivity between all members of a large team.

In disciplines outside of artificial intelligence, including physics, economics, computer science and sociology, networks between people, *social networks*, have been extensively studied. For example, Milgram showed that people were often related into networks with a rich structural property called the *small world property* [15] and recent work has discovered that such structures exist not only between people but in a range of natural and man-made artifacts, including the Internet and electrical grids [7]. In a typical multiagent system, an agent can maintain connections with all other agents at negligible or no cost. This leads to most multiagent systems having completely connected communications graphs, although there is a small number of exceptions [5, 12].

However, the implicit assumption that complete networks are best for facilitating coordination may not be correct.

Coordination involves solving several intertwined problems in a distributed fashion. In this chapter, we first abstracted and isolated four key things that coordination must do and then compared a team's performance at those tasks using seven different types of networks, including a complete network. The four coordination problems were: (1) sharing information; (2) fusing information; (3) allocating tasks and; (4) developing an aggregate picture of the team state. We contend that these problems abstractly capture many of the things that a team must do. The seven types of network were: (1) complete; (2) lattice; (3) loop; (4) hierarchy; (5) random; (6) small world and; (7) scale free. The key result was that for only *one* of the problems did the complete network lead to the best performance. This result suggests that imposing a logical social network with relatively few links on a team of agents, even when a complete network is feasible, can improve performance in a variety of areas. Moreover, no particular network type was best for each of the problems, suggesting that networks might have to be carefully chosen for the situation.

Coordination requires that a team must address several problems in parallel. Our initial results showed that different networks had different performance for the different sub-problems. However, since no particular network was always best, choosing which network to use for overall coordination is not straight forward. In fact it is reasonable to hypothesize that the "best" network for a particular coordination application will depend on the relative frequency of the different coordination sub-problems and the relative importance of different performance metrics. In an abstract coordination simulator we tested networks with problems requiring relatively different emphasis on each sub-problem. In many cases complete networks turned out to be best although they were not best for sub-problems. However, this was largely due to an additional coordination not described above and very suited to complete networks. However when the coordination problem required relatively more information fusion, complete was not best.

In a cooperative team, there is no concern paid to whether one team member does more work than another. However, in practice, equitably sharing the workload can be important, perhaps for the morale or energy requirements of team members, for load balancing reasons or other reasons. While conducting experiments to understand the impact of networks on performance, a correlation was observed between the performance of a particular network and the disparity between the loads on individual agents. More particularly, the scale-free networks were often performing very well, but a small number of nodes with high degree, i.e., nodes with many links were performing a high proportion of the work required to coordinate. The high degree nodes were essentially allowing the team to centralize problem solving, which was leading to good performance even with algorithms designed to be distributed. If an equitable distribution of coordination load is important then the best "performing" network might not be the best choice.

2 Networks

A network N is a pair (A, E), where A is a set of agents, and E is a set of edges between the agents, $E = \{\{u, v\}|u, v \in A\}$. Many useful measurements of networks have been developed [3], but only two are important here: network *width* and the *degree distribution* of the network. The width of the network is the average maximum distance from any agent in the network to any other agent, where distance is the minimum number of edges that need to be traversed to get from one agent to the other. The width gives an indication of the maximum separation of the agents from one another. The number of edges that agent a is involved with is the *degree* of a, which we write $degree(a)$. The degree distribution of the network is the frequency distribution of the number of edges involving each agent.

To understand the basic impact of a social network on coordination, we conducted experiments with seven different networks. The networks were chosen to be be representative of those used in the literature, but diverse enough to uncover interesting properties. Table 1 shows the networks used in the initial experiments and their width and degree distribution. The *Complete* network is most typically used in multiagent systems. The *Lattice* and *Loop* networks might be used when wireless or other communications limitations prevent more highly connected networks from being used. The *Hierarchy* is rarely used within the multiagent community but is a standard communications framework for human organizations. The *Random* network provides a sort of baseline low degree network. Finally, the *Small World* and *Scale Free* networks have been shown to be very common types of networks in nature [16]. Figure 1 shows small examples of each of the network types.

Notice that we are making no assumption about the underlying physical network, which will in many cases be a complete network. In cases where an agent needs to direct by communication to another specific agent whose identity it knows, that communication does not need to go via the logical network described here.

3 Key Coordination Problems

In the following, we formally describe the coordination problem that must be solved by the team and that is the basis for this work.

Agents, $A(t) = \{a_1, \ldots, a_k\}$, are cooperating on a joint goal. Information, $I = \{i_1, \ldots, i_n\}$, are discrete pieces of information. Some, $I_o \subseteq I$, are able to be sensed, at various times, in the environment. The predicate $Observable(i, t)$ returns *true* if it is possible for some agent to observe information i at time t and returns false otherwise. Some information, e.g., location of a fixed resource, will be always observable, while others, e.g., events, will be observable for short periods. There is no implication that something will be sensed, even if it can be. Other information can only be

Name	Description	Degree Dist.	Width
Complete	Each agent is connected to each other agent	All = 1	1
Lattice	Agents are arranged into 2D grid	All = 4, except edges	$\sqrt{|A|}$
Loop	Agents arranged in a circle	All = 2	$\frac{|A|}{2}$
Hierarchy	Traditional tree, branching factor = 3	All = 4	$\log_n |A|$
Random	Each new agent makes 2 connections to other agents	Average is 4	$\log_n |A|$
Small World	Loop plus each agent makes one random connection	Average is 4	$\log_n |A|$
Scale Free	New agents connect to 3 existing with probability proportional to number of connections a node already has	Exponential	$\log_n |A|$

Table 1. The seven different types of network investigated in this chapter.

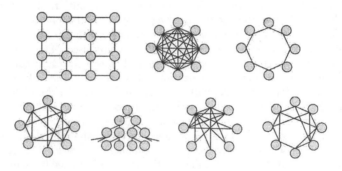

Fig. 1. Examples of the seven network types. Clockwise from top left: Lattice, Complete, Loop, Small World, Scale Free, Hierarchy, Random.

inferred by fusing observable information. i_f can only be inferred from $I_F \subseteq I$ and, hence, must be inferred by some agent $Knows(a,t) \subset I_F$. Some subset of the information is $Knowable(t) \subseteq I$ but an agent might only know some of the information, $Knows(a,t) \subseteq Knowable(t)$. $i_k \in Knowable(t)$ if $\exists t' \leq t, Observable(i_k, t') \vee \forall j \in I_K \exists t' \leq t, Observable(j, t')$

The team $A(t)$ is attempting to achieve a high level goal G, which is broken into discrete sub-tasks $\alpha_1, \ldots, \alpha_n$, typically performed by individuals. A subtask, $alpha_i$ is applicable when the predicate $Applicable(I_{\alpha_i}), I_{\alpha_i}) \subseteq I$ is *true*.

For some tasks, agents require sharable resources. These resources, $R(t) = \{r_1, \ldots, r_m\}$, are a dynamically changing set of available resources. We assume that the resources are discrete and non-consumable. Agent a has exclusive access to $Holds(a) \subseteq R$. $\forall a, b \in A, a \neq b, Holds(a) \cap Holds(b) = \emptyset$.

Agents must perform the individual tasks, when they are applicable for the team to receive reward. The reward received by the team when an agent performs a task is a function of the agent and task as well as what the agent knows and the resources it has. Specifically:

$$Reward(a, \alpha, Knows(a), Holds(a), t) \rightarrow \mathcal{R}$$

Notice that unless $I_{\alpha_i} \subseteq Knowable(t)$, $Reward(\cdot) = 0$, i.e., unless the task is currently applicable the team gets no reward for performing it.

We specifically distinguish between *necessary* and *useful* resources. Define $IR_i \subseteq R$ as a set of substitutable resources. Necessary resources are those where IR_i^* if $Holds(a) \cap IR_i^* = \emptyset$ then $Cap(a, \alpha, Knows(a), Holds(a)) = 0$. Useful resources are those where IR_i^+ if $Cap(a, \alpha, Knows(a), Holds(a)) > Cap(a, \alpha, Knows(a), Holds'(a)) if Holds(a) \cap IR_i^+ \neq \emptyset \cap Holds(a) \cap IR_i^+ = \emptyset$.

The coordination problem is to maximize the reward to the team, while minimizing the *costs of coordination*. The overall reward is simply:

$$\sum_{i=0}^{n} Reward(a, \alpha_i, Knows(a), Holds(a), t)$$

The costs of coordination can be very general and in some cases difficult to define. Here we are specifically concerned with only two elements: the volume of communication and the equity with which the efforts of the team were spread over its team members.

This basic coordination problem imposes a number of coordination requirements on a team. These can be viewed as four abstract problems: (1) sharing information; (2) fusing information; (3) allocating tasks and; (4) gaining a joint perspective. Many of the things that a team must do to coordinate can be mapped to one of these basic activities.

3.1 Information Sharing

The agent that senses an event or situation in the environment will not necessarily be the same agent that needs that information. Moreover, unless it has complete knowledge of a team's activities, it may not know which, if any, agent needs the information. Most solutions to this problem use some sort of information broker or other centralized approach, but Xu [19] presented a distributed approach which we follow here.

To model how information sharing occurs in a social network, we adapt the common technique of using Markov chains [2]. We begin by assuming a randomly distributed selection of source agent, a_s and target agent, a_t. a_s does not know the identity of a_t but it knows the properties of an agent that would be interested in the information and it may know something about its neighbors in the network (see below). The information is passed from agent to agent along links in the network encapsulated in a token.

The state, s_i, is defined to be the situation where the mimimum length path to a_t from the current location of the token is of length i. The special state s_0 corresponds to the token, i.e., information, arriving at the target.

If the token moves randomly from agent to agent, then the structure of the network will govern the behavior of the random walk through the network. Specifically, the structure of the network will define the transition function for the Markov chain. We can write $P(s_i, s_j)$ as the probability of transitioning from s_i to s_j, where only $P(s_i, s_{i-1})$, $P(s_i, s_{i+1})$ and $P(s_i, s_i)$ are possible. We determined values for $P(s_i, s_j)$ empirically, for all the networks considered here. Notice that we average $P(s_i, s_j)$ over each node at distance i, though this will vary from node to node.

Once P is known, the expected time to transition from s_i to s_0, t_i can be calculated in a straightforward way:

$$t_i = \sum_{n=1}^{\infty} n(1 - P(s_i, s_i))P(s_i, s_i)^{(n-1)}(P(s_i, s_{i-1})t_{i-1} \qquad (1)$$

$$+P(s_i, s_{i+1})t_{i+1}) \qquad (2)$$

$$t_i = \frac{1 + P(s_i, s_{i-1})t_{i-1} + P(s_i, s_{i+1})t_{i+1}}{1 - P(s_i, s_i)}$$

Intuitively, in Equation 1, the first term after the sum captures the amount of time the token is expected to stay this distance away from the target and the second captures how long it will take after leaving this distance (and going either closer or further.)

Figures 2 and 3 show the relative rates of $P(s_i, s_{i-1})$, marked "Closer", $P(s_i, s_{i+1})$, marked "Further" and $P(s_i, s_i)$, marked "Same" for Scale Free and Random networks. The x-axis shows the distance of a node to the target node, i.e., the subscript i. Notice that the closer to the target the more likely random movement is to lead further from the target and conversely, the further from the target the more likely random movement will lead the token closer. The figures show that the closer a token is to the target, the easier it is to move away. Moreover, since the figures show different distributions, their information sharing characteristics are likely to be different.

However, in teams, information does not just move randomly from agent to agent. Often the agents will know something (perhaps a lot) about the characteristics of their network neighbors and even the neighbors of their neighbors. Several sociologists have shown how information delivery can be very efficient in human teams with simple models of acquaintances [15, 17] and Xu [19] has effectively illustrated this for multi-agent teams.

To model the fact that the movement is not completely random, but is in fact biased towards the target location, we use a parameter β to make $P(s_i, s_{i-1})$ larger and $P(s_i, s_{i+1})$ smaller. However, this bias should

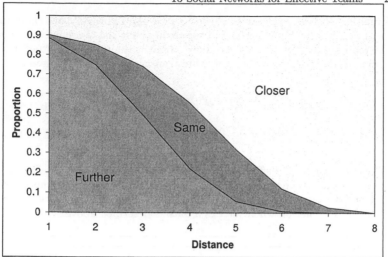

Fig. 2. The relative proportions of links in a Scale Free network that lead closer to, keep the same distance from or move further from some target node, as the distance to the target is varied.

be stronger nearer to the target location, since it is more likely that agents need the target information know what is required to intelligently route the

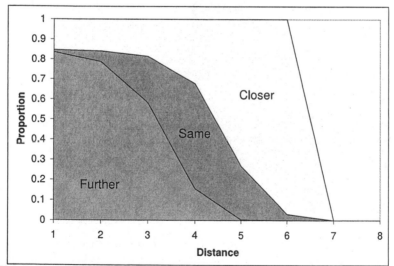

Fig. 3. The relative proportions of links in a Random network that lead closer to, keep the same distance from or move further from some target node, as the distance to the target is varied.

information. We can model this by using $\beta(i) = \frac{1}{e^{\alpha i}}$. Informally, one can think of β as the total learning learning of the team about the team and α as how much more agents "near" an agent know about it than do agents "far" from it. Using α and β the Markov chain state transitions can be rewritten as:

$$\check{P}(s_i, s_{i-1}) = P(s_i, s_{i-1}) + (1 - e^{\beta(i)})P(s_i, s_i) + (1 - e^{2\beta(i)})P(s_i, s_{i+1})$$

$$\check{P}(s_i, s_i) = P(s_i, s_i) - (1 - e^{\beta(i)})P(s_i, s_i)$$

$$\check{P}(s_i, s_{i+1}) = P(s_i, s_{i+1}) - (1 - e^{2\beta(i)})P(s_i, s_{i+1})$$

Figure 4 shows the effect on the scale free distribution from Figure 2. Especially close to the target, i.e., the left of the graph, tokens are much more likely to get closer to the target. Clearly, the result will be more efficient delivery of information when there is a bias as described above.

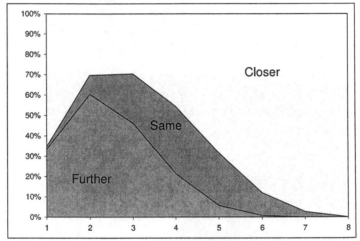

Fig. 4. The relative proportions of links that lead closer to, keep the same distance from or move further from some target node, as the distance to the target is varied. This plus: an α value of 1.5 is used to bias the links towards moving to the target node.

3.2 Information Fusion

In a distributed team, different members of the team may take sensor readings that must be *fused* together to allow action to occur. This basic phenomena can occur for a number of different reasons, including fusing of low confidence sensor readings to get high confidence in the occurrence of some event, for detecting conflicts or synergies between different activities or to know that multiple preconditions for some course of action have all been met. In a co-operative environment, it is typically only necessary for one member of the team to know the sensor readings to be fused and, in the following, we assume there is no preference ordering across agents for who fuses the information.

The probability that an agent a senses the information itself is, *sense*, which is assumed to be uniform across the team. If each piece of sensed information is randomly passed from neighbor to neighbor then the probability that a gets it in a particular step is $\gamma_a = degree(a)/2|E|$. Thus, if the event occurs at $t = 0$ the probability that a knows about it at t is $\Gamma_a(t) = 1 - ((1 - sense)(1 - \gamma)^t)$.

If n of m sensor readings must be "fused" for action to occur, the probability that a can do the fusion at t is:

$$Fuse(a, t) = \sum_{n'=n}^{n'=m} mCn' \times \Gamma_a(t)^{n'} \times (1 - \Gamma_a(t)^{m-n'}) \qquad (3)$$

Informally, this says that the agent can get any combination of sensors readings with equal probability.

$$Fuse(A, t) = 1 - \prod_{a \in A} (1 - Fuse(a, t)) \qquad (4)$$

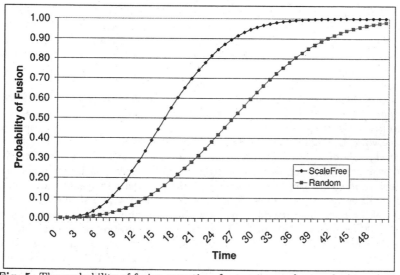

Fig. 5. The probability of fusion over time for two types of network. At each time step five pieces of information move randomly from one agent to another, three must meet for fusion to occur.

Figure 5 shows the probability of fusion for scale free and random networks. The probability of fusion for the scale free network is clearly higher. Examination of Equations 3 and 4, suggests that the advantage of the scale free network is primarily due to the small number of nodes with very many links, since this leads to a high value for γ and consequently Γ. Empirical results bear out this observation (see Section 4).

3.3 Task Allocation

When tasks dynamically arise, capable and available team members need to be found to execute those tasks. There are a large number of different task allocation algorithms that might be used to allocate the tasks. In this chapter, we consider the impact of network structure on LA-DCOP [13] an algorithm that encapsulates a task in a token and moves it around the team until some agent is available and has capability above a *threshold* recorded on the token.

The same formal model developed for the information sharing problem can be applied to understanding the impact of network structure on LA-DCOP. The key is to observe that there are n agents in the team that will accept the task if it reaches them. Each of these n agents can be thought of as a target for the token and we can perform the same analysis as in Equation 2 to determine how far the token will be from an agent that would accept it. Figure 6 shows the relative proportion of different distances to any of n target nodes (on the x-axis) for a random network with 500 agents. The lowest area shows the proportion that are capable of the task themselves, the next area up shows the proportion that is adjacent to an adjacent to an agent capable of the task, the next shows agents 2 links from a capable agent and so on. The figure clearly shows that even if only about 2% of agents are capable, every agent in the network is close to some capable agent.

Task allocation is a specialized coordination task for which it is intuitively important precisely which agent is next to which other. Specifically, it seems intuitive that if agents with different capabilities are close to each other in a network, task allocation can function more effectively, since if the agent is not capable of a task, it is more likely to be able to find someone that is. To evaluate this hypothesis, we configured random networks in two ways: one where random links are created with a strong preference to agents with similar capabilities and one where random links are created with a strong preference for linking to agents with different capabilities. Figure 7 shows an experiment with 500 agents and 2500 tasks. Tasks take some duration to execute and an agent may perform only one task at a time, hence the availability of agents to perform tasks will change over time. As expected the network where agents were connected to others with different capabilities did lead to better results, but the difference was not unexpectedly small.

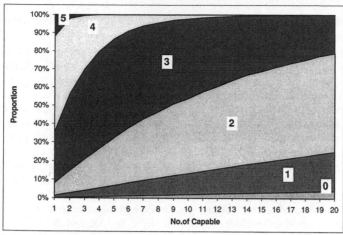

Fig. 6. Percentage of nodes that are of distance 0-5 from an agent capable of performing some task as the number of capable agents is increased. There are 500 agents in all, arranged in a random network.

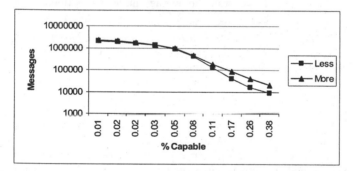

Fig. 7. The number of messages required to allocate 2500 tasks in two random 500 agent networks. One random network arranges agents so those with similar capabilities are near one another (More) and the other arranges agents so that agents with similar capabilities are far from one another (Less).

3.4 Perspective Aggregation

Some coordination algorithms or activities require that each member of the team builds an accurate view of the team's state. For example, the way that an individual agent uses shared resources such as communication bandwidth or fuel should depend on the team's overall need for such resources.

Technically, we can think of every member of the team having a local value for some variable, e.g., their local need for some resource, and needing to know the average value of that variable across the whole team, e.g., the average need for some resource. Formally, each agent has some variable v. The perspective aggregation problem is for each agent to know $\bar{v} = \frac{\sum_{a \in A} v}{|A|}$.

One way of building up a perspective across the team is to have a small number of *propagators* move from agent to agent, taking the current perspective from one agent and adding it to the current perspective at the next agent. If there are multiple propagators simultaneously moving around the team, perspectives build up very quickly. Notice that it is typically infeasible for a propagator to record precisely which agents it has collected values for, since it would need to record all the agent IDs as it moved from agent to agent. In a large team this imposes an unreasonable communication load. However, because the propagator does not know precisely which agents it has visited, some will be visited repeatedly and their values counted repeatedly, distorting the average results.

A simple model of how quickly these perspectives build up can be straightforwardly created by considering how many other values each agent knows about. Before any propagators move, each agent knows only their value. Thus, the average number of values known by each agent $Avg(v, 0)$ is 1. When a propagator moves, one of the agents gets to know 1 new value, hence $Avg(v, 1) = 1 + \frac{1}{|A|}$. In general, the average number of values known by an agent after move t of the propagators is $Avg(v, t) = Avg(v, t-1) + \frac{Avg(v, t-l)}{|A|}$. Because propagators collect information as they move, $Avg(v, t)$ rapidly grows with t. Figure 8, shows $Avg(V, t)$ for a team with 500 members. The x-axis shows the number of propagator moves divided by the number of agents and the y-axis shows $Avg(V, t)$ on a logarithmic scale. The figure suggests that perspective aggregation is not a communication intensive task for a team, even one with relatively few edges.

The average value does not capture two key aspects of the perspective aggregation problem. First, nodes with higher degree will be visited more often by a randomly moving propagator than nodes with lower degree. This effect can be modelled by changing the denominator in $\frac{Avg(v, t-l)}{|A|}$ to be $\rho|A|$, i.e., agents with higher than average degree will have $\rho < 1.0$ and those with lower than average degree will have $\rho > 1.0$. Thus, networks with many nodes with low degree are likely to perform poorly on this task. Second, clearly many of the values an agent gets to know will be repeated. The distortion caused to the agent's perspective by the repeats will be proportional to the relative rates at which repeats occur, i.e., if some values are repeated many times and others are not, the agent's perspective will get very distorted. An agent will likely get to know about another agent's value more often if that agent is close to it in the network than if it is far from it. Thus, networks with higher width are likely to perform poorly on this task.

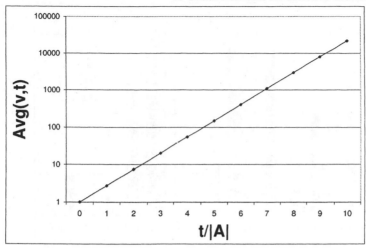

Fig. 8. The average number of samples each agent has (y-axis) after a propagator has moved a fixed number of steps (x-axis). The y-axis has a logarithmic scale.

4 Experiments

To empirically evaluate all the different networks, we developed simple Java programs to simulate each of the algorithms and networks. For each sub-problem we created 100 networks of 500 agents of each type and measured two things: the performance of the algorithm on the network and the standard deviation of the contribution of each agent to the observed performance. For example, for the information fusion problem, we randomly allowed five agents to "sense" five pieces of information and then propagated that information around the network until some agent knew of three of the pieces of information. For this information fusion, we measured the time taken to fuse and the number of times each agent performed the fusion. A summary of the results is shown in Figures 9 (algorithm performance) and 10 (distribution of effort). The results have been normalized for clarity. In Figure 10 the distribution of effort is computed as the magnitude of the standard deviation of effort, based on the idea that higher standard deviation means more variability in effort. The experiments indicate that no network outperforms all others on all tasks and there is often an inverse relationship between performance and distribution of effort.

Finally, to perform a more complete test on the effects of networks on co-ordination, we used an abstract coordination simulator called CoordSim. This simulator is capable of simulating the major aspects of coordination including sensor fusion, plan management, information sharing, task assignment and resource allocation. CoordSim abstracts away the environment, instead just

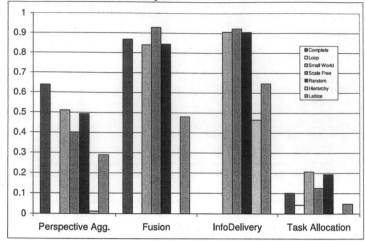

Fig. 9. Relative performance of each network on each abstract problem. Values are normalized and inverted where necessary so that bigger is better.

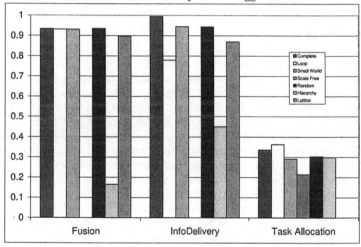

Fig. 10. Relative equitableness of algorithms on different problems. Values are normalized and inverted where necessary so that bigger is better.

simulating its effects on the team. Uncertain sensor readings are received randomly by some agent or agents in the team at a parameterizable rate. Agents cannot "know" anything they do not sense or is not communicated to them from a teammate. Time is simulated and all agents are allowed to "think" and "act" at each step, although the effects of their "actions" are abstractly simulated. Communication is implemented via object passing, making it very fast. Reward is simulated as being received by the team when the agent is allocated the task, the agent's simulated location is at the task location and it has exclusive access to required resources. Reward is received while the agent is simulating the task, which takes one time step.

One key element simulated by CoordSim that is not addressed in the subproblems above is that of deconflicting plan instantiations. In CoordSim, when any agent comes to know that the preconditions for some joint activity are known, it initiates that joint activity and informs *its neighbors in the network*. As tasks are allocated for the joint activity, agents assigned roles in the joint activity also inform their neighbors *in the network* of the joint activity. If any agent gets to know of two activities achieving the same goal, they initiate a process to stop one of the activities. If two activities aimed at achieving the same goal are executed, the team only gets a reward for one.

Figures 11-14 show the results from CoordSim for four different coordination instances. In each case, there are 200 agents and 50 plans. Unless otherwise noted: (i) plans are initiated when two preconditions are true and some agent gets to know both preconditions and starts the plan; (ii) there are between one and four roles per plan; (iii) there are one to five pieces of information available that can improve execution of a task, if the agent performing the task knows of that information; (iv) there are 20 different types of task and each agent can perform between one and five of them and; (v) there are 150 pieces of information that can be known, each must be fused from at least three of five sensor readings randomly distributed to the team. Figure 11 shows a baseline configuration. Figure 12 shows a case where fusion is more important, specifically four preconditions must be known to instantiate a plan and four of five sensor readings are required to fuse a piece of information. Figure 13 shows a case where task allocation is more important, with between five and ten roles per plan, fifty different types of task and each agent only capable of one thing. Finally, Figure 14 shows a case where information sharing is important, because five to ten pieces of information can improve the reward of the agent performing the task.

In all but one case, Figure 12, the complete network performed best. A primary reason for this was that it was never executing conflicting plans, because whenever one agent initiated execution of a plan, it would inform all others and any duplicate instantiations would be immediately removed. The low number of role allocation messages is testemant to this efficiency. This ability to remove conflicting plans overwhelmed any other advantages the other networks had. However, in the configuration in Figure 12 instantiating plans relies on very effective information fusion, both to get the pieces of information from the sensor readings and then to have one agent know four readings and instantiate the plan. Since fusion was so difficult, relatively few duplicate plans were created and hence deconfliction became less important. Instead, the networks more suited to sensor fusion performed relatively better.

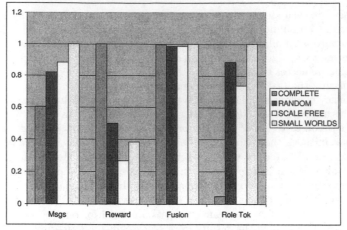

Fig. 11. Coordination experiment baseline case.

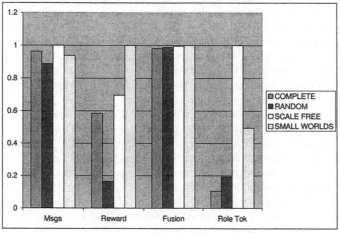

Fig. 12. Coordination experiment with more emphasis on fusion sub-problem.

5 Related Work

Social networks have been an active area of interest since Milgram observed the small world effect nearly 40 years ago [15]. Recent interest in such networks was inspired by models by Barabasi [1] and Watts [16] who observed that similar networks occurred in nature. Over time an amazing array of fields of research have contributed to our understanding of networks [11], from biologists [18], to mathematicians [14] to physicists [10]. Some economists explained such networks by balancing the cost of maintaining an acquaintance against the value of that acquaintance [6]. Sociologists, including Carley, showed the

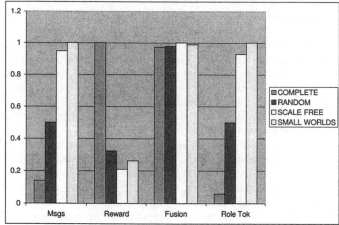

Fig. 13. Coordination experiment with more emphasis on task allocation sub-problem.

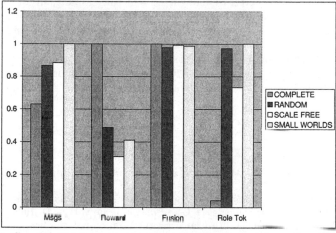

Fig. 14. Coordination experiment with more emphasis on information sharing sub-problem.

rich network structure underlying some organizations and how that network facilitates effective organizational behavior [9].

Computer scientists and agents researchers have also been interested in these network structures. For example, Kleinberg [8] shows some of the information requirements on an agent to replicate Milgram's results in multi-agent systems. Gaston recently published two key papers showing how network structure has an impact on multiagent systems and proposing an algorithm for designing networks for multiagent systems [5, 4].

6 Conclusions

This chapter presented a quantitative investigation of the impact of social networks on multi-agent coordination. Our results support previous work which indicated the importance [5, 4] and utility [20] of social networks. For abstracted coordination tasks, networks with relatively low degree were shown to often significantly outperform completely connected networks. Some of the advantage low degree networks were shown to have was due to a small number of nodes performing a relative large proportion of the work, essentially partially centralizing the coordination. However, we showed that the advantage of social networks disappeared when all coordination tasks were taken into account.

While this chapter advances our understanding of the impact of social networks on coordination, key work is required to utilize this new understanding. Most urgently, we showed that none of the networks we used were best for all coordination problems. A key question is whether there is a particular network that is best in *all* situations. It is possible, or even likely, that changing structure over time is better than any fixed structure. Finally, although initial experiments that looked more carefully at which agent should be adjacent to which other showed little effect, it is likely there are some more significant effects for other adjacencies. Our future work will investigate these questions and apply these social networks to practical multiagent systems.

Another possibility is to have different logical networks for each task in the same team, with networks chosen to be specifically good for the sub-problems they are used for.

References

1. A.-L. Barabasi and E. Bonabeau, Scale free networks, *Scientific American*, pp. 60–69, May 2003.
2. V. Buskens and K. Yamaguchi, A new model for information diffusion in heterogeneous social networks, *Sociological methodology*, Vol. 29(1), 1999.
3. L. Freeman, Centrality in social networks: Conceptual clarification, *Social Networks*, Vol. 1(3), pp. 215–239, 1979.
4. M. Gaston and M. desJardinss, Agent-organized networks for dynamic team formation, in *Proceedings of the Fourth International Joint Conference on Autonomous Agents and Multi-Agent Systems*, Utrecht, Netherlands, 2005.
5. M. Gaston and M. desJardinss, Agent-organized networks for multi-agent production and exchange, in *Proceedings of the Twentieth National Conference on Artificial Intelligenc*, Pittsburgh, PA, 2005.
6. J. Geunes and P. Pardalos, Network optimization in supply chain management and financial engineering: An annotated bibliography, *Networks*, Vol. 42(2), 2003.
7. P. Harrison and W. Knottenbelt, Networks, dynamics, and the small-world phenomenon, *The American Journal of Sociology*, Vol. 105, No. 2:493–527, 1999.

8. J. Kleinberg, The small world phenomenon: An algorithmic perspective, in *Proceedings of Symposium on Theory of Computing*, 200.
9. Z. Lin and K. Carley, Dycorp: A computational framework for examining organizational performance under dynamic conditions, *Journal of Mathematical Sociology*, Vol. 20, 1995.
10. R.M. May and A.L. Lloyd, Infection dynamics on scale-free networks, *Physical Review E.*, 2001.
11. M. Newman, The structure and function of complex networks, *SIAM Review*, Vol. 45(2), 2003.
12. P. Scerri, Y. Xu, E. Liao, J. Lai, and K. Sycara, Scaling teamwork to very large teams, in *Proceedings of AAMAS'04*, 2004.
13. P. Scerri, A. Farinelli, S. Okamoto and M. Tambe, Allocating tasks in extreme teams, in *AAMAS'05*, 2005.
14. C. Topper and K. Carley, A structural perspective on the emergence of network organizations, *Journal of Mathematical Sociology*, Vol. 24(1), 1999.
15. J. Travers and S. Milgram, An experimental study of the small world problem, *Sociometry*, Vol. 32, pp. 425–443, 1969.
16. D. Watts and S. Strogatz, Collective dynamics of small world networks, *Nature*, Vol. 393, pp. 440–442, 1998.
17. D.J. Watts, P.S. Dodds and M.E.J. Newman, Identity and search in social networks, *Science*, Vol. 296(5571), pp. 1302–1305, 2002.
18. Y.I. Wolf, G. Karevand, E.V. Koonin, Scale-free networks in biology: new insights into the fundamentals of evolution? *BioEssays*, 2002.
19. Y. Xu, M. Lewis, K. Sycara and P. Scerri, Information sharing in very large teams, in *AAMAS'04 Workshop on Challenges in Coordination of Large Scale MultiAgent Systems*, 2004.
20. Y. Xu, P. Scerri, B. Yu, S. Okamoto, M. Lewis and K. Sycara, An integrated token-based algorithm for scalable coordination, in *AAMAS'05*, 2005.

16 Optimization Approaches for Vision-Based Path Planning for Autonomous Micro-Air Vehicles in Urban Environments

Michael Zabarankin, Andrew Kurdila, Oleg A. Prokopyev, Ryan Causey, Anukul Goel and Panos M. Pardalos

Summary. A model for optimal path planning for autonomous micro-air vehicles (MAVs), equipped with a vision-based navigation system of limited range, has been proposed. The model is dynamic and deterministic. It assumes no initial information about obstacle locations and uses only local information provided by the camera "vision cone of limited range." To plan a path based on this model, constrained optimization and navigation function approaches have been developed. In the framework of the constrained optimization approach, the system of axioms for kinematic constraints has been suggested, and analytical solutions for a collision free path within a vision cone with obstacles, whose surfaces might be approximated by simple smooth convex hulls, have been obtained. The navigation algorithm has been implemented to solve the associated constrained optimization problem. It has been tested in several numerical examples and has demonstrated robust performance in avoiding obstacles in "reasonable" configurations. The navigation function approach incorporates conditions on obstacle avoidance into an objective function designed to navigate the MAV to a destination point. Although some navigation functions may produce navigation "paradoxes," this approach is less computationally intense than the constrained optimization approach.

1 Introduction

In the last several years, the developments of micro-air vehicles (MAVs) have made technological advances that can potentially be used for a great variety of important practical tasks, including aerial reconnaissance, region mapping, target search, detecting chemical weapons, etc. The main advantage of MAVs is their tiny size. Some of these vehicles have a wing span of less than six inches. Consequently, MAVs have the potential to navigate in complex urban environments without being detected. This feature is essential if MAVs are used for military missions within a hostile territory.

The task of flying autonomous micro-air vehicles in environments confined by buildings and trees requires small vehicles of extreme agility. Moreover, for navigation, MAVs may be equipped with cameras or ultrasonic sensors of limited range. In this case, at each MAV's position we can only consider the information within a "vision cone" of a camera or sensor to be available for making a navigational decision. This type of navigation is referred to as "vision-based navigation." One crucial problem that arises for MAVs is the development of efficient techniques and algorithms for vision-based path planning. In fact, this problem is complex and involves several important aspects, such as vision-based estimation, image processing, optimal path planning algorithms and strategies, optimal control, flight dynamics, etc. Although significant progress has been made in the areas of fabrication, structural design and development of inner-loop controllers using vision-based horizon-tracking algorithms [6, 13], there still remain substantial technical barriers that must be addressed before vehicles are capable of vision-based autonomous flight. The problems of image processing [8], autopilot design and optimal path planning are frequently treated independently. In an ideal case, we assume that an optimal path satisfies all necessary constraints involving path curvature and smoothness. In this case we are able to focus solely on the development of fast optimization algorithm intended for path planning subject to obstacle avoidance. We assume that the autopilot is capable of tracking relatively smooth trajectories.

Vision-based navigation means that at each MAV's position only the information provided by a vision cone is available. The camera installed onboard the MAV captures an image within a spherical vision cone with radius L and field of view (FoV), α (see Figure 1). In practice, L is estimated to be about 100 meters while $\alpha \leq \frac{5\pi}{18}$, see [8]. While there is extensive literature on aerial robotics [1, 4, 5, 7, 10, 11] and algorithms for generating globally optimal trajectories in obstacle-free environments [12], only a few papers develop detailed optimization approaches for autonomous vision-based navigation subject to obstacle avoidance. To focus on the development of optimization approaches to vision-based path planning, we consider neither MAV's dynamic characteristics nor control model for the MAV. In this chapter, we consider only the basic navigation problem: given points A and B, generate a collision-free smooth path from A to B.

Similar problems of optimal path planning for two-dimensional (2D) robots have been widely discussed in the literature [9]. However, the problem of 3D optimal path planning for MAVs using vision-based navigation is much more complicated and computationally intensive due to several factors:

- *The vehicle can not stop and no backward movement is allowed.*
- *There are curvature and smoothness constraints on the trajectory.*
- *There exists a critical distance in avoiding an obstacle.*
- *Information about the region is available only within a vision cone.*

In planning a path from point A to point B with no a priori information about obstacle locations, we also assume that

- *The vision cone is always oriented along trajectory direction.*

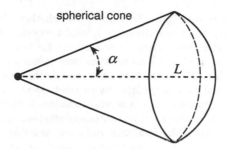

Fig. 1. Spherical vision cone.

- *There is no uncertainty in determining obstacle boundaries within the vision cone.*

and make the following kinematic assumptions:

- *A MAV's trajectory is a piece-wise linear curve in 3D space.*
- *The MAV changes the flight direction only at discrete time moments.*
- *The MAV's velocity is constant.*
- *To avoid sharp turns, the angle between any two joint arcs of MAV's trajectory is constrained.*

Based on these requirements and assumptions, we develop *constrained optimization* and *navigation function* approaches to solve the basic navigation problem with no initial information about the region of consideration. Both approaches rely on "global" and "local" navigation functions. But, they use them differently. Both approaches consider the distance between a MAV's current position and point B to be a "global" navigation function.

The constrained optimization approach minimizes the global navigation function subject to several local constraints intended for obstacle avoidance. "Passive" and "active" local constraints are proposed to control the distance from the MAV to an obstacle. This approach is developed in discrete and continuous cases. In a discrete case, the obstacles are represented by collections of feature points, while in the continuous case, the surfaces of obstacles are approximated by smooth convex hulls. In the continuous case, the smoothness of hulls allows one to employ techniques of differential geometry for formulating and handling kinematic constraints, and the convexity of hulls significantly simplifies calculation of the distance from current MAV's position to an obstacle. In this case of the approach, we suggest the system of axioms for determining kinematic constraints to avoid obstacles, and obtain analytical solutions for a collision free path within the spherical vision cone with obstacles, whose surfaces are approximated by either planes or spheres. In the discrete case of the constrained optimization approach, we present the navigation algorithm capable of navigating the MAV from A to B in an environment with obstacles, whose surfaces cannot be satisfactorily approximated by simple smooth convex hulls. Depending on whether a global network is used for a whole region of consideration, a local network within a vision cone may be a part of the global network or may use its own network structure. The difference in generating a local network determines a modification of the navigation algorithm. The variant which uses its own local structure is less computationally costly and automatically satisfies the smoothness constraint. We test the navigation algorithm in several examples with respect to running time and robustness. It proves to be extremely efficient in navigating among obstacles having "reasonable" locations and configurations. Ideally, optimization algorithms for vision-based navigation should be capable of handling a relatively complex geometry of an urban environment, including tunnels and closed 3D constructions.

The navigation function approach minimizes a single linear combination of "global" and "local" navigation functions. In the framework of this approach, three navigation functions are considered. Two of them are shown to have drawbacks in some situations, while the third one, being a combination of the previous two, is free from these flaws. Still, this approach requires further investigation, which, probably, calls for a separate publication.

An important issue, which remains open, is how the geometry of a trajectory depends on the quantity of local information provided by a vision cone. For instance, if the volume of the vision cone is too small compared to obstacle size, and the geometry of obstacle locations is complicated then, obviously, the MAV can easily be trapped or lost causing a pathological trajectory. Addressing this question is beyond the scope of this chapter.

The remainder of the chapter is organized as follows. Section 2 formulates a basic navigation problem. Section 3 develops the constrained optimization approach in the discrete case. Section 4 develops the constrained optimization approach in the continuous case. It suggests a system of axioms for determining kinematic constraints and obtains analytical solutions for a collision free path in a vision cone with obstacles, whose surfaces are approximated either by planes or spheres. Section 5 considers techniques for approximating obstacle surfaces by smooth convex hulls. Section 6 presents the navigation function approach and discusses three examples of navigation functions. Section 7 considers two modifications of the navigation algorithm and demonstrates algorithm performance with several numerical examples. Finally, Section 8 summarizes the results of the chapter.

2 Problem Formulation

This section develops optimization approaches to the navigation problem with no initial information about the region of consideration. We formulate and solve the dynamic model for obstacle avoidance using local information provided by a vision-based system of limited range.

In planning a path from point A to point B with no a priori information about obstacle locations, we make the following assumptions

- *The vehicle travels in 3D space.*
- *The current coordinates of the MAV and coordinates of point B are known.*
- *The Euclidean metric for any two points A_1 and A_2 with coordinates $A_1(x_{A_1}, y_{A_1}, z_{A_1})$ and $A_2(x_{A_2}, y_{A_2}, z_{A_2})$ is denoted*

$$\rho(A_1, A_2) = \sqrt{(x_{A_2} - x_{A_1})^2 + (y_{A_2} - y_{A_1})^2 + (z_{A_2} - z_{A_1})^2}.$$

- *The camera installed onboard the MAV captures an image within a spherical vision cone with radius L and angle α (see Figure 1). At each point of a MAV's trajectory, only information within the vision cone is available.*
- *The vision cone is always oriented along trajectory direction.*
- *Image processing identifies only finite number of points on the curve that defines an obstacle boundary. That is, visible boundary of an obstacle is presented by a finite set of nodes.*
- *There is no uncertainty in determining the coordinates of nodes within the vision cone described above.*
- *Any two nodes O_1 and O_2, identifying obstacle boundaries with $\rho(O_1, O_2) \leq h$, belong to the boundary of the same obstacle. The parameter h defines a minimal characteristic dimension associated with an obstacle.*

Consequently, the navigation problem based on a vision system with a limited vision range is formulated as: *navigate from point A to specified point B avoiding obstacles, where the boundaries are observable only within the vision cone and identified by finite number of feature points.*

The case of an obstacle free region. In the case when a region of consideration has no obstacles, the solution to the formulated navigation problem is trivial. Namely, at point A, a MAV orients towards point B and moves to such point A' within the vision cone that minimizes distance to point B. When point A' is reached, the next vision cone will be automatically oriented toward point B and the MAV moves to such point A'' within the new vision cone that again minimizes the distance to point B. This procedure is repeated until point B is reached. Obviously, in this case, the MAV's trajectory is the straight line between points A and B.

"Global" and "local" navigation principles. In general, solving the navigation problem subject to obstacle avoidance is based on two principles: the *"global navigation"* principle, which navigates the MAV towards point B; and the *"local navigation"* principle, which prevents collisions with obstacles. For instance, the "global navigation" principle, used in generating the trajectory from point A to point B in obstacle free region, minimizes distance between MAV current position $A^{(k)}$ and the point of destination B. Lumelsky and Stepanov [9] showed that in a 2D region with obstacles, this principle may not always lead to a solution. For example, a 2D robot can be trapped or move around point B never reaching it. Nevertheless, being perfectly aware about such an issue, we consider minimization of $\rho(A^{(k)}, B)$ to embody the "global navigation" principle for the case of a 3D region with obstacles, and focus on development of "local navigation" principles.

Trivial solution to the navigation problem subject to obstacle avoidance. Assume that MAV flight occurs in an unbounded 3D region containing finite size obstacles. There is usually a trivial solution to the formulated problem of navigation subject to obstacle avoidance in an outdoor urban environment. We can always generate a trajectory within the plane going through points A and B, simply overflying the urban obstacles (Figure 2). Nevertheless, there are at least two reasons why we are not satisfied with this solution. First, such a trajectory may not be feasible from a MAV's control point of view. Indeed, we can always complicate obstacle geometry or impose additional constraints on a path, so that this trivial path can be the only geometrically feasible one. However, in this case we need to prove that another feasible path is really not available. A second reason concerns satisfaction of the "global navigation" principle. For instance, without imposing additional constraints, the trajectory may not minimize the distance between the next trajectory's point and point B.

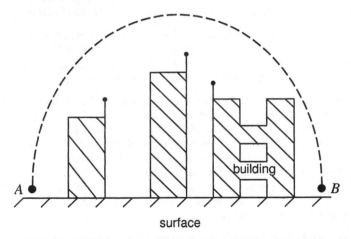

Fig. 2. Trivial solution to navigation problem subject to obstacle avoidance.

We suggest *constrained optimization* and *navigation function* approaches to planning a collision free path. The constrained optimization approach minimizes the distance between MAV's current position and point B subject to constraints on distance and direction to approaching obstacles. This approach is considered in discrete and continuous cases. The navigation function approach develops a single function, which encodes the goals for approaching point B and avoiding obstacles, and is considered only in the discrete case. The constrained optimization approach in the discrete case and navigation function approach are based on the following ideas:

- *The space within every vision cone is represented by the finite set of points including those which identify obstacle boundaries. Based on this set of points, a local network $\mathcal{G}_l = (\mathcal{N}_l, \mathcal{A}_l)$ is introduced, where $\mathcal{N}_l$ and $\mathcal{A}_l$ denote sets of nodes and arcs in $\mathcal{G}_l$, respectively.*
- *The MAV can be directed toward way points selected from the set of local network nodes.*
- *Three sets of nodes are considered: $\mathcal{F}$ is the set of free nodes (no-obstacle), $\mathcal{O}$ is the set of nodes identifying obstacle boundaries, and $\Im$ is the dynamic set of accessible nodes such that $\Im \subset (\mathcal{N}_l \cap \mathcal{F})$.*
- *A MAV's trajectory is approximated by a piece-wise linear curve with a finite number of segments $\langle E_{k-1}, E_k \rangle$ where $E_0 = A$ and $E_n = B$. Point $E_k \in \Im_k$ is the "exit" node in local network $\mathcal{G}_{lk}$, generated within the corresponding vision cone with the vertex at $E_{k-1} \in \Im_{k-1}$.*

3 The Constrained Optimization Approach: Discrete Case

There are several constraints which would force a MAV to avoid an obstacle. A "passive" constraint assigns a critical distance $\rho_{\min}$ between the MAV and an obstacle. For example, for all $O \in \mathcal{O}$ we require $\rho(E, O) \geq \rho_{\min}$. This constraint will be incorporated in the definition of the set of accessible nodes $\Im_k$. However, it will not turn the MAV away from the obstacle. For this purpose, we use an "active" constraint in the following form

$$\mathbf{e}_{E_{k-1}E} \cdot \mathbf{e}_{EO_E^*} \leq \cos\beta, \quad \cos\beta = 1 - \rho_{\min}/\rho(E, O_E^*), \tag{1}$$

where O_E^* is the obstacle node closest to E within a vision cone, i.e. $O_E^* = \arg\min_{O \in \mathcal{N}_{lk} \cap \mathcal{O}} \rho(E, O)$, and angle β is a variable, depending on $\rho(E, O_E^*)$ (see Figures 3 and 4).

Figure 4 provides geometric interpretation of constraint (1) and variable β. Consider triangle EFO_E^* shown in Figure 4–b. FD is the height perpendicular to the base EO_E^* and $\rho(E, F) = \rho(E, O_E^*)$. Then, according to the definition of β and construction of the triangle EFO_E^*, $\cos\beta = \frac{\rho(E,O_E^*) - \rho(D,O_E^*)}{\rho(E,O_E^*)} = \frac{\rho(E,D)}{\rho(E,O_E^*)} = \frac{\rho(E,D)}{\rho(E,F)}$. For example, let $\rho(E', O_E^*) = 2\rho_{\min}$ (see Figure 4–b). The MAV is going to approach an obstacle at double the critical distance. According to the introduced "active" constraint, we obtain $\cos\beta' = \frac{1}{2}$ or $\beta' = \frac{\pi}{3}$. This condition means that at the current position, the MAV can approach the obstacle at double the critical distance only if it turns relative to the obstacle at an angle greater than $\frac{\pi}{3}$. Moreover, if $\rho(E, O_E^*) \to \rho_{\min}$ then $\beta \to \frac{\pi}{2}$.

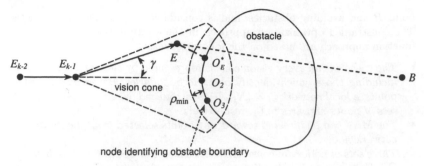

Fig. 3. Navigation avoiding obstacles.

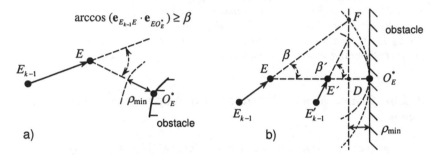

Fig. 4. Meaning of obstacle avoidance constraints.

Thus, at the critical distance from an obstacle, the MAV can fly tangent to the boundary only.

To determine the "exit" point E_k, constrained optimization with global objective function $\rho(E, B)$ is proposed and the following three constraints are considered:

- *Feasible exits are characterized by $E \in \mathfrak{I}_k$, where $\mathfrak{I}_k$ consists of all $O \in \mathcal{O}$ such that $\rho(E, O) \geq \rho_{\min}$.*
- *The trajectory must satisfy the smoothness condition with parameter γ, which is active if $\gamma < \alpha$ (see Figure 3).*
- *A feasible selection must satisfy constraint (1).*

Consequently, E_k solves the following optimization problem

$$E_k = \arg \min_{E \in \mathfrak{I}_k} \rho(E, B)$$
$$\text{s.t. } \mathbf{e}_{E_{k-1}E} \cdot \mathbf{e}_{EO_E^*} \leq 1 - \frac{\rho_{\min}}{\rho(E, O_E^*)}, \qquad (2)$$
$$\mathbf{e}_{E_{k-2}E_{k-1}} \cdot \mathbf{e}_{E_{k-1}E} \geq \cos \gamma.$$

Note that (2) is a nonlinear optimization problem. However, due to relatively small size of the local network $\mathcal{G}_{lk}$, associated with the vision cone, we confine ourselves to the *enumeration* approach. For discrete optimization methods, the reader may refer to [2, 3]. However, in this case, enumeration proves to be fast.

4 The Constrained Optimization Approach: Continuous Case

As in in the discrete case, we suppose that at each MAV's position, only the information provided by a vision cone is available. We introduce a global cartesian system of coordinates (OX_1, OX_2, OX_3). Let $x = (x_1, x_2, x_3)$, $y = (y_1, y_2, y_3)$, and $d(x, y)$ be the distance between the points x and y. Let x_{k-2}, x_{k-1} and x denote previous, current and next-unknown MAV's positions, respectively. To navigate the MAV toward its destination, point B, we minimize the distance between the MAV's next position and the destination point, i.e. $d(x, x_B)$. Because of the delay in image processing and computation of the next position, we impose a constraint on a minimal move distance $d(x_{k-1}, x) \geq \rho_0$. Since the vision cone is bounded, the MAV can move only within the range of camera visibility. This means that $d(x_{k-1}, x) \leq L$. The MAV's trajectory is modelled by a piece-wise linear curve.

Suppose that the boundary of any obstacle is a *convex smooth* surface, determined in the global system of coordinates by equation $F(x) = 0$. Let z be a point on the obstacle surface closest to the next MAV position x, i.e.

$$z = \arg \min_{F(y)=0} d(x, y). \tag{3}$$

Note that the uniqueness of point z is guaranteed by convexity of the obstacle surface. Another reason for the assumption of the obstacle surface to be convex is to avoid situations when nonconvex obstacles can trap the MAV at the next vision cone. We impose a *safe-distance constraint* $d(x, z) \geq \rho_{\min}$. In order to avoid sharp turns in the trajectory, we constrain the angle between any two consecutive arcs in the trajectory, i.e. a constraint on trajectory smoothness is given in the form

$$\frac{(x_{k-1} - x_{k-2}) \cdot (x - x_{k-1})}{d(x_{k-2}, x_{k-1}) d(x_{k-1}, x_k)} \geq \cos \gamma,$$

where we assume that $\gamma \leq \alpha$. The next step is to develop kinematic constraints intended for obstacle avoidance. The form of the kinematic constraint depends on the geometry of the obstacle surface. Not to confine the model to a particular geometry of obstacle surface, we define the kinematic constraint axiomatically. Let e_x define the unit vector of the direction of MAV's motion, i.e. $x - x_{k-1}$, and let n_z define the normal vector of the surface at point z, i.e. n_z is a normalized gradient $\nabla F(z)$. We assume that the normal n_z is oriented inside the obstacle. The kinematic constraint accounts for the distance from the next MAV's position to the obstacle surface and the angle at which the MAV approaches the obstacle surface. The essence of the kinematic constraint may be expressed as: *the closer the MAV is to the obstacle surface, the greater the angle, at which the MAV approaches the surface, is.* Stating the kinematic constraint in a general form $f(x, z, e_x, n_z) \geq 0$, a problem of navigating from point x_{k-1} toward point B subject to obstacle avoidance is formulated as

$$\min_x d(x, x_B)$$

$$\text{s.t. } \rho_0 \leq d(x_{k-1}, x) \leq L,$$

$$\frac{(x_{k-1}-x_{k-2})\cdot(x-x_{k-1})}{d(x_{k-2},x_{k-1})d(x_{k-1},x_k)} \geq \cos\gamma,$$

$$z = \arg\min_{F(y)=0} d(x, y), \tag{4}$$

$$d(x, z) \geq \rho_{\min},$$

$$f(x, z, e_x, n_z) \geq 0.$$

A solution x^* to the optimization problem (4) is the next MAV's position, i.e. $x_k = x^*$.

The geometry of vision cone suggests introducing a local spherical system of coordinates. Let (r, φ, θ) be coordinates of point x in the local spherical system associated with the vision cone. The origin of the local system coincides with the vertex of the vision cone and the plane (r, φ) is orthogonal to the axis of revolution of the cone. A triad $(e_r, e_\varphi, e_\theta)$ defines the basis of the local system of coordinates. Assuming that coordinates of the point x in the local system are (r, φ, θ), we have the following relation $r e_r = x - x_{k-1}$. Similarly, if coordinates of point B in the local system are $(r_B, \varphi_B, \theta_B)$, then $r_B e_B = x_B - x_{k-1}$. Let

$$\cos\zeta(\varphi, \theta) = e_r \cdot e_B = \sin\theta \sin\theta_B \cos(\varphi - \varphi_B) + \cos\theta \cos\theta_B, \tag{5}$$

and

$$\cos\omega(\varphi, \theta) = e_r \cdot n_z = \sin\theta \sin\theta_z \cos(\varphi - \varphi_z) + \cos\theta \cos\theta_z. \tag{6}$$

Figure 5 illustrates the local spherical system of coordinates and depicts angles $\zeta(\varphi, \theta)$ and $\omega(\varphi, \theta)$.

Denoting the difference between $d(x, z)$ and the critical distance $\rho_{\min}$ by

$$\delta(\varphi, \theta) = d(x, z) - \rho_{\min}, \tag{7}$$

we express the kinematic constraint in the following form

$$f(r, \omega(\varphi, \theta)) \geq 0.$$

The kinematic constraint is defined by a system of axioms

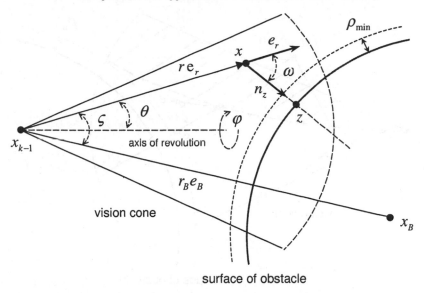

Fig. 5. Local spherical system of coordinates and angles $\zeta(\varphi, \theta)$ and $\omega(\varphi, \theta)$.

Axioms 1 (kinematic constraint)

(K1)   If $f(r, \omega) \geq 0$ then $\delta \geq 0$.

(K2)   If $f(r, \omega) \geq 0$ and $r > 0$ then $\omega > 0$.

(K3)   If $f(r, \omega) = 0$ and δ decreases then ω increases on $(0, \frac{\pi}{2}]$, in particular, if $f(r, \omega) = 0$ and $\delta = 0$ then $\omega = \frac{\pi}{2}$.

Figure 6 provides an intuitive insight for the axioms defining the kinematic constraint. The shaded area is the set of MAV's feasible positions determined by $f(r, \omega(\varphi, \theta)) \geq 0$.

Although, several forms of f will satisfy this system of axioms for a given geometry of obstacle surface, we are interested in the form, which would be simplest from the optimization point of view. We will consider plane and sphere as "generic" geometries of obstacle surfaces.

Note that now we do not include the safe-distance constraint, since by definition of the kinematic constraint, it should be satisfied automatically. The navigation problem (4) is rewritten as

$$\min_{r, \varphi, \theta} r^2 - 2rr_B \cos \zeta(\varphi, \theta) + r_B^2$$

$$\text{s.t. } r \in [\rho_0, L], \ \theta \in [0, \gamma], \tag{8}$$

$$f(r, \omega(\varphi, \theta)) \geq 0.$$

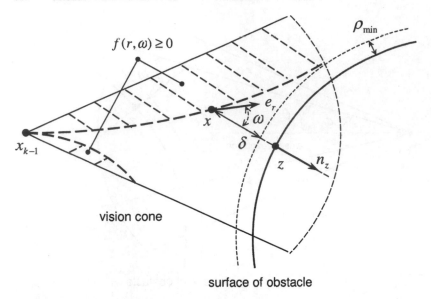

Fig. 6. The set of MAV's feasible positions determined by $f(r, \omega(\varphi, \theta)) \geq 0$.

where $\zeta(\varphi, \theta)$, $\omega(\varphi, \theta)$ and $\delta(\varphi, \theta)$ are determined by (5), (6) and (7), respectively, and coordinates of point z needed for computing all these characteristics are given by (3).

Principle 1 *If the segment $x_{k-1}x_B$ intersects the surface of the obstacle within the vision cone then the kinematic constraint for obstacle avoidance is always active, i.e., $f(r, \omega(\varphi, \theta)) = 0$.*

Figure 7 provides geometrical interpretation on why the kinematic constraint is active if the segment $x_{k-1}x_B$ intersects the surface of the obstacle within the vision cone.

In the case with no obstacles in the vision cone, the navigation problem reduces to

$$\min_{r, \varphi, \theta} r^2 - 2rr_B(\sin\theta \sin\theta_B \cos(\varphi - \varphi_B) + \cos\theta \cos\theta_B) + r_B^2$$

$$\text{(9)}$$

$$\text{s.t. } r \in [\rho_0, L], \ \theta \in [0, \gamma].$$

Since in (9), φ is a free variable, the optimal value of φ is $\varphi^* = \varphi_B$ and the objective function in (9) reduces to $r^2 - 2rr_B\cos(\theta - \theta_B) + r_B^2$. Then an optimal value of θ is determined $\theta^* = \min\{\theta_B, \gamma\}$. Note that the objective function in (9) is quadratic with respect to r. Consequently, optimal r is given by

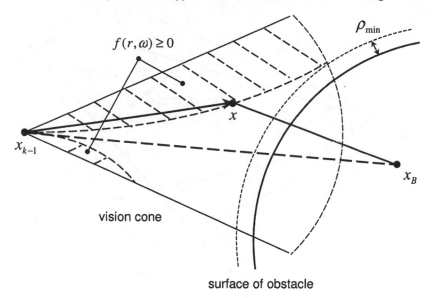

Fig. 7. Geometrical interpretation of Principle 1.

$$r^* = \begin{cases} \rho_0, & r_B \cos(\min\{0, \gamma - \theta_B\}) \le \rho_0 \\ r_B \cos(\min\{0, \gamma - \theta_B\}), & r_B \cos(\min\{0, \gamma - \theta_B\}) \in [\rho_0, L] \\ L, & r_B \cos(\min\{0, \gamma - \theta_B\}) \ge L \end{cases}$$

4.1 Obstacle with Plain Surface

Let the surface of an obstacle be a plane determined by $F(y) = e_h \cdot y - d_0 = 0$, where e_h is the plane normal satisfying $e_r \cdot e_h \ge 0$ (see Figure 8). Then for any point x, which is in the same half-space with the cone vertex (the origin of local system of coordinates), the distance to the plane is given by $d(x, z) = d_0 \quad re_r \cdot e_h$, where

$$\cos \omega(\varphi, \theta) = e_r \cdot e_h = \sin \theta \sin \theta_h \cos(\varphi - \varphi_h) + \cos \theta \cos \theta_h. \tag{10}$$

Consequently, $d(x, z) = d_0 - r \cos \omega(\varphi, \theta)$. Since the obstacle surface is represented by the hyperplane, we suggest the kinematic constraint in the following form

$$f(r, \omega(\theta, \varphi)) = (d_0 - \rho_{\min}) \tan \omega(\theta, \varphi) - r \ge 0. \tag{11}$$

Let us verify that the axioms K1–K3 for kinematic constraints are satisfied. To verify that the safe-distance constraint is satisfied, we have

$$\begin{aligned} d(x, z) &= d_0 - r \cos \omega(\varphi, \theta) \\ &\ge d_0 - (d_0 - \rho_{\min}) \sin \omega(\varphi, \theta) \\ &= \rho_{\min} + (d_0 - \rho_{\min})(1 - \sin \omega(\varphi, \theta)). \end{aligned} \tag{12}$$

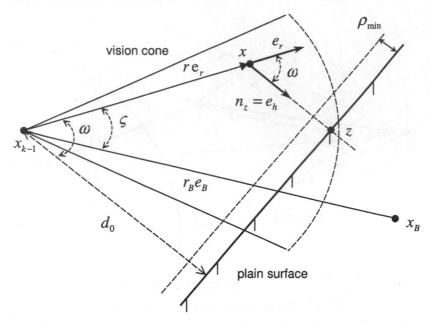

Fig. 8. Navigation subject to avoidance of an obstacle with plain surface.

Hence, $d(x,z) \geq \rho_{\min}$ since $(d_0 - \rho_{\min})(1 - \sin \omega(\varphi, \theta) \geq 0$. To verify the second axiom, note that $\tan \omega(\theta, \varphi) \geq \frac{r}{d_0 - \rho_{\min}} > 0$. This means that $\omega(\varphi, \theta) > 0$. Now suppose that $f(r, \omega(\varphi, \theta)) = 0$ and δ decreases. From $\delta = d(x,z) - \rho_{\min}$ we have $r = \frac{d_0 - \rho_{\min} - \delta}{\cos \omega(\varphi, \theta)}$. Substituting this expression into $f(\omega, \delta) = 0$, we obtain $\delta = (d_0 - \rho_{\min})(1 - \sin \omega(\varphi, \theta))$. Consequently, if δ decreases, then $\omega(\varphi, \theta)$ increases. Obviously, if $\delta = 0$, then $\omega(\varphi, \theta) = \frac{\pi}{2}$.

In the case when the segment $x_{k-1} x_B$ intersects the hyperplane within the vision cone, by Principle 1 the kinematic constraint for obstacle avoidance is active. Consequently, the navigation problem (8) with the plain obstacle reduces to

$$\min_{r, \varphi, \theta} r^2 - 2 r r_B \cos \zeta(\varphi, \theta) + r_B^2$$

$$\text{s.t. } r = (d_0 - \rho_{\min}) \tan \omega(\theta, \varphi), \tag{13}$$

$$r \in [\rho_0, L], \ \theta \in [0, \gamma].$$

Note that because of condition $\rho_0 \leq r \leq L$, variable r can not be replaced by expression $(d_0 - \rho_{\min}) \tan \omega(\theta, \varphi)$ in the objective function of (13), unless we determine a domain of (φ, θ) such that any pair of (φ, θ) from this set satisfy condition $(d_0 - \rho_{\min}) \tan \omega(\theta, \varphi) \in [\rho_0, L]$. Instead, we consider r and θ

to be free variables and express variable φ from $f(r, \omega(\varphi, \theta)) = 0$. From (10) and $f(r, \omega(\varphi, \theta)) = 0$, we have

$$\cos(\varphi - \varphi_h) = \frac{1}{\sin\theta \sin\theta_h} \left(\frac{d - \rho_{\min}}{\sqrt{r^2 + (d - \rho_{\min}^2)}} - \cos\theta \cos\theta_h \right).$$

Note that $\cos(\varphi - \varphi_h) = \cos(2\pi - (\varphi - \varphi_h))$. Since the angle between any two vectors is within the range $[0, \pi]$, i.e. $\min\{\varphi - \varphi_h,\ 2\pi - (\varphi - \varphi_h)\} \in [0, \pi]$ we have

$$\sin(\varphi - \varphi_h) = \sqrt{1 - \cos^2(\varphi - \varphi_h)} \geq 0.$$

Using the relation $\cos(\varphi - \varphi_B) = \cos(\varphi - \varphi_h)\cos(\varphi_h - \varphi_B) - \sin(\varphi - \varphi_h)\sin(\varphi_h - \varphi_B)$ the objective function in (13) is given by

$$
\begin{aligned}
g_h(r, \theta) &= r^2 - 2rr_B \cos\zeta(\varphi, \theta) + r_B^2 \\
&= r^2 - 2rr_B \left(\frac{\sin\theta_B}{\sin\theta_h} \left(\cos(\varphi_h - \varphi_B) \left(\frac{d - \rho_{\min}}{\sqrt{r^2 + (d - \rho_{\min}^2)}} - \cos\theta \cos\theta_h \right) \right. \right. \\
&\quad \left. - \sin(\varphi_h - \varphi_B) \sqrt{\sin^2\theta \sin^2\theta_h - \left(\frac{d - \rho_{\min}}{\sqrt{r^2 + (d - \rho_{\min}^2)}} - \cos\theta \cos\theta_h \right)^2} \right) \\
&\quad + \cos\theta \cos\theta_B \bigg) + r_B^2.
\end{aligned}
$$

$$(14)$$

Thus, optimization problem (13) is reduced to

$$\min_{r, \theta}\ g_h(r, \theta)$$

$$(15)$$

$$\text{s.t.}\ r \in [\rho_0, L],\ \theta \in [0, \gamma].$$

To solve the problem (13), we represent the set $[\rho_0, L] \times [0, \gamma]$ by the grid with nodes (r_i, θ_j), where $r_i = \rho_0 + \frac{i}{n_r}(L - \rho_0)$ and $\theta_j = \frac{j}{n_\theta}\gamma$, $0 \leq i \leq n_r$, $0 \leq j \leq n_\theta$, and find an optimal pair (r_{i^*}, θ_{j^*}) by enumeration.

4.2 Obstacle with Spheroidal Surface

Let the surface of an obstacle be a sphere with the center at x_s and radius R, i.e. $F(y) = d(y, x_s) - R = 0$. If center x_s is defined by $(r_s, \varphi_s, \theta_s)$ in the local system of coordinates, then $r_s e_s = x_s - x_{k-1}$, where e_s is the sphere normal satisfying $e_r \cdot e_s \geq 0$ (see Figure 9). The distance from any point x to the sphere is given by $d(x, z) = d(x, x_s) - R = \sqrt{r_s^2 - r^2 \sin^2\omega(\theta, \varphi)} - r\cos\omega(\theta, \varphi) - R$, where

$$\cos\omega(\theta, \varphi) = e_r \cdot n_z = e_r \cdot e_s = \sin\theta \sin\theta_s \cos(\varphi - \varphi_s) + \cos\theta \cos\theta_s. \quad (16)$$

The safe-distance constraint reduces to

$$\delta = \sqrt{r_s^2 - r^2 \sin^2\omega(\varphi, \theta)} - r\cos\omega(\varphi, \theta) - R - \rho_{\min} \geq 0.$$

Since the obstacle surface is represented by the sphere, we suggest the kinematic constraint in the following form

$$f(r, \omega(\varphi, \theta)) = \sqrt{r_s^2 - (R + \rho_{\min})^2} \sin \omega(\varphi, \theta) - r \geq 0. \qquad (17)$$

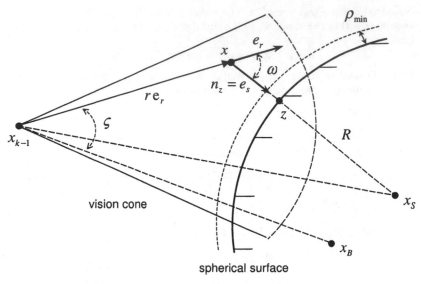

Fig. 9. Navigation subject to avoidance of an obstacle with plain surface.

To verify that the safe-distance constraint is satisfied, it is sufficient to consider only the case of $\omega \in (0, \frac{\pi}{2}]$. Under condition $\omega \in (0, \frac{\pi}{2}]$, the safe-distance constraint is equivalent to

$$r \leq (R + \rho_{\min}) \cos \omega + \sqrt{r_s^2 - (R + \rho_{\min})^2 \sin^2 \omega}.$$

However, if kinematic constraint (17) holds, then the inequality above is satisfied automatically, since $\sqrt{r_s^2 - (R + \rho_{\min})^2} \sin \omega \leq \sqrt{r_s^2 - (R + \rho_{\min})^2 \sin^2 \omega}$ and $(R + \rho_{\min}) \cos \omega > 0$ for $\omega \in (0, \frac{\pi}{2}]$. Obviously, if (17) holds, and $r > 0$ then $\omega > 0$. Now suppose that $f(r, \omega) = 0$, i.e. $r = \sqrt{r_s^2 - (R + \rho_{\min})^2} \sin \omega$. Then we have

$$\delta(\omega) = \sqrt{r_s^2 - (r_s^2 - (R + \rho_{\min})^2) \sin^4 \omega} - \frac{1}{2}\sqrt{r_s^2 - (R + \rho_{\min})^2} \sin 2\omega - (R + \rho_{\min}).$$

From this expression, it is obvious that if $\delta(\omega)$ decreases then ω increases. Moreover, if $\delta = 0$, then $\omega = \frac{\pi}{2}$.

In the case when the segment $x_{k-1}x_B$ intersects the sphere within the vision cone, by Principle 1, the kinematic constraint for obstacle avoidance is active. Thus, the navigation problem (8) with the obstacle having spheroidal surface is reduced to

$$\min_{r,\varphi,\theta} r^2 - 2rr_B \cos \zeta(\varphi,\theta) + r_B^2$$

$$\text{s.t. } r = \sqrt{r_s^2 - (R + \rho_{\min})^2} \sin \omega(\varphi,\theta), \tag{18}$$

$$r \in [\rho_0, L], \ \theta \in [0, \gamma].$$

To solve (18), we use the same technique as for the problem (13). Namely, from $f(r, \omega(\varphi,\theta)) = 0$ and (16), we express $\cos(\varphi - \varphi_s)$ and $\sin(\varphi - \varphi_s)$ through r and θ

$$\cos(\varphi - \varphi_s) = \frac{1}{\sin \theta \sin \theta_s} \left(\sqrt{\frac{r_s^2 - r^2 - (R + \rho_{\min})^2}{r_s^2 - (R + \rho_{\min})^2}} - \cos \theta \cos \theta_s \right).$$

Using the same arguments and repeating all the transformations as in solving (13), we obtain

$$\begin{aligned}
g_s(r,\theta) &= r^2 - 2rr_B \cos \zeta(\varphi,\theta) + r_B^2 \\
&= r^2 - 2rr_B \left(\frac{\sin \theta_B}{\sin \theta_s} \left(\cos(\varphi_s - \varphi_B) \left(\sqrt{\frac{r_s^2 - r^2 - (R+\rho_{\min})^2}{r_s^2 - (R+\rho_{\min})^2}} - \cos \theta \cos \theta_s \right) \right. \right. \\
&\quad \left. - \sin(\varphi_s - \varphi_B) \sqrt{\sin^2 \theta \sin^2 \theta_s - \left(\sqrt{\frac{r_s^2 - r^2 - (R+\rho_{\min})^2}{r_s^2 - (R+\rho_{\min})^2}} - \cos \theta \cos \theta_s \right)^2} \right) \\
&\quad + \cos \theta \cos \theta_B) + r_B^2.
\end{aligned} \tag{19}$$

Thus, optimization problem (18) reduces to

$$\min_{r,\theta} g_s(r,\theta)$$

$$\text{s.t. } r \in [\rho_0, L], \ \theta \in [0, \gamma]. \tag{20}$$

To solve the problem (20), we represent the set $[\rho_0, L] \times [0, \gamma]$ by the grid with nodes (r_i, θ_j), where $r_i = \rho_0 + \frac{i}{n_r}(L - \rho_0)$ and $\theta_j = \frac{j}{n_\theta}\gamma$, $0 \le i \le n_r$, $0 \le j \le n_\theta$, and find an optimal pair (r_{i^*}, θ_{j^*}) by enumeration.

5 Approximation of Obstacle Surfaces

Output of image processing is usually given as a collection of "clouds" of feature points [8]. Further by "cloud" we will mean the cloud of feature points.

To plan a path subject to obstacle avoidance, in this section we suggested a kinematic constraint, which for each cloud within a vision cone requires the knowledge of the feature point closest to the next MAV position. Because "clouds" are generally non-convex, finding the distance to a cloud is computationally intense and is often uses complete enumeration. To reduce computational time in this procedure, we suggest to approximate the boundary of a cloud within the vision cone by a smooth convex hull, see Figure 10-a. If the vision cone embraces two obstacles in the form of two disjoint clouds then those clouds are approximated by two different smooth convex hulls. A problem of approximation is to minimize the total deviation error with respect to parameters of a smooth convex hull. For instance, the total error may be defined as the sum of squared distances from feature points to the hull.

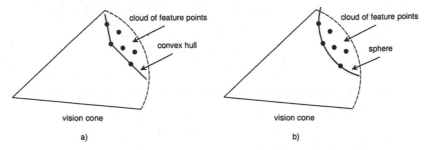

Fig. 10. Approximation of the boundary of feature point cloud by convex hulls.

Let a smooth convex hull be determined by equation $F(x, p) = 0$, where $x = (x_1, x_2, x_3)$ and p is a vector of parameters. Let C be the set of feature points of the part of the cloud boundary within the vision cone, and let $N = |C|$ be the number of feature points in the set C, i.e. $C = \{y_i\}_{i=1}^N$. A problem of approximating the set C by the surface $F(x, p) = 0$ is formulated as

$$\min_{p} \sum_{y \in C} d^2(y, z)$$
$$\text{s.t. } z = \arg \min_{F(x,p)=0} d(x, y).$$

However, since the hull is smooth and convex, condition $z = \arg \min_{F(x,p)=0} d(x, y)$ reduces to solving two equations $[(y - z) \times \nabla F(z, p)] = 0$ and $F(z, p) = 0$, and the problem above is rewritten as

$$\min_{p} \sum_{y \in C} d^2(y, z)$$
$$\text{s.t. } F(z, p) = 0, \tag{21}$$
$$[(y - z) \times \nabla_z F(z, p)] = 0.$$

We consider approximation of the set C by a plane and a sphere (see Figure 10-b). For example, for a set C we may use such an approximation that has the least total error. Note that instead of $d^2(x, y)$ we can use different error measures assigning greater or smaller weights for larger positive deviations.

5.1 Plain Surface

A plane is determined by a normal e_h, i.e. unite vector e_h orthogonal to the plane, and a scalar d_0. Any point y belonging to this plane satisfy a linear equation

$$e_h \cdot y - d_0 = 0.$$

Since the distance from any point x to the plane equals $|e_h \cdot x - d_0|$, problem (21) reduces to

$$\min_{e_h, d_0} \sum_{i=1}^{N} (e_h \cdot y_i - d_0)^2 \tag{22}$$
$$\text{s.t.} \quad e_h^2 = 1.$$

Let $y_C = \frac{1}{N} \sum_{i=1}^{N} y_i$, $\xi_i = y_i - y_C$, $1 \leq i \leq N$, and matrix $\Xi_{3 \times N}$ be defined

$$\Xi_{3 \times N} = \begin{bmatrix} \xi_{11} & \xi_{21} & \cdots & \xi_{N1} \\ \xi_{12} & \xi_{22} & \cdots & \xi_{N2} \\ \xi_{13} & \xi_{23} & \cdots & \xi_{N3} \end{bmatrix}$$

then the unit vector e_h, which solves problem (22), is an eigen-vector of the matrix $\Xi_{3 \times N} \times \Xi_{3 \times N}^T$, and the optimal d_0 is given by $d_0 = e_h \cdot y_C$.

5.2 Spherical Surface

For a sphere, determined by its center x_s and radius R, the problem (21) is equivalent to

$$\min_{x_s, R} \sum_{i=1}^{N} (d(y_i, x_s) - R)^2. \tag{23}$$

Further, problem (23) reduces to a system of algebraic equations for finding x_s and R

$$R = \frac{1}{N} \sum_{i=1}^{N} d(y_i, x_s), \tag{24}$$

and

$$\sum_{i=1}^{N} (y_i - x_s) \left(1 - \frac{R}{d(y_i, x_s)} \right) = 0. \tag{25}$$

Note that (25) is a vectorial equation. Equation (25) subject to (24) is solved numerically.

6 The Navigation Function Approach

A function accounting for a global objective, such as navigation from point A to point B, and capable of finding a collision free path among obstacles is called a *navigation* function. Although, it may seem quite appealing to use a single navigation function, its construction is not a trivial task. For illustration purpose, we consider two *"intuitively obvious"* navigation functions

$$L_1(E, \lambda) = \rho^2(E, B) - \lambda \sum_{O_i \in (\mathcal{N}_{ik} \cap \mathcal{O})} \rho^2(E, O_i), \qquad (26)$$

and

$$L_2(E, \mu) = \rho^2(E, B) - \mu \, \rho^2(E, O_E^*). \qquad (27)$$

In these expressions $\lambda > 0$ and $\mu > 0$ are weight coefficients, which may be updated at each step. The notation O_E^* has the same meaning introduced earlier $O_E^* = \arg \min_{O \in \mathcal{N}_{ik} \cap \mathcal{O}} \rho(E, O)$. Note that in the case when the vision cone identifies no obstacle, the navigation functions (26) and (27) are reduced to $\rho^2(E, B)$. The squared distance between points E and B. Given any arbitrary navigation function $L(E)$, "exit" node E_k is determined by

$$E_k = \arg \min_{E \in \mathfrak{I}_k} L(E).$$

Navigation functions (26) and (27) resemble Lagrange functions in constrained optimization. Indeed, both functions are linear combinations of the global objective $\rho^2(E, B)$, which should be minimized, and expressions accounting for proximity to obstacles. While function L_1 accounts for the whole identified geometry of the obstacle boundary within vision cone (Figure 11), function L_2 simply controls the minimal distance to the obstacle boundary (Figure 12). However, each function has its drawbacks. In some cases functions may be incapable of distinguishing obviously different obstacle shapes or making a correct decision regarding the next move. Consider some examples illustrating these situations.

We show that with respect to obstacle geometry, the function L_1 controls only the distance between the current position E and the *center of gravity* of the nodes along the obstacle boundary within the vision cone. Assuming these nodes to be equally "weighted," their center of gravity C is defined by the radius vector $\mathbf{r}_C = \frac{1}{n} \sum_{O_i \in (\mathcal{N}_{ik} \cap \mathcal{O})} \mathbf{r}_{O_i}$, where n is the number of nodes in $\mathcal{N}_{lk} \cap \mathcal{O}$. The center of gravity has the following property: $\sum_{O_i \in (\mathcal{N}_{ik} \cap \mathcal{O})} CO_i = 0$, where CO_i defines the vector with length $\rho(C, O_i)$ and has direction from C to O_i. By definition,

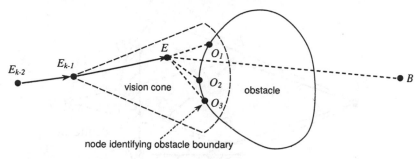

Fig. 11. Navigation function L_1 accounts for the whole visible geometry of an obstacle.

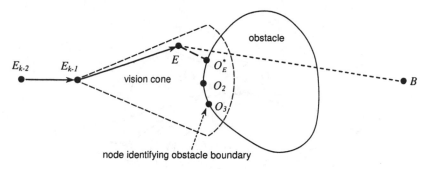

Fig. 12. Navigation function L_2 controls minimal distance to the obstacle.

$$L_1(E, \lambda) = EB^2 - \lambda \sum_{O_i \in (\mathcal{N}_{lk} \cap \mathcal{O})} EO_i^2$$
$$= EB^2 - \lambda \sum_{O_i \in (\mathcal{N}_{lk} \cap \mathcal{O})} (EC + CO_i)^2$$
$$= EB^2 - n\lambda \, EC^2 - \lambda \sum_{O_i \in (\mathcal{N}_{lk} \cap \mathcal{O})} CO_i^2.$$

Since term $\sum_{O_i \in (\mathcal{N}_{lk} \cap \mathcal{O})} CO_i^2$ does not depend on E, minimization of func-
tion (26) is equivalent to minimization of function $L_1'(E, \lambda') = \rho^2(E, B) - \lambda' \rho^2(E, C)$ with $\lambda' = n\lambda$. Figure 13 illustrates an example when $L_1'(E, \lambda')$
remains the same after rotation of the obstacle boundary around its center
of gravity C. Given only two accessible points E and E', minimization of
function $L_1'(E, \lambda')$ leads to the same decision in cases a) and b) despite the
obvious difference in geometries of obstacle boundaries. Note that the choice
of E in Figure 13a might agree with our intuition in that it avoids the obstacle.
However, the choice of E in Figure 13b does not.

Figure 14 shows that minimization of function $L_2(E, \mu)$, which accounts
only for the minimal distance to an obstacle, leads to a wrong decision when a
vision cone is orthogonal to a wall. Again, given only two accessible points E

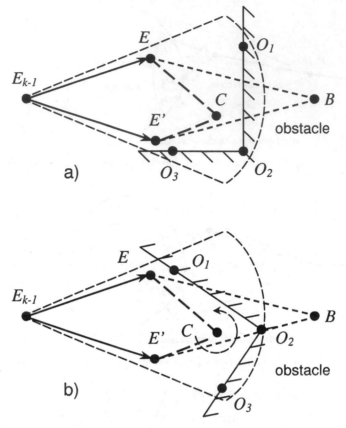

Fig. 13. Navigation function L_1 remains the same after rotation of the obstacle boundary around its center of gravity C.

and E' with the same distance to the wall, the navigation function $L_2(E, \lambda)$ is less at the point which is closer to the line of cone symmetry. This fact contradicts our intuitive notion of what constitutes a correct choice.

Concluding the discussion of the navigation function approach, we propose another navigation function, which is a combination of L_1' and L_2. We define

$$L(E, \lambda, \mu) = \rho^2(E, B) - \lambda \rho^2(E, C) - \mu \rho^2(E, O_E^*), \qquad \lambda > 0, \ \mu > 0. \quad (28)$$

Minimization of function (28) with appropriate a priori selection of λ and μ solves both aforementioned examples correctly. However, in this case, we face the following issues:

- *There is complex interplay of λ and μ depending on different geometries of obstacle boundaries.*

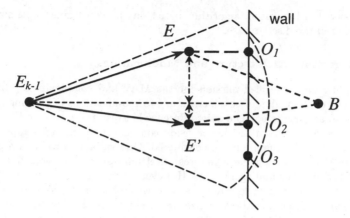

Fig. 14. Navigation function L_2 is lesser at the point that is closer to the line of cone symmetry.

- *We must update λ and μ at each step based on new information about boundary geometry.*

A thorough investigation of these issues as well as other forms for the navigation functions merits further study. Although the navigation function approach has a great potential and requires further development, this chapter is focused on solving the constrained optimization problem (2).

Remark. It seems that the Lagrange function for optimization problem (2) may also be considered as a navigation function. However, there are two crucial distinctions between the Lagrange function for (2) and a "true" navigation function as exemplified in (26)–(28). First, all terms in a "true" navigation function are expressed in the same "metric" units simplifying geometrical interpretation and analysis, while Lagrange function combines an objective function and constraints of different "nature" adjusting their dimensions by multipliers and, second, these multipliers or dual variables can not arbitrarily be assigned as in a navigation function, since they are found from solving a corresponding dual problem. Although, if we assume that constraint parameters in optimization problem (2) may be updated at each step, two approaches should formally be equivalent.

7 The Navigation Algorithm

Lumelsky and Stepanov [9] suggested two simple algorithms for navigation of a robot in a 2D space with no initial information about the obstacles. Accounting for the difference in navigation of a MAV and a 2D-robot, we have developed the navigation algorithm based on the constrained optimization approach in the discrete case.

7.1 Algorithm Architecture and Modifications

Depending on a specified mission for the MAV and complexity of obstacle geometry in the region of consideration, two modifications of the underlying algorithm are suggested. The only difference between these two modifications is how the local network within a vision cone is generated. Two approaches to local network generation are considered. One variant introduces a global network $\mathcal{G}$ representing the whole region of interest. We consider $\mathcal{G}$ to be a 3D grid with axis and diagonal arcs. However, in the dynamic case, we do not know which arcs and nodes are associated with obstacles. Thus, no arcs and nodes can be deleted at the initialization of the graph. In this case, all nodes of global network $\mathcal{G}$ within a vision cone form a local network $\mathcal{G}_l = (\mathcal{N}_l, \mathcal{A}_l)$ (Figure 15). A second variant assumes no global network and generates a local network $\mathcal{G}_l$ within the vision cone at each step of MAV movement. In this case, nodes are generated along radial rays starting at the vertex of the vision cone (Figure 16).

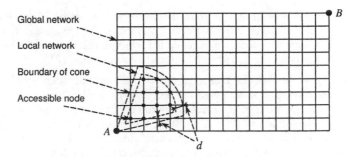

Fig. 15. Local network as a part of the global network.

Both approaches to local network generation use three sets of nodes: $\mathcal{F}$ is the set of free nodes (no-obstacle), $\mathcal{O}$ is the set of nodes identifying obstacle boundaries (so-called feature points), and $\Im$ is the set of accessible nodes. Note that a node free from an obstacle does not mean that it is accessible. The definition of the set of accessible nodes accounts for several conditions and constraints. For instance, all nodes on the boundary of a vision cone should not be accessible. If the MAV reaches such a point, the search of the next cone can identify the boundary of an obstacle. As a result, the MAV will be unable to avoid a collision. Moreover, to reserve space for MAV maneuvering, we consider all nodes within a distance d from the boundary of a cone to be inaccessible (see Figures 15, 16 and 17). As noted earlier, nodes within the

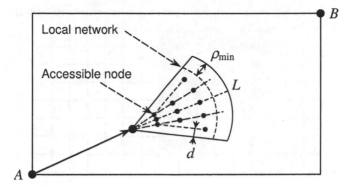

Fig. 16. Local network without introducing a global network.

minimum distance $\rho_{\min}$ from any node along an obstacle boundary are also inaccessible. The set of accessible nodes $\Im$ is formally defined by

$$\Im := \{\, i \in \mathcal{N}_l \mid \rho(i,a) > d \text{ for all } a \in \partial \mathcal{G}_l \text{ and } \rho(i,j) > \rho_{min} \text{ for all } j \in \mathcal{O} \,\},$$
(29)

where $\partial \mathcal{G}_l$ denotes the boundary of the vision cone. In this case, $\partial \mathcal{G}_l$ is a continuous set.

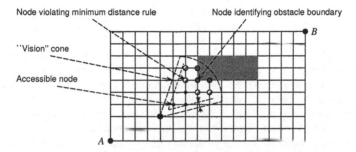

Fig. 17. Node status in the vicinity of an obstacle.

So far, we have considered conditions for obstacle avoidance without using the third dimension explicitly. For instance, in 2D space, these conditions do not guarantee that the MAV will not be trapped as it is shown in Figure 18. However, since we consider 3D space, the MAV can always move to higher altitude to avoid such "traps" (we do not consider tunnels or closed 3D constructions). Moreover, suggested constraints should automatically utilize the third dimension for avoiding complicated 2D geometry. When a solution to the optimization problem (2) is not unique, then the "exit" point with the highest altitude is chosen.

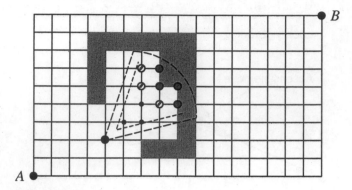

Fig. 18. Escape to higher altitude is suggested to avoid "traps."

Comparing the two approaches to local network generation, we note that neither is superior. Each of them is efficient in different aspects. For instance, the introduction of a global network essentially facilitates region analysis. Indeed, there is only a finite number of locations where the MAV can move in the entire region, since distances to obstacles are discrete. Moreover, if v and V are volumes of a vision cone and global network, respectively, then the whole region may be mapped in V/v steps. The information about the region is stored in sets $\mathcal{F}$ and $\mathcal{O}$. At each step, these sets are updated and used in making decision for the MAV's next move. Thus, this approach is efficient in target search, region mapping and navigation in a region with complicated obstacle geometry. However, identification of nodes for a local network complicates the computational process. In turn, an approach that assumes no global network is more realistic. It generates a local network of simple structure, easily incorporates a smoothness constraint and minimum distance rule. One disadvantage is that there can be a difficulty in collecting information about a region. Even overlapped vision cones have different nodal locations. As a consequence, we must be able to collect almost "continuous" information with no fixed, prior geometrical structure. Coupling with an efficient procedure for collecting and processing information about a region of consideration, this approach would be preferable to one which uses a global network. The fundamental algorithm integrates both modifications and is presented in pseudo-code form below.

The Navigation Algorithm

Step 0:   INITIALIZE $\mathcal{F} := \emptyset$; $\mathcal{O} := \emptyset$; $\mathfrak{I} := \emptyset$;
Vision cone is initially oriented in the direction of point B.

Step 1:   Based on chosen approach to local network generation, DEVELOP $\mathcal{G}_l = (\mathcal{N}_l, \mathcal{A}_l)$;

Let $\mathcal{O}_l \subset \mathcal{G}_l$ be the set of nodes identifying obstacle boundaries in $\mathcal{G}_l$;

UPDATE $\mathcal{F}$, $\mathcal{O}$ and $\mathfrak{S}$, correspondingly, i.e.

$\mathcal{O} := \mathcal{O} \cup \mathcal{O}_l$;
$\mathcal{F} := \mathcal{F} \cup \{\mathcal{N}_l \backslash \mathcal{O}\}$;
$\mathfrak{S} :=$ dynamic set, defined by (29).

Step 2:   SOLVE optimization problem (2) for $\mathcal{G}_l$ and UPDATE "exit" point E;
if several optimal E are available, CHOSE the one with the highest altitude;
if there is no feasible E, **then** STOP – there is no feasible trajectory.

Step 3:   MOVE from point A to point E;
if $E = B$ **go to** End, else $A = E$ and **go to** Step 1.

End.

We note that while moving from A to E, camera intermediate observations may also be made. In this case, $\mathcal{F}$, $\mathcal{O}$ and $\mathfrak{S}$ should be updated after each intermediate observation.

7.2 Results of Testing The Navigation Algorithm with Global Network

Algorithm performance is tested with three examples. All three examples use the same data for a global network, vision cone, smoothing condition and obstacle configuration. These examples differ only in the coordinates of points A and B. Because of scaling, all size parameters are integer. Every obstacle is a parallelepiped determined by the coordinates of its left-lower and right-upper vertices. Tables 1 and 2 show the obstacle data and coordinates of points A and B, respectively. In all examples, the vision cone and smoothness parameters are $L = 5$, $\alpha = \frac{5}{18}\pi$ and $\gamma = \frac{\pi}{6}$.

Figures 19–24 illustrate several MAV's trajectories corresponding to Examples 1, 2 and 3. Comparing Examples 1 and 2, it should be noted that only the start and end points are switched. However, the trajectories are not the same.

8 Conclusions

We have developed constrained optimization and navigation function approaches for vision-based trajectory planning for autonomous MAVs subject to obstacle avoidance. The navigation model assumes no initial information about obstacle locations. Only local information provided by a vision cone of

Table 1. Obstacle data.

obstacle	left-lower vertex	right-upper vertex
1	$(17, 9, 0)$	$(24, 12, 8)$
2	$(17, 14, 0)$	$(24, 17, 8)$
3	$(17, 19, 0)$	$(24, 22, 8)$
4	$(6, 11, 0)$	$(10, 18, 8)$
5	$(6, 17, 0)$	$(10, 24, 8)$
6	$(22, 2, 0)$	$(29, 5, 8)$
7	$(19, 15, 3)$	$(22, 21, 6)$
8	$(8, 14, 1)$	$(19, 17, 4)$
9	$(6, 16, 1)$	$(10, 19, 4)$

Table 2. Coordinates of start and end points in each example.

point	example 1	example 2	example 3
A	$(4, 20, 4)$	$(29, 11, 4)$	$(18, 4, 4)$
B	$(29, 11, 4)$	$(4, 20, 4)$	$(18, 27, 4)$

limited range is used. The constrained optimization approach is considered in discrete and continuous cases. In discrete case, obstacles are represented by collections of feature points, while in the continuous case, surfaces of the obstacle are approximated by smooth convex hulls. Analytical solutions have been obtained for a collision free path within a vision cone with obstacles, whose surfaces were approximated by either planes or spheres. In contrast to the constrained optimization approach, the navigation function approach incorporates conditions on obstacle avoidance into an objective function designed to navigate the MAV to a destination point. This approach has been developed only in the discrete case. Both approaches incorporate a constraint on trajectory smoothness. Based on the constrained optimization approach, we have developed the navigation algorithm in two modifications. The first modification relies on the existence of a global network, which covers a whole region of consideration. The second modification generates a local network within a vision cone, which reduces computational time, and automatically satisfies the smoothness constraint. The fundamental algorithm has proved to be computationally fast. It has also demonstrated excellent performance and flexibility in navigation among obstacles of "reasonable" configurations. In all test examples, the MAV has successfully avoided all the obstacles and found locally optimal paths to targets. Moreover, the algorithm can effectively handle several constraints, such as trajectory smoothness, maximum turn angle, and critical distance for obstacle avoidance.

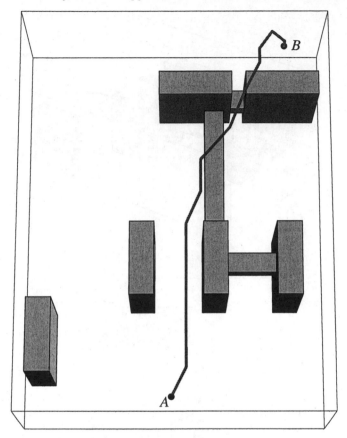

Fig. 19. Example 1: view from above.

References

1. P.K. Agarwal, T. Biedl, S. Lazard, S. Robbins, S. Suri and S. Whitesides, *Curvature-constrained Shortest Paths in a Convex Polygon,* Proceedings of 14th Annual ACM Symposium on Computational Geometry, pp. 392–401, 1998.
2. D.P. Bertsekas, *Network Optimization: Continuous and Discrete Models,* presented at Athena Scientific, Belmont, 1998.
3. D.P. Bertsekas, *Dynamic Programming and Optimal Control,* presented at Athena Scientific, Belmont, 2001.
4. W.B. Dunbar and R.M. Murray, *Model Predictive Control of Coordinated Multi-Vehicle Formations,* in Proceedings of the 41st IEEE International Conference on Decision and Control, 2002.

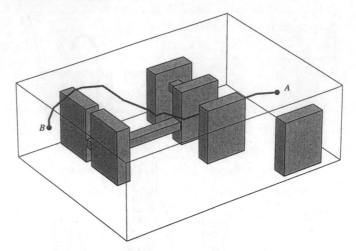

Fig. 20. Example 1: side view.

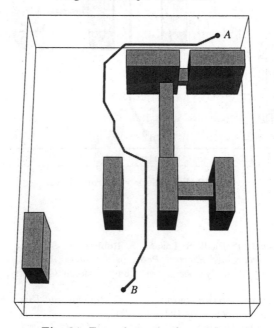

Fig. 21. Example 2: view from above.

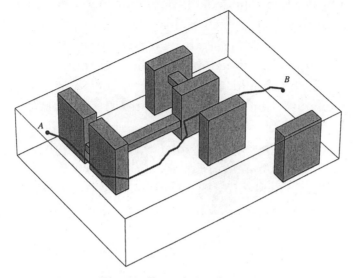

Fig. 22. Example 2: side view.

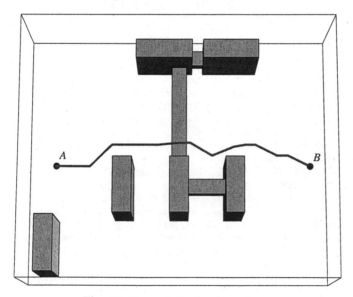

Fig. 23. Example 3: view from above.

342   Michael Zabarankin et al.

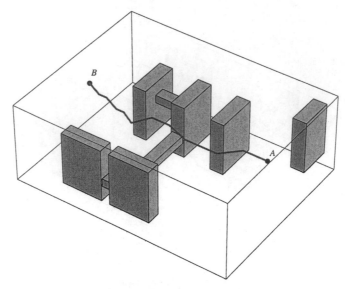

Fig. 24. Example 3: side view.

5. N. Faiz, S. Agrawal and R.M. Murray, *Trajectory Planning of Differentially Flat Systems with Dynamics and Inequalities,* Journal of Guidance, Control, and Dynamics, Vol. 24, No. 2, pp. 219–226, 2001.
6. P.G. Ifju, S. Ettinger, D.A. Jeinkins, Y. Lian, W. Shyy and M.R. Waszak, *Flexible Wing-Based Micro Air Vehicles,* presented at 40th AIAA Aerospace Sciences Meeting and Exhibit (2002-0705), Reno, 2002.
7. M. Jun, A.I. Chaudhry and R. D'Andrea, *The Navigation of Autonomous Vehicles in Uncertain Dynamic Environments: a Case Study,* SIEEE Conference on Decision and Control, Las Vegas, Nevada, December, 2002.
8. T. Kanade, O. Amidi and Q. Ke, *Real-Time and 3D Vision for Autonomous Small and Micro Air Vehicles,* Invited paper, IEEE Conference on Decision and Control (CDC 2004), December, pp. 1655–1662, 2004.
9. V. Lumelsky and A. Stepanov, *Dynamic Path Planning for a Mobile Automation with Limited Information on the Environment,* IEEE Transactions on Automatic Control, Vol. Ac-31, No. 11, November, pp. 1058–1063, 1986.
10. A. Richards and J.P. How, *Aircraft Trajectory Planning with Collision Avoidance using Mixed Integer Linear Programming,* presented at Proceedings of the American Control Conference, 2002.
11. A. Richards, T. Schouwenaars, J.P. How and E. Feron, *Spacecraft Trajectory Planning with Collision and Plume Avoidance Using Mixed-Integer Linear Programming,* AIAA Jornal of Guidance, Control and Dynamics, July, 2002.
12. J.N. Tsitsiklis, *Efficient Algorithms for Globally Optimal Trajectories,* IEEE Transactions on Automatic Control, Vol. 40, No. 9, pp. 1528–1538, 1995.
13. M.R. Waszak, J.B. Davidson and P.G. Ifju, *Simulation and Flight Control of an Aeroelastic Fixed Wing Micro Aerial Vehicle,* presented at AIAA Atmospheric Flight Mechanics Conference (2002-4875), Monterey, 2002.

17 Cooperative Stabilization and Tracking for Linear Dynamic Systems

Yi Guo

Summary. A key problem in cooperative control is the convergence to a common value of the systems, which is called the consensus or agreement problem. Among recent results in cooperative control of multi-agent dynamic systems, the method used can be categorized in either graph theory or matrix theory based techniques, and the agent dynamics is either a single/double integrator or a linear system transformable to a canonical form. We design cooperative stabilization and tracking control in the framework of Lyapunov theorem for general linear dynamic systems. Decentralized control laws are explicitly constructed for individual systems with inter-system communications. Simulations show asymptotically tracking of a time-varying trajectory with pre-designated formation for a group of dynamic agents.

1 Introduction

Cooperative control has been an active research area due to its application importance in robotics, networked systems, and biological systems. A key problem in cooperative control is the convergence to a common value of the systems, which is called the consensus or agreement problem. Since Jadbabaie, Lin and Morse [1] present the first analytic results for analyzing cooperative behaviors of networked agents, excellent work has appeared in the literature [8]. It is revealed [2, 4, 9] that the sufficient and necessary condition for a networked system to achieve consensus is that the information exchange topology has a spanning tree. Olfati-Saber and Murray [5] relate information flow and communication structure to the stability property of the group of agents. Local control laws are presented to achieve formation stability [10, 3]. Recently, Qu [6, 7] presents a comprehensive study of consensus problem using matrix-theory-based framework, where cooperative control for dynamic vehicles in their canonical forms is solved explicitly.

Among existing cooperative control results, most focuses on single or double integrator dynamics or a particular vehicle dynamics, except that Qu [6, 7] provides solutions to general, higher dimensional linear systems which are in their canonical forms. Though the dynamics of many agent systems can be transformed into one of the above structures, the extension to general linear dynamic systems is needed from both theoretical and practical points of view. From another perspective, the method used in most existing consensus work is either graph-theory or matrix theory, and the core interest has been the convergence of consensus protocol in association with the communication topology. While such a problem is well understood, the problem of applying the consensus principle to general dynamic systems is naturally the next to explore.

We present cooperative stabilization and tracking control in the chapter. A group of dynamic systems is designed to cooperatively get to their equilibrium points or to track a set of desired trajectories. Lyapunov theorem based methods are used in proving the convergence of the system state or the error (between the system and its reference). Decentralized feedback control laws for individual systems are explicitly constructed with inter-system communications. Simulation results are provided which demonstrate satisfactory performances. Besides the formation stability demonstrated, the method used in the chapter sets up a connection between Lyapunov stability with consensus protocol, and provides a theoretical tool to study cooperative control for uncertain dynamic systems or systems with uncertain communication channels.

2 Problem Statement

We assume the ith vehicle has the following linear model:

$$\dot{x}_i = A_i x_i + B_i u_i \tag{1}$$

where $x_i \in \Re^m, u_i \in \Re^k$, $A_i \in \Re^{m \times m}$, and $B_i \in \Re^{m \times 1}$.

For a group of vehicles $i = 1, \ldots, N$ to achieve cooperative behaviors such as formation, we need to design control law

$$u_i = u_i(x_i, x_j), \quad j \in \mathcal{N}_i \tag{2}$$

where x_j is the state of the jth vehicle, and $\mathcal{N}_i$ denotes the set of the neighbors of the ith vehicle. It can be seen that the control law is decentralized in the sense that it uses feedback from its own and its neighbors' states only.

Depending on the control objectives, we define the cooperative stabilization and cooperative tracking control problems as follows:

Cooperative Stabilization: *For a group of multi-vehicle systems (1) with $i = 1, \ldots, N$, given communication structure of the group forming a connected undirected graph, design decentralized control law*

$$u_i = u_i(x_i, x_j) \quad j \in \mathcal{N}_i, \tag{3}$$

such that the vehicles return to their own equilibriums and keep formation. That is,

$$x_i = 0 \quad as \quad t \to \infty$$
$$x_i = x_j \quad as \quad t \to \infty \tag{4}$$

for all $i, j = 1, 2, \ldots, N$.

Cooperative Tracking: *Given a set of formation trajectories described by:*

$$\dot{x}_{id} = A_{id} x_{id} + B_{id} u_{id}, \tag{5}$$

design decentralized control law

$$u_i = u_i(x_i, x_j) \quad j \in \mathcal{N}_i, \tag{6}$$

such that the ith vehicle asymptotically tracks its reference trajectory and keep formation. That is,

$$e_i = 0 \quad as \quad t \to \infty$$
$$e_i = e_j \quad as \quad t \to \infty \tag{7}$$

for all $i, j = 1, 2, \ldots, N$, where $e_i = x_i - x_{id}$.

3 Preliminaries from Graph and Matrix Theory

We use some notations from graph theory to describe the communications structure of the group. A graph consists of a pair $(\mathcal{V}, \mathcal{E})$, where $\mathcal{V}$ is a nonempty set of nodes and $\mathcal{E} \subset \mathcal{V}^2$ is a set of pairs of nodes, called edges. $\mathcal{E}$ is unordered (or ordered) for a undirect (or direct) graph. A (undirect or direct) path is a sequence of (unordered or ordered) edges connecting two distinct vertices. A graph is called connected if there is a path between any distinct pair of nodes. A tree is a graph where every node, except the root, has exactly one parent node. A spanning tree is a tree formed by graph edges that connect all the nodes of the graph. A graph has a spanning tree if there exists a spanning tree that is a subset of the graph.

We use the Laplacian matrix, L_G, to describe the connectivity of the nodes in a graph. $L_G = D - A$, where A is the adjacent matrix with diagonal entries 0 and off-diagonal entries $a_{ij} = 1$ if $(j, i) \in \mathcal{E}$; D is the degree matrix with diagonal entries $d_{ii} = \{j \in \mathcal{V} : (j, i) \in \mathcal{E}\}$ and off-diagonal entries 0. By definition, L_G is a zero row sum matrix. L_G is positive semi-definite for an undirected graph.

Definition 1. *A matrix $E \in \Re^{r \times r}$ is said to be reducible if the set of its indices, $\mathcal{I} \triangleq \{1, 2, \ldots, r\}$, can be divided into two disjoint nonempty set $\mathcal{S} \triangleq \{i_1, i_2, \ldots, i_\mu\}$ and $\mathcal{S}^c \triangleq \mathcal{I} \backslash \mathcal{S} = \{i_1, i_2, \ldots, i_\nu\}$ (with $\mu + \nu = r$) such that $e_{i_\alpha j_\beta} = 0$, where $\alpha = 1, \ldots, \mu$ and $\beta = 1, \ldots, \nu$. A matrix is said to be irreducible if it is not reducible.*

It is easy to see that the Laplacian matrix of a connected undirected graph is irreducible since its adjacent matrix does not have two disjoint vertex sets.

We'll need the following properties which was proposed by Wu [12, 11] for zero row sum matrices.

Lemma 1. *Let the set W consists all zero row sum matrices which have only non-positive off-diagonal elements. If A is a symmetric matrix in W, then A can be decomposed as $A = M^T M$ where M is a matrix such that row i consists of zeros and exactly one entry α_i and one entry $-\alpha_i$ for some nonzero α_i. Furthermore, if A is irreducible, then the graph associated with M is connected.*

The matrix M can be constructed as follows ([11]): For each nonzero row of A, we generate several rows of M if the same length: for the ith row of A, and for each $i < j$ such that $A_{ij} = -\alpha$ for some $\alpha > 0$, we add a row to M with the ith element being $\sqrt{\alpha}$, and the jth element being $-\sqrt{\alpha}$. This matrix M satisfies $A = M^T M$.

4 Cooperative Control Design

4.1 Cooperative Stabilization for One-Dimensional Vehicle Model

First, we illustrate the idea of the design using a one-dimensional vehicle system. The ith vehicle system model is:

$$\dot{x}_i = -x_i + u_i \tag{8}$$

where x_i, u_i are scalars. It's obvious that the equilibrium of each vehicle is at 0. We show next that the consensus protocol,

$$u_i = -\sum_{j \in \mathcal{N}_i} \alpha_{ij}(x_i - x_j), \tag{9}$$

achieves cooperative stabilization using Lyapunov theory.

Define Lyapunov candidate

$$V = \sum_{i=1}^{N} \frac{1}{2} x_i^2. \tag{10}$$

Its time derivative along the system dynamics is

$$\dot{V} = \sum_{i=1}^{N} (-x_i^2 + x_i u_i)$$

$$= -\sum_{i=1}^{N} x_i^2 - \sum_{i=1}^{N} x_i \sum_{j \in N_i} \alpha_{ij}(x_i - x_j)$$

$$= -\|x\|^2 - x^T L x \tag{11}$$

where $x = [x_1, x_2, \ldots, x_N]^T$, and L is the positive semi-definite Laplacian matrix defined in Section 3.

From (11), we obtain:

$$\dot{V} \leq -\|x\|^2 \tag{12}$$

which is negative definite. From Lyapunov theory, we conclude that each vehicle system goes to its equilibrium, 0, as $t \to \infty$.

From (11), we also obtain:

$$\dot{V} \leq -x^T L x = - \sum_{i,j \in \mathcal{N}_i} a_{ij}(x_i - x_j)^2 \tag{13}$$

where a_{ij} are real constants. From LaSalle's invariance theorem, we know that the system gets to the invariance set $\sum_{i,j} a_{ij}(x_i - x_j)^2 = 0$. Since the graph is connected, we get $x_i = x_j$ for all i, j as $t \to \infty$.

4.2 Cooperative Stabilization for Higher Dimensional Vehicle Model

In this subsection, we take the vehicle model as the general linear form in (1). First, rewrite the system dynamics in the following compact form:

$$\dot{x} = Ax + Bu \tag{14}$$

where

$$x = [x_1^T, x_2^T, \ldots, x_N^T]^T, \quad A = I_N \otimes A_i, \quad B = I_N \otimes B_i,$$

and I_N is the N-dimensional identity matrix.

Choose Lyapunov candidate

$$V = x^T P x \tag{15}$$

where

$$P = I_N \otimes P_i$$

and P_i are positive definite matrices to be chosen later.

Taking time derivatives of V along the vehicles' dynamics, we get

$$\dot{V} = x^T(PA + A^T P)x + x^T PBu. \tag{16}$$

Let

$$u = u_i + u_j. \tag{17}$$

That is, we divide the control into two separate parts: one for stabilizing its own states, u_i, called stabilizing control law; the other for the group coordination, u_j, called coordination control law.

We choose the stabilizing control law as

$$u_i = -(I_N \otimes \varepsilon_i B_i^T P_i)x \tag{18}$$

where ε_i is a positive constant. Substituting it into (16), we get

$$\dot{V} = \sum_{i=1}^{N}\{x_i^T(P_iA_i + A_i^TP_i - 2\varepsilon_iP_iB_iB_i^TP_i)x_i\} + 2x^TPBu_j. \tag{19}$$

Choose P_i to solve the following algebraic Riccati equation

$$P_iA_i + A_i^TP_i - 2\varepsilon_iP_iB_iB_i^TP_i + Q_i = 0 \tag{20}$$

where Q_i is a positive definite matrix. Then we obtain

$$\dot{V} = -x^TQx + 2x^TPBu_j \tag{21}$$

where

$$Q = I_N \otimes Q_i.$$

To design the coordination control law, we need to make use of the Laplacian matrix of the group, L_G. Choose

$$u_j = -FLx \tag{22}$$

where

$$F = I_N \otimes F_i, \quad L = L_G \otimes I_m,$$

and $F_i \in \Re^{1 \times m}$ to be chosen later. Recall that m is the dimension of the vehicle state x_i. Simplify BFL as follows:

$$\begin{aligned}
BFL &= (I_N \otimes B_i)(I_N \otimes F_i)(L_G \otimes I_m) \\
&= (I_N \otimes B_iF_i)(L_G \otimes I_m) \\
&= L_G \otimes B_iF_i.
\end{aligned} \tag{23}$$

Substituting u_j into the second term of (21), we have

$$\begin{aligned}
x^TPBu_j &= -x^TPBFLx \\
&= -x^T(I_n \otimes P_i)(L_G \otimes I_m)x \\
&= -x^T[L_G \otimes (P_iB_iF_i)]x
\end{aligned} \tag{24}$$

From Lemma 1, we know that L_G can be decomposed as $L_G = C^TC$ where C is a matrix such that row i consists of zeros and exactly one entry α_i and one entry $-\alpha_i$ for some nonzero α_i. Therefore we have

$$\begin{aligned}
x^TPBu_j &= -x^T[(C^TC \otimes (P_iB_iF_i)]x \\
&= -x^T(C^T \otimes I_m)[C \otimes (P_iB_iF_i)]x \\
&= -\sum_{i,j}\alpha_{ij}^2(x_i - x_j)^T(P_iB_iF_i)(x_i - x_j)
\end{aligned} \tag{25}$$

Since the graph associated with C is connected, all $i, j = 1, 2, \ldots, N$ are included in the above equation.

Substitute the above equation into (21), we obtain

$$\dot{V} = -x^T Q x - \sum_{i,j} 2\alpha_{ij}^2 (x_i - x_j)^T (P_i B_i F_i)(x_i - x_j) \tag{26}$$

Choose F_i such that $P_i B_i F_i$ is positive definite, then the second term of the above equation is negative definite. We conclude stability for each individual vehicle since

$$\dot{V} \leq -x^T Q x. \tag{27}$$

Also,

$$\dot{V} \leq -\sum_{i,j} 2\alpha_{ij}^2 (x_i - x_j)^T (P_i B_i F_i)(x_i - x_j) \tag{28}$$

The right hand side of the equation is 0 only when $x_i = x_j$ for all i, j. From the LaSalle's invariant theorem, we conclude that the group achieves cooperative stabilization.

Note that if $P_i B_i F_i$ is positive semi-definite, which is a weaker condition, then $x_i - x_j$ does not guarantee to go to zero but a small neighborhood of zero by (28).

We have the following theorem to summarize the result of cooperative stabilization:

Theorem 1. *Given a strongly connected communication graph, the cooperative stabilization problem is solvable using the decentralized control law (17), (18), (22) if there exists positive definite matrices P_i, Q_i to solve (20), and a feedback control matrix, F_i, such that $P_i B_i F_i$ is positive definite.*

4.3 Cooperative Tracking

Define the error state as follows:

$$e_i = x_i - x_{id} \tag{29}$$

We have the error dynamics

$$\dot{e}_i = A_i e_i + B_i(u_i - u_{id}). \tag{30}$$

Rewrite it in the compact form

$$\dot{E} = AE + BV \tag{31}$$

where

$$A = I_N \otimes A_i, \quad B = I_N \otimes B_i,$$
$$E = [e_1^T, e_2^T, \ldots, e_N^T]^T, \quad V = [v_1^T, v_2^T, \ldots, v_N^T]^T$$

and $v_i = u_i - u_{id}$ for $i = 1, 2, \ldots, N$.

Now the cooperative tracking design for the original system (1) turns into cooperative stabilization design for the error system (30). Since (31) is in the same form as (14), it follows the same procedure to design v_i as that described in 4.2. After we obtain decentralized control v_i, we can easily get $u_i = v_i + u_{id}$.

In the case that we do not have the reference model (5), but only a reference trajectory x_{id} and its derivatives $\dot{x}_{id}$, we can define the dynamics of the error systems as

$$\dot{e}_i = A_i e_i + (A x_{id} + B_i u_i - \dot{x}_{id}) \overset{\text{def}}{=} A_i e_i + v_i \tag{32}$$

By defining the new virtual control input v_i, the cooperative tracking problem is solved.

5 Simulations

Consider the following two-dimensional vehicle model:

$$x_{i1} = x_{i2} + u_{i1}$$
$$x_{i2} = u_{i2} \tag{33}$$

where $i = 1, 2, 3, 4$. Correspondingly,

$$A_i = \begin{bmatrix} 0 & 1 \\ 0 & 0 \end{bmatrix}, \quad B_i = \begin{bmatrix} 1 \\ 1 \end{bmatrix}.$$

The communication structure of the vehicles is shown as in Figure 1. Its Laplacian matrix is

$$L_G = \begin{bmatrix} 1 & -1 & 0 & 0 \\ -1 & 3 & -1 & -1 \\ 0 & -1 & 1 & 0 \\ 0 & -1 & 0 & 1 \end{bmatrix} \tag{34}$$

Choose $Q_i = I_2$, i.e., a 2-dimensional identity matrix, and $\varepsilon_i = 0.5$. The solution to the Riccati equation (20) is

$$P_i = \begin{bmatrix} 0.9102 & 0.4142 \\ 0.4142 & 1.28720 \end{bmatrix}.$$

The decentralized control law is:

$$u_i = -\varepsilon_i B_i^T P_i x_i - (FLx)_i, \tag{35}$$

Fig. 1. Communication topology of a four-robot team

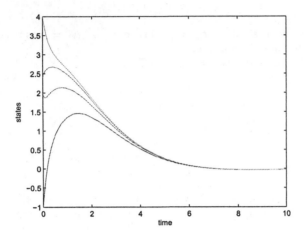

Fig. 2. The time history of the first states of four robots

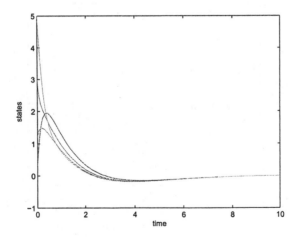

Fig. 3. The time history of the second states of four robots

where $(FLx)_i$ denotes the ith row element of FLx. The simulation results are shown in Figures 2-3.

To demonstrate the performance of cooperative tracking, we show the states of the four-robot team following a unit circle in a rectangular formation. The geometry of the formation is illustrated in Figure 4. The origin of the formation frame tracks a unit circle. To set up the reference trajectories appropriately, we define a set of coordinations, $d_i, i = 1, 2, \ldots, N$, on a moving frame (h_1, h_2), see Figure 5. The reference trajectory is:

$$x_i^d = P_i + d_{i1}h_1 + d_{i2}h_2 \tag{36}$$

where $P_i = [\cos t \ \sin t]^T$, $h_1 = [-\sin t \ \cos t]^T$, $h_2 = [\cos t \ \sin t]^T$, and $d_i, i = 1, 2, 3, 4$ are given in Figure 4.

Define the error states $e_i = x_i - x_i^d$. Then

$$\dot{e}_{i1} = e_{i2} + v_{i1}$$
$$e_{i2} = v_{i2} \tag{37}$$

where

$$v_{i1} = u_{i1} + x_{i2}^d - \dot{x}_{i1}^d$$
$$v_{i2} = u_{i2} - \dot{x}_{i2}^d. \tag{38}$$

We design the new control input (v_{i1}, v_{i2}) following the procedure described in Section 4.3. The state histories of the four robots are shown in Figure 6. It can be seen that cooperative formation with formation is achieved.

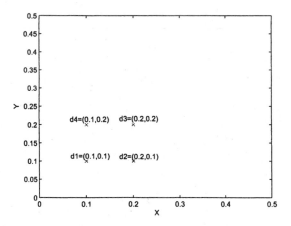

Fig. 4. Rigid formation of four robots

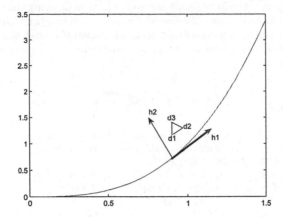

Fig. 5. Formation of a three-robot team in a moving coordinate frame

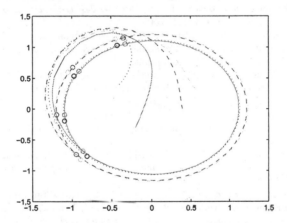

Fig. 6. The time history of four robots tracking a unit circle with rigid formation

6 Conclusions

We have designed a cooperative stabilization and a cooperative tracking control for a group of dynamic linear systems in their general state space representations. Decentralized control laws are explicitly constructed for each system with information exchange between them. Making use of the existing results on consensus protocol, we combined the vehicle-level control and rendered the group tracking a desired time-varying trajectory with pre-designate formation. Lyapunov theorem based method is used in deriving the control laws, and simulation results are illustrated for a four-robot group circling a unit circle. Future research includes the study of cooperative control for uncertain dynamic systems and systems with uncertain communication channels under the present framework.

References

1. A. Jadbabaie, J. Lin, and A. S. Morse, Coordination of groups of mobile autonomous agents using nearest neighbor rules, *IEEE Trans. on Automatic Control*, Vol. 48(6), pp. 988–1001, 2003.
2. G. Lafferriere, J. Caughman and A. Williams, Graph theoretic methods in the stability of veicle formations, in *Proceedings of American Control Conference*, pp. 3729–3734, Boston, June 2004.
3. Z. Lin, B. Francis and M. Maggiore, Local control strategies for groups of mobile autonomous agents, *IEEE Transactions on Automatic Control*, Vol. 49(4), pp. 622–629, 2004.
4. Z. Lin, B. Francis and M. Maggiore, Necessary and sufficient graphical conditions for formation control of unicycles, *IEEE Transactions on Automatic Control*, Vol. 50(1), pp. 121–127, 2005.
5. R. Olfati-Saber and R. M. Murray, Consensus problems in networks of agents with switching topology and time-delays, *IEEE Transactions on Automatic Control*, Vol. 49(9), pp. 1520–1533, 2004.
6. Z. Qu, J. Wang and R. Hull, Leadless cooperative formation control of autonomous mobile robots under limited communication range constraints, in *Coorperative Control and Optimization*, Gainesville, FL, January 2005.
7. Z. Qu, J. Wang and R. A. Hull, Cooperative control of dynamical systems with application to autonomous vehicles, *IEEE Transactions on Automatic Control*, to appear May 2008.
8. W. Ren, R. W. Beard and E. M. Atkins, A survey of consensus problems in multi-agent coordination, in *American Control Conference*, pp. 1859–1864, Portland, OR, June 2005.
9. W. Ren, R. W. Beard and T. W. McLain, Coordination variable and consensus building in multiple vehicle systems, in V. Kumar, N. E. Leonard and A. S. Moore (eds), *Cooperative Control, Springer-Verlag Series, Lecture Notes in Control and Information Sciences*, pp. 171–188, 2004.
10. H. G. Tanner, A. Jadbabaie and G. J. Pappas, Flocking in teams of nonholonomic agents, in V. Kumar, N. E. Leonard, and A. S. Moore (eds), *Cooperative Control, Springer-Verlag Series, Lecture Notes in Control and Information Sciences*, pp. 229–239, 2004.
11. C. W. Wu, *Synchronization in Coupled Chaotic Circuits and Systems*, World Scientific, Singapore, 2002.
12. C. W. Wu and L. O. Chua, Synchronization in an array of linearly coupled dynamical systems, *IEEE Transactions on Circuits and Systems*, Vol. 42(8), pp. 430–447, 1995.

Index

ATR, *see* automatic target recognition
auction, 271
auctioneer, 16
automatic target recognition, 209
autonomous underwater vehicle, 81
AUV, *see* autonomous underwater
vehicle

banana function, 89
belief state, 8
bid
two-level, 271
biologically inspired algorithm, 83
blocked connection, 108
Brownian motion, 85

CCPM, *see* cooperative communication
problem
center of gravity, 330
collaborative search, 88
complex system, 147, 171
global symmetry of, 162
optimization of, 171
concave function, 171
Conditional Value-at-Risk, 107
connectivity index, 104
problem, 105
consensus, 30, 63, 66
building, 67
controllability, 27
controllable, 30
problem, 27
state, 30
constrained shortest path problem, 113

construction phase, 192
contractive algorithm, 124
σ_x, 128
σ_y, 128
complexity of, 126
domain-optimal, 129
optimal, 128
range-optimal, 130
cooperative communication problem,
190
cooperative search, 281
cooperative stabilization, 344
cooperative tracking, 345
coordination, 1
market-based, 16
token-based, 3
CoordSim, 13
coupling constraint, 47
active, 47
CPLEX, 56
CSPP, *see* constrained shortest path
problem
CVaR, *see* Conditional Value-at-Risk

Dijkstra algorithm, 116
Dirichlet zeta function, 151
dual problem, 115
duality gap, 116
dynamic sensor coverage problem, 253

eavesdropping, 101
edge, 28, 64
head, 28
tail, 28

extended shortest path algorithm, 118
  correctness of, 121
external feedback, 63, 68

formation control problem, 68

geometric representation, 148
global information, 30
graph, 4
  connected, 64
  directed, 28, 64
  order, 28
  strongly connected, 29, 64
  symmetric, 28
  undirected, 29, 64
  weakly connected, 29
GRASP, see greedy randomized
    adaptive search procedure
greedy randomized adaptive search
    procedure, 191
guides, 196
guiding solution, 196

hierarchical structure, 151, 152, 173

ICW, see task allocation
improvement phase, 193
in-degree, 29
information feedback, 63, 67
information flow, 63, 67
information fusion, 298
information sharing, 295
integer alphabet, 148, 172
intelligence, surveillance and reconnais-
    sance, 209
intensification phase, 192
irreducible theory, 147
ISR, see intelligence, surveillance and
    reconnaissance
iterative algorithm, 121

jammed communication node, 108
jamming, 101
  effectiveness, 102
  under complete uncertainty, 107

Kalman filter, 254
Kelley's cutting plane algorithm, 117
kinematic constraint, 320
Kronecker product, 37, 66

Lagrange multiplier, 50
levy flight, 81
linear time invariant system, 254
LTI system, see linear time invariant
    system
Lyapunov function, 68

Mackay Transport Technology, 185
majorization principle, 172
Markov chain
  ergodic, 254
Markov Decision Process, 3
matrix
  adjacency, 29, 65, 66
  confusion, 212
  diagonal, 30
  distance, 178
  identity, 66
  Laplacian, 30, 65
  permutation, 34
  quadratic trace, 179
  strategy, 176
  trace, 178
  Vandermonde, 150, 173
  variance-covariance, 179
  weighted adjacency, 29
MAV, see micro-air vehicle
MDP, see Markov Decision Process
merchandize, 16
micro-air vehicle, 311
minimum
  global, 86
  local, 86

Nash equilibrium, 49, 53
navigation
  algorithm, 333
  function, 330
negotiation parameter, 50
neighbor, 29, 47, 67
network
  acquaintance, 4
  ad hoc, 188
  complete, 293
  degree distribution, 293
  hierarchy, 293
  lattice, 293
  loop, 293
  mobile ad hoc, 188

random, 293
scale free, 293
small world, 4, 293
social, 4
width, 293

obstacle surface
approximation of, 327
plain, 329
spherical, 329
optimal network covering, 103
optimality condition, 147, 167, 178
optimization
concave, 171
constrained, 317
cooperative, 45, 49
decentralized, 45
global, 45
local, 48, 51
of complex system, 171
robot swarm, 83
trajectory, 45
out-degree, 29

Pareto optimal solution, 58
Partially Observable Markov Decision
    Process, 7
path
directed, 64
planning, 45, 46
shortest, 113
strong, 29
vision-based planning, 312
weak, 29
weight feasible, 115
path-relinking, 196
perspective aggregation, 301
plan template, 3
POMDP, see Partially Observable
    Markov Decision Process
power law, 96
prime integer relation, 147, 172
programming
mixed-integer, 48
quadratic, 48
Prouhet-Thue-Morse sequence, 176
PTM sequence, see Prouhet-Thue-
    Morse sequence

quantum algorithm, 171, 184

random walk, 85
RCL, see restricted candidate list
reference state, 63, 74
restricted candidate list, 192
reward, 3
Riemann zeta function, 151
root, 34, 64

SAV, see small air vehicle
scatter search, 259
search algorithm
creeping line, 282
random, 283
zone-allocation based, 283
search score, 284
self-organization process, 147, 172
sequential inspection, 214
small air vehicle, 209
strong component, 34
root, 35
structural complexity, 147, 171
sub-goal, 3
surveillance metric, 284
swarm, 81
swarm gathering, 94

task allocation, 300
iterative Clarke-Wright, 273
market-based, 271
sorted cheapest insertion, 273
team, 1
heterogeneous, 17
team reward, 6, 13
token, 1, 296
auction, 16
coordination, 5
information, 5
resource, 5
role, 5
similarity, 11
TraderBots, 16
traveling salesman problem, 171
tree
rooted directed spanning, 34, 64
TSP, see traveling salesman problem

UAS, see unmanned aircraft system

UAV, *see* unmanned aerial vehicle
UGV, *see* unmanned ground vehicle
unmanned aerial vehicle, 5, 27, 45
unmanned aircraft system, 269
unmanned ground vehicle, 45

Value-at-Risk, 107
VaR, *see* Value-at-Risk

vehicle routing problem, 270
vertex, 28
  parent, 29
vision cone, 312
vision-based navigation, 312
VRP, *see* vehicle routing problem

wireless network jamming problem, 101